W0107747

Lectures on Soft Computing
and Fuzzy Logic

Advances in Soft Computing

Editor-in-chief
Prof. Janusz Kacprzyk
Systems Research Institute
Polish Academy of Sciences
ul. Newelska 6
01-447 Warsaw, Poland
E-mail: kacprzyk@ibspan.waw.pl
http://www.springer.de/cgi-bin/search-bock.pl?series=4240

Esko Turunen
Mathematics Behind Fuzzy Logic
1999. ISBN 3-7908-1221-8

Robert Fullér
Introduction to Neuro-Fuzzy Systems
2000. ISBN 3-7908-1256-0

Robert John and Ralph Birkenhead (Eds.)
*Soft Computing Techniques
and Applications*
2000. ISBN 3-7908-1257-9

Mieczysław Kłopotek, Maciej Michalewicz
and Sławomir T. Wierzchoń (Eds.)
Intelligent Information Systems
2000. ISBN 3-7908-1309-5

Peter Sinčák, Ján Vaščák, Vladimír Kvasnička
and Radko Mesiar (Eds.)
The State of the Art in Computational Intelligence
2000. ISBN 3-7908-1322-2

Bernd Reusch and Karl-Heinz Temme (Eds.)
Computational Intelligence in Theory and Practice
2001. ISBN 3-7908-1357-5

Robert John and Ralph Birkenhead (Eds.)
Developments in Soft Computing
2001. ISBN 3-7908-1361-3

Mieczysław A. Kłopotek, Maciej Michalewicz
and Sławomir T. Wierzchoń (Eds.)
Intelligent Information Systems 2001
2001. ISBN 3-7908-1407-5

Antonio Di Nola
Giangiacomo Gerla
Editors

Lectures on Soft Computing and Fuzzy Logic

With 37 Figures
and 4 Tables

Physica-Verlag

A Springer-Verlag Company

Prof. Antonio Di Nola
Prof. Giangiacomo Gerla
Università di Salerno
Dipartimento di Matematica e Informatica
Via S. Allende
84081 Baronissi (SA)
Italy
dinola@unina.it
gerla@matna2.dma.unina.it

ISSN 1615-3871
ISBN 978-3-7908-1396-8

Cataloging-in-Publication Data applied for
Die Deutsche Bibliothek – CIP-Einheitsaufnahme
Lectures on soft computing and fuzzy logic; with 4 tables / Antonio DiNola; Giangiacomo Gerla ed. – Heidelberg; New York: Physica-Verl., 2001
    (Advances in soft computing)
    ISBN 978-3-7908-1396-8   ISBN 978-3-7908-1818-5 (eBook)
    DOI 10.1007/978-3-7908-1818-5

Physica-Verlag Heidelberg New York
a member of BertelsmannSpringer Science+Business Media GmbH

© Physica-Verlag Heidelberg 2001

Hardcover Design: Erich Kirchner, Heidelberg

SPIN 10831940   88/2202-5 4 3 2 1 0 – Printed on acid-free paper

Preface

The present volume collects selected papers arising from lectures delivered by the authors at the School on Fuzzy Logic and Soft Computing held during the years 1996/97/98/99 and sponsored by the Salerno University.

The authors contributing to this volume agreed with editors to write down, to enlarge and, in many cases, to rethink their original lectures, in order to offer to readership, a more compact presentation of the proposed topics. The aim of the volume is to offer a picture, as a job in progress, of the effort that is coming in founding and developing soft computing's techniques.

The volume contains papers aimed to report on recent results containing genuinely logical aspects of fuzzy logic. The topics treated in this area cover algebraic aspects of Lukasiewicz Logic, Fuzzy Logic as the logic of continuous t-norms, Intuitionistic Fuzzy Logic. Aspects of fuzzy logic based on similarity relation are presented in connection with the problem of flexible querying in deductive database. Departing from fuzzy logic, some papers present results in Probability Logic treating computational aspects, results based on indishernability relation and a non commutative version of generalized effect algebras. Several strict applications of soft computing are presented in the book. Indeed we find applications ranging among pattern recognition, image and signal processing, evolutionary agents, fuzzy cellular networks, classification in fuzzy environments. The volume is then intended to serve as a reference work for foundational logico-algebraic aspect of Soft Computing and for concrete applications of soft computing technologies.

The editors are grateful to Maria Sessa for helping them to edit the present volume and wish to thank their University for supporting the seminars on Soft Computing originating this volume.

Salerno,
December 2000

Antonio Di Nola
Giangiacomo Gerla

Contents

A Natural Deduction System for Intuitionistic Fuzzy Logic

Matthias Baaz[1] *, Agata Ciabattoni[1]**, and Christian G. Fermüller[2]

[1] Institut für Algebra und Computermathematik E118.2
Technische Universität Wien
A–1040 Vienna, Austria,
[2] Institut für Computersprachen E185, Technische Universität Wien
A–1040 Vienna, Austria {baaz, agata, chrisf}@logic.at

Abstract. Intuitionistic fuzzy logic IF was introduced by Takeuti and Titani. This logic coincides with the first-order Gödel logic based on the real unit interval $[0, 1]$ as set of truth-values. We present a natural deduction system **NIF** for IF. **NIF** is defined by suitably translating a first-order extension of Avron's hypersequent calculus for Gödel logic. Soundness, completeness and normal form theorems for **NIF** are provided.

1 Introduction

Intuitionistic fuzzy logic IF was defined by Takeuti and Titani [18] as the logic of the complete Heyting algebra over the real unit interval $[0, 1]$. IF turned out to coincide with the first-order Gödel logic based on the truth-value set $[0, 1]$.

The finite-valued propositional versions of this logic were introduced by Gödel in 1933 to show that intuitionistic logic does not admit a characteristic finite matrix [11]. Dummett later generalized these to the infinite set of truth-values $[0, 1]$, and showed that the set of its tautologies LC is axiomatized as intuitionistic logic extended by the linearity axiom $(A \supset B) \vee (B \supset A)$.

Together with Lukasiewicz and product logic, LC is also one of the most important formalizations of fuzzy logic [12].

Cut-free sequent calculi for LC have been defined in [16,AFM99,8]. However these calculi make use of rules with an arbitrary number of premises. A cut-free calculus for LC which does not have this drawback has been introduced in [2]. This calculus, called **GLC**, is based on *hypersequents* — a simple and natural generalization of Gentzen's sequents to *multisets* of sequents (see [3] for an overview). Contrary to the other calculi for LC, in **GLC** the rules introducing the connectives are exactly those of Gentzen's sequent calculus **LJ** for intuitionistic logic. In [7] the hypersequent calculus $\mathbf{GLC}^{\forall\exists}$ was defined by extending **GLC** with rules for quantifiers. It was shown that $\mathbf{GLC}^{\forall\exists}$ is sound and complete for IF.

* Research supported by the Austrian Science Fund under grant P–12652 MAT
** Research supported by EC Marie Curie fellowship HPMF–CT–1999–00301

In this paper we introduce a natural deduction system **NIF** for *IF*. **NIF** is defined by suitably translating **GLC**$^{\forall\exists}$. **NIF** is based on a natural deduction system for intuitionistic logic — called **NJ'** — that is different from the usual one. Albeit equivalent to Gentzen's original system **NJ** for intuitionistic logic [10], **NJ'** is more perspicuously related to the sequent calculus **LJ** than **NJ**.

We prove that any cut-free proof in **GLC**$^{\forall\exists}$ translates into a normal natural deduction derivation in **NIF**, where "normal" essentially means that for no branch of the derivation tree an elimination of a formula follows an introduction of the same formula. Moreover we show that any derivation in **NIF** can be translated into a proof in **GLC**$^{\forall\exists}$. These results, in particular, imply that **NIF** is sound and complete for *IF*.

In future work we will investigate the possibility of strong normalization for **NIF**. That is we aim at a non-deterministic procedure that stepwise converts any given derivation into a unique normal form. We also plan to investigate the relationships between *LC* and parallel computation (see [2] for some hints). The analysis of **NIF** can be seen as a first step to achieve these research goals.

2   Syntax and Semantic of Intuitionistic Fuzzy Logic

The language of Gödel logics is the same as that of classical and intuitionistic logic. More precisely, we use the binary *connectives* $\wedge$, $\vee$, and $\supset$ and the *truth constant* $\bot$ and abbreviate $A \supset \bot$ by $\neg A$. *Object variables* are denoted by x, y, ...; the usual existential and universal *quantifiers*, $\forall$ and $\exists$, refer to these variables. *Bound* and *free* occurrences of variables are defined as usual. Moreover, for every $n \geq 0$, there is an infinite supply of n-ary *predicate symbols* and *function symbols*. Constants are considered as 0-ary function symbols. *Terms* and *formulas* are inductively defined in the usual way. Propositional variables are identified with predicate symbols of arity 0.

In this work we consider intuitionistic fuzzy logic *IF*, that is the first-order Gödel logic defined over the real unit interval $[0, 1]$. In general, the truth-values of Gödel logic can be taken from a set W such that $\{0, 1\} \subseteq W \subseteq [0, 1]$. Note that whereas for propositional Gödel logic, the tautologies coincide for all infinite W, uncountably many different Gödel logics are induced by different infinite W. (This is proved in [6].)

An *interpretation* $\mathcal{I}$ consists of a non-empty *domain* D and a *valuation function* val$_{\mathcal{I}}$ that maps constants and object variables to elements of D and n-ary function symbols to functions from D^n into D. val$_{\mathcal{I}}$ extends in the usual way to function mapping all terms of the language to an element of the domain. Moreover, val$_{\mathcal{I}}$ maps every n-ary predicate symbol P to a function from D^n into $[0, 1]$. The truth-value of an atomic formula $A \equiv P(t_1, \ldots, t_n)$ is thus defined as

$$\mathrm{val}_{\mathcal{I}}(A) = \mathrm{val}_{\mathcal{I}}(P)(\mathrm{val}_{\mathcal{I}}(t_1), \ldots, \mathrm{val}_{\mathcal{I}}(t_n)).$$

For the truth constant $\perp$ we have $\mathrm{val}_\mathcal{I}(\perp) = 0$.

The semantics of propositional connectives is given by

$$\mathrm{val}_\mathcal{I}(A \supset B) = \begin{cases} 1 & \text{if } \mathrm{val}_\mathcal{I}(A) \leq \mathrm{val}_\mathcal{I}(B) \\ \mathrm{val}_\mathcal{I}(B) & \text{otherwise,} \end{cases}$$

$$\mathrm{val}_\mathcal{I}(A \wedge B) = \min(\mathrm{val}_\mathcal{I}(A), \mathrm{val}_\mathcal{I}(B)),$$
$$\mathrm{val}_\mathcal{I}(A \vee B) = \max(\mathrm{val}_\mathcal{I}(A), \mathrm{val}_\mathcal{I}(B)).$$

To assist a concise formulation of the semantics of quantifiers we define the *distribution* of a formula A and a free variable x with respect to an interpretation $\mathcal{I}$ as $\mathrm{Distr}_\mathcal{I}(A(x)) = \{\mathrm{val}_{\mathcal{I}'}(A(x)) \mid \mathcal{I}' \sim_x \mathcal{I}\}$, where $\mathcal{I}' \sim_x \mathcal{I}$ means that $\mathcal{I}'$ is exactly as $\mathcal{I}$ with the possible exception of the domain element assigned to x. The semantics of quantifiers is given by the infimum and supremum of the corresponding distribution:

$$\mathrm{val}_\mathcal{I}((\forall x)A(x)) = \inf \mathrm{Distr}_\mathcal{I}(A(x)),$$

$$\mathrm{val}_\mathcal{I}((\exists x)A(x)) = \sup \mathrm{Distr}_\mathcal{I}(A(x)).$$

A formula A is a *tautology* iff for all $\mathrm{val}_\mathcal{I}$, $\mathrm{val}_\mathcal{I}(A) = 1$. Moreover A is a *logical consequence* of a set of formulas Γ (in symbols $\Gamma \models_{IF} A$) iff, for all $\mathrm{val}_\mathcal{I}$, $\min\{\mathrm{val}_\mathcal{I}(\gamma) \mid \gamma \in \Gamma\} \leq \mathrm{val}_\mathcal{I}(A)$. Note that, for Gödel logics, this condition is equivalent to the alternative definition: for all $\mathrm{val}_\mathcal{I}$, either there is a formula $\gamma \in \Gamma$ such that $\mathrm{val}_\mathcal{I}(\gamma) < 1$ or $\mathrm{val}_\mathcal{I}(A) = 1$.

Remark 1. Equivalent semantics for LC which stress the close relationship with intuitionistic logic are provided by linearly ordered Heyting algebras [13] and by the class of all rooted linearly ordered Kripke models [9].

A Hilbert style calculus for LC is obtained by extending the *intuitionistic propositional calculus*

I1 $A \supset (B \supset A)$

I2 $(A \wedge B) \supset A$

I3 $(A \wedge B) \supset B$

I4 $A \supset (B \supset (A \wedge B))$

I5 $A \supset (A \vee B)$

I6 $B \supset (A \vee B)$

I7 $(A \wedge \neg A) \supset B$

I8 $(A \supset \neg A) \supset \neg A$

I9 $\perp \supset A$

I10 $A \supset \top$

I11 $(A \supset (B \supset C)) \supset ((A \supset B) \supset (A \supset C))$

I12 $((A \supset C) \wedge (B \supset C)) \supset ((A \vee B) \supset C)$

with the following axiom (see [9]) which expresses the linearity of the ordering of truth-values or linear Kripke models, respectively:

$$(Linearity) \quad (A \supset B) \vee (B \supset A).$$

A Hilbert style calculus for *IF* is obtained by adding to the above calculus the following axioms and rules for quantifiers:

$$\forall x A(x) \supset A(t) \qquad\qquad A(t) \supset \exists x A(x)$$

$$\vdash B \supset A(x) \Longrightarrow \vdash B \supset \forall x A(x)$$

$$\vdash A(x) \supset B \Longrightarrow \vdash \exists x A(x) \supset B$$

together with axiom:

$$(\forall\text{-}shift) \quad \forall x(A(x) \vee B) \supset \forall x A(x) \vee B$$

The axioms and rules for quantifiers are subject to the usual variable conditions: t is free for x and x does not occur free in B.

Remark 2. The above Hilbert style axiomatization for *IF* was introduced by Horn [13]. Takeuti and Titani axiomatized *IF* in a different way, see [18]. Indeed they added to the sequent calculus for first-order intuitionistic logic a number of extra axioms as well as the following inference rule:

$$\frac{\Gamma \Rightarrow A \vee (C \supset p) \vee (p \supset B)}{\Gamma \Rightarrow A \vee (C \supset B)}$$

where p is a propositional variable that does not occur in the lower sequent. This rule expresses the fact that the set of truth-values is densely ordered.

3   A Hypersequent Calculus for *IF*

In this section we describe Avron's hypersequent calculus **GLC** for *LC* and its first-order version **GLC**$^{\forall\exists}$.

Definition 1. A *hypersequent* is a multiset

$$\Gamma_1 \Rightarrow \Delta_1 \mid \Gamma_2 \Rightarrow \Delta_2 \mid \cdots \mid \Gamma_n \Rightarrow \Delta_n$$

where every $\Gamma_i \Rightarrow \Delta_i$ is an ordinary sequent, called *component* of the hypersequent. We say that a hypersequent is *single-conclusioned* if for all $i = 1, \ldots, n$, Δ_i consists of at most one formula.

The intended interpretation of the symbol " $\mid$ " is disjunctive.

Henceforth we will only consider single-conclusioned hypersequents. As a matter of convenience, we follow [17] in assuming the left hand sides of sequents to be *sets* instead of *sequences* of formulas.

Like in ordinary sequent calculi, in a hypersequent calculus there are axioms and two types of rules: *logical rules* and *structural rules*. The logical rules are essentially the same as those in sequent calculi, the only difference being the presence of dummy contexts G and G' called *side hypersequents* which are used as meta-variables for (possibly empty) hypersequents. The structural rules are divided into *internal* and *external rules*. The former deal with formulas within components. If they are present, they are the same as in ordinary sequent calculi. The external rules manipulate whole components within a hypersequent. By the presence of these structural rules, *multiplicative* and *additive* versions of the rules are interderivable (see, e.g., [19] for this terminology). The connection to natural deduction as presented in Section 4, however, is more direct using additive logical rules with multiplicative side hypersequents.

We first display a hypersequent calculus for propositional intuitionistic logic. Note that the underlying sequent calculus is a variant of Gentzen's calculus **LJ**.

Axioms

$$(id) \quad A \Rightarrow A \quad (\bot) \quad \bot \Rightarrow A$$

Cut Rule

$$\frac{G \mid \Gamma \Rightarrow A \quad G' \mid A, \Gamma \Rightarrow C}{G \mid G' \mid \Gamma \Rightarrow C} \quad (cut)$$

Internal Structural Rule

$$\frac{G \mid \Gamma \Rightarrow C}{G \mid \Gamma, B \Rightarrow C} \quad (w, l)$$

External Structural Rules

$$\frac{G \mid \Gamma \Rightarrow C}{G \mid \Gamma \Rightarrow C \mid \Gamma' \Rightarrow D} \quad (EW) \qquad \frac{G \mid \Gamma \Rightarrow C \mid \Gamma \Rightarrow C}{G \mid \Gamma \Rightarrow C} \quad (EC)$$

Logical Rules

$$\frac{G \mid \Gamma \Rightarrow A \quad G' \mid \Gamma, B \Rightarrow C}{G \mid G' \mid \Gamma, A \supset B \Rightarrow C} \quad (\supset, l) \qquad \frac{G \mid \Gamma, A \Rightarrow B}{G \mid \Gamma \Rightarrow A \supset B} \quad (\supset, r)$$

$$\frac{G \mid \Gamma, A \Rightarrow C \quad G' \mid \Gamma, B \Rightarrow C}{G \mid G' \mid \Gamma, A \vee B \Rightarrow C} \quad (\vee, l) \qquad \frac{G \mid \Gamma \Rightarrow A_i}{G \mid \Gamma \Rightarrow A_1 \vee A_2} \quad (\vee_i, r)_{i=1,2}$$

$$\frac{G \mid \Gamma, A_i \Rightarrow C}{G \mid \Gamma, A_1 \wedge A_2 \Rightarrow C} \quad (\wedge_i, l)_{i=1,2} \qquad \frac{G \mid \Gamma \Rightarrow A \quad G' \mid \Gamma \Rightarrow B}{G \mid G' \mid \Gamma \Rightarrow A \wedge B} \quad (\wedge, r)$$

Remark 3. The "hyperstructure" of the above calculus is redundant, in the sense that a hypersequent $\Gamma_1 \vdash A_1 | \ldots | \Gamma_k \vdash A_k$ is derivable iff for some $i \in \{1, \ldots, k\}$, $\Gamma_i \vdash A_i$ is derivable.

In hypersequent calculi it is possible to define further structural rules which simultaneously act on several components of one or more hypersequents. It is this type of rule which increases the expressive power of hypersequent calculi compared to ordinary sequent calculi.

Avron's calculus **GLC** for LC is obtained by adding to the above hypersequent calculus for intuitionistic logic the following two rules (see [2]):

$$\frac{G \mid \Pi, \Gamma \Rightarrow A}{G \mid \Pi \Rightarrow A \mid \Gamma \Rightarrow A} \quad (S_I) \qquad \text{and} \qquad \frac{G \mid \Gamma_1 \Rightarrow A \quad G' \mid \Gamma_2 \Rightarrow B}{G \mid G' \mid \Gamma_2 \Rightarrow A \mid \Gamma_1 \Rightarrow B} \quad (com)$$

Remark 4. As pointed out by Avron, a hypersequent can be thought of as a specification of a multi-process. With respect to this interpretation the (com) rule models the exchange of information within multi-processes.

Alternative cut-free hypersequent calculi for LC have been presented in [2] (see also [3]). These calculi are obtained by extending the hypersequent calculus for intuitionistic logic with either the rule

$$\frac{G \mid \Gamma_1, \Gamma_2 \Rightarrow A \quad G \mid \Gamma_1, \Gamma_2 \Rightarrow B}{G \mid \Gamma_1 \Rightarrow A \mid \Gamma_2 \Rightarrow B} \quad (com')$$

or

$$\frac{G \mid \Gamma_1, \Gamma_1' \Rightarrow A \quad G \mid \Gamma_2, \Gamma_2' \Rightarrow B}{G \mid \Gamma_1, \Gamma_2 \Rightarrow A \mid \Gamma_1', \Gamma_2' \Rightarrow B} \quad (com'')$$

It is not hard to show that all these calculi are equivalent.

As an example, we display a proof of the linearity axiom in **GLC**. Recall that this axiom is not valid intuitionistic logic.

$$\frac{\dfrac{A \Rightarrow A \quad B \Rightarrow B}{\dfrac{A \Rightarrow B \mid B \Rightarrow A}{\dfrac{A \Rightarrow B \mid \Rightarrow B \supset A}{\dfrac{\Rightarrow A \supset B \mid \Rightarrow B \supset A}{\dfrac{\Rightarrow A \supset B \mid \Rightarrow (A \supset B) \vee (B \supset A)}{\dfrac{\Rightarrow (A \supset B) \vee (B \supset A) \mid \Rightarrow (A \supset B) \vee (B \supset A)}{\Rightarrow (A \supset B) \vee (B \supset A)} (EC)} (\vee_1, r)} (\vee_2, r)} (\supset, r)} (\supset, r)} (com)}$$

A cut-free calculus for IF is defined by adding to **GLC** the following rules for quantifiers (see [4])

$$\frac{G \mid \Gamma \Rightarrow F(x)}{G \mid \Gamma \Rightarrow \forall x F(x)} \ (\forall, r) \qquad \frac{G \mid \Gamma, F(t) \Rightarrow C}{G \mid \Gamma, \forall x F(x) \Rightarrow C} \ (\forall, l)$$

$$\frac{G \mid \Gamma \Rightarrow F(t)}{G \mid \Gamma \Rightarrow \exists x F(x)} \ (\exists, r) \qquad \frac{G \mid \Gamma, F(x) \Rightarrow C}{G \mid \Gamma, \exists x F(x) \Rightarrow C} \ (\exists, l)$$

where x does not occur as a free variable in the lower hypersequents of $(\forall, r)$ and $(\exists, l)$.

Henceforth we shall refer to this calculus as $\mathbf{GLC}^{\forall\exists}$.

Theorem 1. *[7]* $\mathbf{GLC}^{\forall\exists}$ *is sound and complete for IF.*

Theorem 2 (Cut-elimination). *[4,7] If a hypersequent H is derivable in* $\mathbf{GLC}^{\forall\exists}$ *then it is derivable in* $\mathbf{GLC}^{\forall\exists}$ *without using the cut rule.*

4   Natural Deduction

In his seminal paper [10] Gentzen formulated the system $\mathbf{NJ}$ of natural deduction for intuitionistic logic. In correspondence with the sequent calculus $\mathbf{LJ}$ for intuitionistic logic, where the right hand side of a sequent is restricted to at most one formula, $\mathbf{NJ}$ deals with inference patterns ("Schlußweisen") of *one* conclusion from a set of assumptions. At the application of rules, assumptions of a certain form can be *cancelled* in parts of the proof. A proof of a formula is a derivation tree where all assumptions have been cancelled. A natural deduction system $\mathbf{NK}$ for classical logic is obtained by either adding to $\mathbf{NJ}$ some rule or axiom corresponding to *tertium non datur*, or by dropping the restriction to one formula in the conclusion and allowing multi-conclusions, i.e., sets of formulas as nodes of a derivation tree, instead. In [5] it has been shown how to generalize the multi-conclusion version of $\mathbf{NK}$ to natural deduction systems for each finite-valued logic.

Here, we first describe a version of natural deduction for intuitionistic logic called $\mathbf{NJ'}$. $\mathbf{NJ'}$ is equivalent to the Gentzen's original system $\mathbf{NJ}$, but more perspicuously related to the sequent calculus $\mathbf{LJ}$ than $\mathbf{NJ}$. In a second step we build on $\mathbf{NJ'}$ to define a system of natural deduction $\mathbf{NIF}$ that is sound and complete for *IF*. $\mathbf{NIF}$ is defined by suitably translating $\mathbf{GLC}^{\forall\exists}$.

4.1   An Alternative Natural Deduction System for Intuitionistic Logic

As usual, a *derivation* in $\mathbf{NJ'}$ consists in an upward rooted tree, where the inner nodes are formulas and the leaf nodes are multisets of formulas called *assumptions*. We say that A is derivable from assumptions Γ and write

$$\Gamma$$
$$\vdots$$
$$A$$

if there is a derivation ν with root A such that the union of all assumptions of ν is a subset of the set Γ.

Derivations in **NJ'** are generated inductively as follows:

1. *Initial derivations* are of the form:

$$\frac{\alpha}{A}$$

where A is any formula and α (i.e., the assumptions of this derivation) is any multiset of formulas containing A or $\bot$.

2. New derivations can be generated from given ones using the following rules for the *elimination* and *introduction* of *connectives*:

$$\frac{\begin{array}{cc}\Gamma & \Gamma \\ \vdots & \vdots \\ A & B\end{array}}{A \wedge B}\,(i\,\wedge) \qquad\qquad \frac{\begin{array}{cc}\Gamma & \Gamma,[A_j] \\ \vdots & \vdots \\ A_1 \wedge A_2 & C\end{array}}{C}\,(e\,\wedge_j)\ j=1,2$$

$$\frac{\begin{array}{c}\Gamma \\ \vdots \\ A_j\end{array}}{A_1 \vee A_2}\,(i\,\vee_j)\ j=1,2 \qquad \frac{\begin{array}{ccc}\Gamma & \Gamma,[A] & \Gamma,[B] \\ \vdots & \vdots & \vdots \\ A \vee B & C & C\end{array}}{C}\,(e\,\vee)$$

$$\frac{\begin{array}{c}\Gamma,[A] \\ \vdots \\ B\end{array}}{A \supset B}\,(i\supset) \qquad \frac{\begin{array}{ccc}\Gamma & \Gamma & \Gamma,[B] \\ \vdots & \vdots & \vdots \\ A \supset B & A & C\end{array}}{C}\,(e\supset)$$

$$\frac{\bot}{A}\,(e\,\bot)$$

The application of a rule is to be understood as follows: The formula below the line is appended as a new root to the derivations rooted in the formulas indicated above the line. In writing

$$\begin{array}{c}\Gamma,[F] \\ \vdots \\ G\end{array}$$

for the premise of a rule we mean that the formula F may be (but need not be) *cancelled* from some or all the assumptions (leaf nodes) of the derivation of G from $\Gamma \cup \{F\}$ when applying the rule. Cancelled formulas are marked as such in the leaf nodes and cannot be used anymore in later applications of rules.

3. The *elimination* and *introduction* rules of **NJ'** for *quantifiers* are:

$$
\frac{\begin{array}{c}\Gamma\\ \vdots\\ A(x)\end{array}}{\forall x A(x)}\;(i\,\forall)
\qquad\qquad
\frac{\begin{array}{cc}\begin{array}{c}\Gamma\\ \vdots\\ \forall x A(x)\end{array} & \begin{array}{c}\Gamma,[A(t)]\\ \vdots\\ C\end{array}\end{array}}{C}\;(e\,\forall)
$$

where in $(i\,\forall)$ x neither occurs free in any assumption that $A(x)$ depends on, nor free in $\forall x A(x)$. Moreover:

$$
\frac{\begin{array}{c}\Gamma\\ \vdots\\ A(t)\end{array}}{\exists x A(x)}\;(i\,\exists)
\qquad\qquad
\frac{\begin{array}{cc}\begin{array}{c}\Gamma\\ \vdots\\ \exists x A(x)\end{array} & \begin{array}{c}\Gamma,[A(x)]\\ \vdots\\ C\end{array}\end{array}}{C}\;(e\,\exists)
$$

where in $(e\,\exists)$ x neither occurs free in any assumption that C depends on except $A(x)$ nor free in $\exists x A(x), C$.

The application of these rules is defined as for the case of connectives above.

In an elimination rule — i.e., $(e\,\wedge_i)$, $(e\,\vee)$, $(e\,\supset)$, $(e\,\exists)$, or $(e\,\forall)$ — the premise containing the occurrence of the connective or quantifier being eliminated is called *major premise*; all the other premise(s) are called *minor premise(s)* of the rule. In the case of $(e\,\supset)$, we also refer to the exhibited occurrence of A in

$$
\begin{array}{c}\Gamma\\ \vdots\\ A\end{array}
$$

as the *left minor premise*.

Remark 5. In order to be able to recognize effectively wellformed derivations one should not only label the inner nodes of the derivation tree with the names of the applied rules as indicated above, but also point to the formulas that are cancelled at the application of the rules. This can be achieved, e.g., by labelling the formulas in assumptions and referring to these labels in denoting applications of rules. In our context it is not necessary to make this more explicit. We rather refer to [19] and [14] for more exact definitions.

Remark 6. The above system is sound and complete for first-order intuitionistic logic. Note that the $(e\,\wedge_i)$, $(e\,\supset)$ and $(e\,\forall)$ rules differ from those given by Gentzen for **NJ**. We follow [5] in the definition of them. They are more close to the corresponding rules of **LJ** than that of **NJ**. Moreover, as we shall see, the **NJ'**-rules allow for a direct translation of cut-free **LJ**-proof into normal derivations. This transfers also to the level of hypersequents and hyperderivations, respectively, as shown in Theorem 4, below.

Is easy to check that **NJ** and **NJ'** are equivalent. For instance, the usual version for the elimination rules for the conjunction connective are

$$
\frac{\begin{array}{c}\Gamma\\\vdots\\A_1 \wedge A_2\end{array}}{A_1}\ (e\wedge_1)' \qquad \text{and} \qquad \frac{\begin{array}{c}\Gamma\\\vdots\\A_1 \wedge A_2\end{array}}{A_2}\ (e\wedge_2)'
$$

These rules are obtained from $(e\wedge_1)$ and $(e\wedge_2)$ by simply instantiating C with A_1 and A_2, respectively, and disregarding the redundant minor premise.

On the other hand, an application of $(e\wedge_i)$ can be replaced by an application of $(e\wedge_i)'$ by substituting each assumption (i.e., each leaf node) in

$$
\begin{array}{c}\Gamma \cup \{A_i\}\\\vdots\\C\end{array}
$$

that contains A_i, by an instance of a derivation

$$
\frac{\begin{array}{c}\Gamma\\\vdots\\A_1 \wedge A_2\end{array}}{A_i}\ (e\wedge_i)'
$$

The equivalence of the rules $(e\supset)$ and $(e\forall)$ to the corresponding formulations in **NJ** is handled similarly.

Remark 7. As usual, in **NJ'** we have no rules corresponding to the (w,l) rule of **GLC**$^{\forall\exists}$. This is in accordance to the traditional understanding of natural deduction: A derivation of a formula A from assumptions Γ is, by definition, also a derivation of A from assumptions *containing* Γ.

As usual in natural deduction we are interested in *normal proofs*; i.e., proofs that do not contain redundancies corresponding to cuts and enjoying a properly defined subformula property (see [14,19]).

Definition 2. A *maximum segment* or *cut sequence* in a derivation ν of **NJ'** is a sequence of consecutive occurrences $A^1, \ldots, A^n$ of a formula A in ν such that A^1 is the conclusion of an introduction rule and A^n is the major premise of an elimination rule.

Note that while in **NJ**, for $i < n$, A^i can only be a minor premise of an elimination rule for either $\vee$ or $\exists$, in **NJ'** it can also be a minor premise of the rules $(e\wedge_i)$, $(e\supset)$ or $(e\forall)$.

Definition 3. An **NJ'**-derivation ν is called *normal* when it contains no maximum segment.

It is not hard to see that both the proof transformations described in Remark 6 — that from **NJ** to **NJ′** and that from **NJ′** to **NJ** — preserve normality.

To define paths in **NJ′**, we suitably adapt the notion of *tracks* in [19].

Definition 4. A *path* in a derivation ν of **NJ′** is a sequence of occurrences of formulas $A_0, \ldots, A_n$ such that

1. A_0 is a formula in an assumption (leaf node) of ν that is not cancelled by an application of an elimination rule in ν;
2. A_i for $i < n$ is not the left minor premise of an instance of $(e \supset)$, and *either*
    (a) A_i is not the major premise of an application of an elimination rule and A_{i+1} is the conclusion for which A_i is a premise, *or*
    (b) A_i is the major premise of an application of an elimination rule, and A_{i+1} is a formula in an assumption cancelled by this rule application;
3. A_n is *either*
    (a) the left minor premise of an instance of $(e \supset)$, *or*
    (b) the conclusion of ν, *or*
    (c) the major premise of an application of an elimination rule, and no formulas are cancelled by this rule application.

Definition 5. A *path of order* 0 in a normal derivation ν of **NJ′** is a path ending in a conclusion of ν. A *path of order* $n + 1$ is a path ending in the left minor premise of $(e \supset)$, with major premise belonging to a path of order n.

Theorem 3 (Subformula Property). *Let ν be a normal derivation in* **NJ′** *of A from Γ. Then each formula in ν is a subformula of a formula in* $\Gamma \cup \{A\}$.

Proof. By an easy induction on the order of paths in **NJ′**.

4.2   Extending natural deduction to the hyper-level

In this section we extend **NJ′** to a natural deduction system for *IF* in analogy to the extension of **LJ** to **GLC**$^{\forall\exists}$. To this aim we have to generalize natural deduction derivations to operate on *multisets* of derivation trees called *hyperderivations*. We call the resulting calculus **NIF**. We use the symbol $*$ to separate the components (i.e., the derivation trees) which a hyperderivation consists of.

Naturally, the definition of **NIF**-hyperderivations refers to that for **NJ′**-derivations above. In fact, as long as no rules corresponding to the external structural rules of **GLC**$^{\forall\exists}$ is applied, the constructed hyperderivation will consist of a single component which is an ordinary **NJ′**-derivation.

1. Initial **NJ′**-derivations are also initial **NIF**-hyperderivations.

2. The rules for introducing and eliminating connectives and quantifiers are the same as for **NJ'**. However, their interpretation is slightly changed in the context of **NIF**. Applying, e.g., the rule $(e\,\exists)$ here means that the two hyperderivations $N_1 * \nu_1$ and $N_2 * \nu_2$ are replaced by the hyperderivation $N_1 * N_2 * \nu$, where N_1 and N_2 denote arbitrary *side hyperderivations* and ν denotes the derivation resulting from the application of $(e\,\exists)$ to ν_1 and ν_2, where the eigenvariable conditions refer to the whole hyperderivation. (Analogously for all the other rules of **NJ'**).

3. The additional rules of **NIF** correspond to the external structural rules (EW), (EC), (S_I), and (com) of $\mathbf{GLC}^{\vee\exists}$. In their formulation the double bar shall indicate that the upper hyperderivation is *replaced* by the lower one. Let N and N' be metavariables for hyperderivations.

   (a) For *external weakening* we have the following rule:

   $$\frac{N}{N * \dfrac{\Gamma}{A}} \; (EW)$$

   Observe that the new component is a tree with exactly two nodes: a leaf node consisting in a finite multiset of formulas Γ and a root node consisting in the formula A.

   (b) For *external contraction* we have:

   $$\frac{N * \begin{matrix} \Gamma \\ \vdots \\ A \end{matrix} * \begin{matrix} \Gamma' \\ \vdots \\ A \end{matrix}}{N * \dfrac{\begin{matrix} \Gamma & \Gamma' \\ \vdots & \vdots \\ A & A \end{matrix}}{A}} \; (EC)$$

   Applying this rule means the following: Given a hyperderivation in which there are two components that end with occurrences of the same formula A we combine these two components into one by appending another occurrence of A as a common root node to the components.

   (c) In applying the *splitting* rule

   $$\frac{N * \begin{matrix} \Gamma,\Gamma' \\ \vdots \\ A \end{matrix}}{N * \dfrac{\begin{matrix} \Gamma \\ \vdots \\ A \end{matrix}}{A} \; (S_I) \; * \; \dfrac{\begin{matrix} \Gamma' \\ \vdots \\ A \end{matrix}}{A} \; (S_I)}$$

the set of formulas occurring in leaf nodes of the indicated component of the upper hyperderivation is split into two (not necessarily disjoint) subsets Γ and Γ'. The application itself can be described in two steps. First, the indicated component of the upper derivation is duplicated. Second, we remove formulas not occurring in Γ' from the assumptions of the first copy, and formulas not occurring in Γ from the assumptions of the second copy of the original component, respectively.

(d) In applying the *communication* rule

$$
\frac{N * \begin{array}{c}\Gamma \\ \vdots \\ A\end{array} \qquad N' * \begin{array}{c}\Gamma' \\ \vdots \\ B\end{array}}{N * N' * \dfrac{\begin{array}{c}\Gamma \\ \vdots \\ A\end{array}}{B}\ (com) \quad * \quad \dfrac{\begin{array}{c}\Gamma' \\ \vdots \\ B\end{array}}{A}\ (com)}
$$

two hyperderivations — one containing a derivation of A from Γ, the other one containing a derivation of B from Γ' — are replaced by a hyperderivation consisting of the two indicated components with their conclusions interchanged, together with the union of the respective side hyperderivations.

As an example we give a proof of the linearity axiom in **NIF**. It starts by taking the two initial derivations:

$$
\frac{\{A\}}{A} \qquad \text{and} \qquad \frac{\{B\}}{B}
$$

and applying (*com*) to obtain the hyperderivation

$$
\frac{\{A\}}{\dfrac{A}{B}}\ (com) \quad * \quad \frac{\{B\}}{\dfrac{B}{A}}\ (com)
$$

In both components we now apply $(i \supset)$, thereby discarding the assumptions:

$$
\frac{\dfrac{\{[A]\}}{\dfrac{A}{B}}\ (com)}{A \supset B}\ (i \supset) \quad * \quad \frac{\dfrac{\{[B]\}}{\dfrac{B}{A}}\ (com)}{B \supset A}\ (i \supset)
$$

We then introduce $\vee$ in both components and obtain

$$
\frac{\dfrac{\dfrac{\{[A]\}}{\dfrac{A}{B}}\ (com)}{A \supset B}\ (i \supset)}{(A \supset B) \vee (B \supset A)}\ (i \vee_1) \quad * \quad \frac{\dfrac{\dfrac{\{[B]\}}{\dfrac{B}{A}}\ (com)}{B \supset A}\ (i \supset)}{(A \supset B) \vee (B \supset A)}\ (i \vee_2)
$$

Since the two components end with the same formula we can apply external contraction (EC) to obtain the following proof (i.e., a hyperderivation with a single component in which all assumptions are discarded).

$$\cfrac{\cfrac{\cfrac{\cfrac{\{[A]\}}{\cfrac{A}{B}\ (com)}}{A \supset B}\ (i \supset)}{(A \supset B) \vee (B \supset A)}\ (i \vee_1)\qquad \cfrac{\cfrac{\cfrac{\{[B]\}}{\cfrac{B}{A}\ (com)}}{B \supset A}\ (i \supset)}{(A \supset B) \vee (B \supset A)}\ (i \vee_2)}{(A \supset B) \vee (B \supset A)}\ (EC)$$

Remark 8. The rules (S_I) and (com) allow to move assumptions and formulas, respectively, between different components. Thus the components of a hyperderivation need not be **NJ'**-derivations generated according to the rules of Section 4.1.

A similar remark holds for the definition of a *maximum segment*. In the communication rule, as formulated above, the lower occurrence of B is *not* to be understood as "consecutive" to the upper occurrence of A but rather is defined to be consecutive to the upper occurrence of B. (The same also holds for A and B interchanged, of course.) Given this modification, the definition of *normality* for **NIF**-hyperderivations is the same as for **NJ'**-derivations (see Definition 3).

Definition 6. A *path* in an **NIF**-hyperderivation is defined exactly as a path for **NJ'** (Definition 4) except for keeping track with the exchange of assumptions and formulas between different components. More exactly, let A be a premise of (com), i.e., A is moved from one component ν of a hyperderivation to another component ν' by an application of (com). Then any path containing the indicated occurrence of A crosses from ν to ν' at the position of A. I.e., the predecessor of A (occurring in ν', by the application of (com)) in the path is the occurrence of the premise (possible assumption) in ν that was used to deduce A. Similarly, for (S_I).

Another small modification concerns external contraction. In any path $A_0, \ldots, A_n$ the successor A_{i+1} of A_i for $i < n$ may also be the conclusion in the new component arising from an application of (EC), where A_i is the conclusion of one of the combined components.

Theorem 3 (subformula property) also holds for **NIF**.

Definition 7. The *height* of a derivation in $\mathbf{GLC}^{\forall \exists}$ is the maximal number of consecutive internal nodes (i.e., inferences) in it.

Theorem 4. *Every cut-free* $\mathbf{GLC}^{\forall \exists}$-*proof of a hypersequent*

$$\Gamma_1 \Rightarrow A_1 \mid \ldots \mid \Gamma_m \Rightarrow A_m$$

can be translated into a normal hyperderivation in **NIF** *of the form*

$$
\begin{array}{ccc}
\Gamma_1 & & \Gamma_m \\
\vdots & * \ldots * & \vdots \\
A_1 & & A_m
\end{array}
$$

Proof. We describe the construction of an appropriate hyperderivation π^* by induction on the height h of the cut-free proof π of $\Gamma_1 \Rightarrow A_1 \mid \ldots \mid \Gamma_m \Rightarrow A_m$.

If $h = 0$, then π is an axiom either of the form $A \Rightarrow A$ or of the form $\perp \Rightarrow A$. The corresponding hyperderivations π^* in **NIF** are the initial derivations of A from the assumptions $\{A\}$ and $\{\perp\}$, respectively.

For $h > 0$ we distinguish the following cases according to the last inference rule applied in π:

1. π ends with an internal weakening rule:

$$
\cfrac{\begin{array}{c} \vdots\; \pi_1 \\ G \mid \Gamma \Rightarrow C \end{array}}{G \mid \Gamma, B \Rightarrow C} \;(w,l)
$$

By the induction hypothesis there exists a normal hyperderivation π_1^* corresponding to the **GLC**$^{\vee\exists}$-derivation π_1 of $G \mid \Gamma \Rightarrow C$. The hyperderivation π^* is then generated by adding (everywhere) B to the assumptions in

$$
\begin{array}{c}
\Gamma \\
\vdots \\
C
\end{array}
$$

which, by definition, is a component of π_1^*. Obviously π^* is normal, too.

2. π ends in a right hand side rule for either a connective or a quantifier. We present the case for conjunction. The other cases are analogous. Suppose π ends with an application of the $(\wedge r)$ rule:

$$
\cfrac{\begin{array}{cc} \vdots\; \pi_1 & \vdots\; \pi_2 \\ G \mid \Gamma \Rightarrow A & G' \mid \Gamma \Rightarrow B \end{array}}{G \mid G' \mid \Gamma \Rightarrow A \wedge B} \;(\wedge, r)
$$

By the induction hypothesis we have the normal hyperderivations π_1^* and π_2^* of the form

$$
N_G * \begin{array}{c} \Gamma \\ \vdots \\ A \end{array} \qquad \text{and} \qquad N_{G'} * \begin{array}{c} \Gamma \\ \vdots \\ B \end{array}
$$

corresponding to π_1 and π_2, respectively. By applying the **NIF**-rule $(i \wedge)$ to these two hyperderivations we obtain π^* as

$$
N_G * N_{G'} * \cfrac{\begin{array}{cc} \Gamma & \Gamma \\ \vdots & \vdots \\ A & B \end{array}}{A \wedge B} \;(i \wedge)
$$

Since π^* is defined by adding an introduction rule at the end of two normal derivations π^* is normal.

3. π ends in a left hand side rule for either a connective or a quantifier. Then π^* is obtained by applying the corresponding elimination rule to the hyperderivation(s) corresponding to the $\mathbf{GLC^{\forall\exists}}$-proofs of the premise(s) of this rule application. Again, we illustrate the case for conjunction. Suppose that π is of the form

$$\begin{array}{c} \vdots \ \pi_1 \\ G \mid \Gamma, A_1 \Rightarrow C \\ \hline G \mid \Gamma, A_1 \wedge A_2 \Rightarrow C \end{array} \ (\wedge_1, l)$$

Then, by induction hypothesis there is a normal hyperderivation $\pi_1{}^*$ of the form

$$N_G \ * \ \begin{array}{c} \Gamma, A_1 \\ \vdots \\ C \end{array}$$

By applying the rule $(e \wedge_1)$ to $\pi_1{}^*$ and the initial derivation

$$\frac{\{A_1 \wedge A_2\}}{A_1 \wedge A_2}$$

we obtain π^* as

$$N_G \ * \ \frac{\dfrac{\{A_1 \wedge A_2\}}{A_1 \wedge A_2} \quad \begin{array}{c} \Gamma, [A_1] \\ \vdots \\ C \end{array}}{C} \ (e\wedge)$$

Since the elimination rule is applied immediately below the initial derivation, π^* is normal.

4. π ends with one of the rules: (EC), (EW), (S_I), or (com). Like in the cases above, π^* is obtained by applying the corresponding $\mathbf{NIF}$-rule to the hyperderivation(s) that correspond to the $\mathbf{GLC^{\forall\exists}}$-proof(s) of the premise(s) of this application of the rule. It is easy to see that π^* is normal.

Theorem 5. *Every hyperderivation ν in* **NIF**

$$\begin{array}{ccc} \Gamma_1 & & \Gamma_m \\ \vdots & * \ \dots \ * & \vdots \\ A_1 & & A_m \end{array}$$

can be translated into a $\mathbf{GLC^{\forall\exists}}$-*proof $\sigma(\nu)$ of the hypersequent*

$$\Gamma_1 \Rightarrow A_1 \mid \dots \mid \Gamma_m \Rightarrow A_m$$

Proof. We define $\sigma(\nu)$ by translating each inference step of ν into a corresponding inference of $\mathbf{GLC}^{\forall\exists}$. More exactly, the proof is by induction on the number of inference steps applied to construct the derivation ν.

1. If ν is an initial derivation

$$\frac{\alpha}{A}$$

   then either A or $\perp$ occurs in α. In the first case $\sigma(\nu)$ starts with axiom $A \Rightarrow A$; in the second case with axiom $\perp \Rightarrow A$. The other formulas of α are then added to the left hand side of these sequents by internal weakenings.
2. If the last inference step in deriving ν is the application of an introduction rule (for either a connective or a quantifier). Then $\sigma(\nu)$ simply ends with the corresponding right hand side rule in $\mathbf{GLC}^{\forall\exists}$.
3. If the last rule applied to derive ν is an elimination rule we combine the corresponding left hand side rule of $\mathbf{GLC}^{\forall\exists}$ with an application of cut to define $\sigma(\nu)$. We illustrate this for the case for implication. The other cases are analogous. ν is of the form

$$\frac{\begin{array}{ccc} \Gamma & \Gamma & \Gamma,[B] \\ \vdots\, \nu_1 & \vdots\, \nu_2 & \vdots\, \nu_3 \\ A \supset B & A & C \end{array}}{C}(e\supset)$$

The corresponding derivation $\sigma(\nu)$ in $\mathbf{GLC}^{\forall\exists}$ is

$$\cfrac{\vdots\,\sigma(\nu_1) \qquad \cfrac{\vdots\,\sigma(\nu_2) \qquad \vdots\,\sigma(\nu_3)}{\Gamma \Rightarrow A \qquad \Gamma,B \Rightarrow C}(\supset,l)}{\begin{array}{cc}\Gamma \Rightarrow A \supset B & \Gamma,A \supset B \Rightarrow C\end{array}}(cut)$$
$$\Gamma \Rightarrow C$$

4. If the last rule applied to derive ν is (EC), (EW), (S_I) or (com), then $\sigma(\nu)$ simply ends with the corresponding external structural rule in $\mathbf{GLC}^{\forall\exists}$.

Corollary 1. NIF *is sound and complete for IF.*

Proof. Since $\mathbf{GLC}^{\forall\exists}$ is sound and complete for *IF*, the claim follows from Theorems 4 and 5.

Corollary 2 (Normal Form Property). *For every derivation in* **NIF**, *there exists a normal derivation in* **NIF** *of the same formula from the same assumptions.*

Proof. If there exists a derivation in **NIF** of A from assumptions Γ, then, by Theorem 5 there exists a proof in $\mathbf{GLC}^{\forall\exists}$ of $\Gamma \Rightarrow A$. By Theorem 2 there exists also a cut-free proof in in $\mathbf{GLC}^{\forall\exists}$ of $\Gamma \Rightarrow A$ whose translation according to Theorem 4 yields a normal derivation of A from assumptions Γ.

References

1. Avellone, A., Ferrari, M. and Miglioli P.: Duplication-free tableau calculi together with cut-free and contraction free sequent calculi for the interpolable propositional intermediate logics. *Logic Journal of the IGPL*, 7(4) (1999), 447–480.

2. Avron, A.: Hypersequents, logical consequence and intermediate logics for concurrency. *Annals for Mathematics and Artificial Intelligence*, 4 (1991), 225–248.

3. Avron, A.: The method of hypersequents in the proof theory of propositional nonclassical logics. In *Logic: from Foundations to Applications, European Logic Colloquium*, Oxford Science Publications. Clarendon Press. Oxford, 1996, 1–32.

4. Baaz, M., Ciabattoni A., Fermüller, C.G., Veith, H.: On the Undecidability of Some Sub-classical First-Order Logics. Proceedings of FSTTCS'99. LNCS 1738, 1999, 258–268.

5. Baaz, M., Fermüller, C.G., Zach, R.: Systematic Construction of Natural Deduction Systems for Many-valued logics. In: *Proc. 23th International Symposium on Multiple-Valued Logics*. Los Gatos, CA. IEEE Press. 1993, 208-213.

6. Baaz, M., Veith, H.: An Axiomatization of Quantified Propositional Gödel Logic Using the Takeuti-Titani Rule. Proc. Logic Colloquium 1998, Lecture Notes in Logic, ASL, 2000.

7. Baaz, M., Zach, R.: Hypersequents and the Proof Theory of Intuitionistic Fuzzy Logic. *Proc. CSL'2000*, 2000, 187–201.

8. Dyckhoff, R.: A Deterministic Terminating Sequent Calculus for Gödel-Dummett logic. *Logic Journal of the IGPL*, 7 (1999), 319–326.

9. Dummett, M.: A propositional calculus with denumerable matrix. *J. Symbolic Logic*, 24 (1959), 97–106.

10. Gentzen, G.: Untersuchungen über das logische Schliessen I, II. *Mathematische Zeitschrift*, 39 (1934-35), 176–210, 405–431.

11. Gödel, K.: Zum intuitionistischen Aussagenkalkül. *Anz. Akad. Wiss. Wien*, 69 (1932), 65–66.

12. Hájek, P.: *Metamathematics of Fuzzy Logic*. Kluwer, 1998.

13. Horn, A.: Logic with truth values in a linearly ordered Heyting algebra. *J. Symbolic Logic*, 27 (1962), 159–170.

14. Prawitz, D.: *Natural Deduction: A Proof Theoretical Study*. Almquist and Wiksell. 1965.

15. Prawitz, D.: Ideas and results in proof theory. In *Proceedings of the Second Scandinavian Logic Symposium*, North-Holland, Amsterdam, 1971, 235–307.

16. Sonobe, O.: A Gentzen-type Formulation of Some Intermediate Propositional Logics. *J. of Tsuda College*, 7 (1975), 7–14.

17. Tait, W.W.: Normal derivability in classical logic. In: *The Syntax and Semantics of infinitary Languages*. J. Barwise ed., *Lectures Notes in Mathematics*, Springer, Berlin. 72 (1968), 204–236.

18. Takeuti, G., Titani, S.: Intuitionistic fuzzy logic and intuitionistic fuzzy set theory. *J. of Symbolic Logic*, 49 (1984), 851–866.

19. Troelstra, A.S., Schwichtenberg, H.: *Basic Proof Theory*. Cambridge Tracts in Theoretical Computer Science 43. Cambridge, 1996.

Minimal Ideals and the Socle in MV-algebras[*]

Peter L. Belluce[1] and Salvatore Sessa[2]

[1] Department of Mathematics, University of British Columbia
V6T 1Y2, Vancouver, B.C. Canada
[2] Istituto di Matematica, Facoltá di Architettura, Universitá di Napoli
Via Monteoliveto 3, 80134 Napoli, Italy - sessa@.unina.it

Abstract. We borrow from ring theory the well known notions of essential ideal and socle reversing them profitably in a MV-algebra A and proving that the socle is the l.w.b. of the minimals and the g.l.b. of the essentials in the lattice of the ideal of A. We also characterize non-essential prime ideals of A as the isolated points in the hull-kernel topology of the minimals. The above concepts are in particular studied in the MV-algebra of continuous functions from a compact Hausdorff topological space into $[0, 1]$.

1 Preliminaries

MV-algebras [6] and [7] are a non-lattical generalization of Boolean algebras that have their origin in the study of the infinite-valued logic of Lukasiewicz . MV-algebras also appear in relation to abelian l-groups and AF C^*-algebras [14], fuzzy set theory [5] and regular rings [3].

From an axiomatic point of view, MV-algebra + idempotency = Boolean algebra. Unlike Boolean algebras, MV-algebras have no known equivalent ring structure. Nevertheless, being a system with an addition, multiplication and a relative subtraction, we may formally import from ring theory certain useful notions. We intend in this work to exam some of these notions, namely the socle and ideals minimal with respect to set inclusion.

Minimal ideals are also those ideals that have zero intersection with all ideals that do not contain them. This leads us to the dual notion of essential or large ideal, that is to those ideals which have non-zero intersection with all non-zero ideals. Wedged between these two types of ideals is the socle. The socle is the upper bound of the minimals and the greatest lower bound of the essentials in the lattice of the ideals of an MV-algebra.

We shall study how the behavior of the socle relates to the behavior of the algebra.

The socle of an MV-algebra had been introduced in [4] and [11]. The latter work defines the socle in terms of the essential ideals and does not relate it to the minimal ideals. We define the socle [4] in terms of the minimal ideals and show the equivalence of the two definitions.

[*] Research under the auspices of C.N.R. - G.N.S.A.G.A. (Italy).

We assume some familiarity with MV-algebras [5], [6], [14]. We give here some facts concerning usage in this paper. Recall that an MV-algebra is a system $< A, +, \cdot, -, 0, 1 >$, where we have $< A, +, 0 >$ a commutative monoid such that

$$a \cdot b = \overline{(\bar{a} + \bar{b})}, \quad a + \bar{a} = 1, \quad \bar{\bar{a}} = a, \quad \bar{0} = 1,$$

and $a + \bar{a} \cdot b = b + \bar{b} \cdot a$ for all $a, b \in A$.

We write ab for $a \cdot b$ and if n is a positive integer, we put

$$na = a + \ldots + a \ (n \text{ times})$$
$$a^n = a \cdot \ldots \cdot a \ (n \text{ times}).$$

By defining $a \vee b = a + \bar{a}b$, and $a \wedge b = a(\bar{a} + b)$, the induced system $< A, \vee, \wedge, 0, 1 >$ becomes a bounded distributive lattice.

We then have that $< A, +, \leq, 0 >$ becomes an ordered commutative monoid, where $a \leq b$ iff $a \wedge b = a$. If this ordering is linear, A is said linearly ordered.

Spec A will denote the set of prime ideals of A. On Spec A is given the Stone-Zariski topology, where the sets $U(x) = \{P \in \text{Spec } A \mid x \notin P\}$, $x \in A$, form a basis of compact open sets. Max A, Min A denote the subspaces of maximal ideals and minimal primes respectively.

From [7], $\cap \text{Spec } A = \{0\}$. For $M \in \text{Max } A$, the quotient algebra A/M is locally finite, that is, a subalgebra of the MV-algebra $[0, 1]$. For any $P \in \text{Spec}(A)$, A/P is always a linearly ordered algebra.

BA is the subalgebra of idempotents of A; BA is always a Boolean algebra, $U(x)$ is clopen iff $x \in BA$ [5]. At(BA) designates the set of atoms of BA, Rad A is the intersection of all the maximal ideals of A. A is semisimple if Rad $A = \{0\}$, A is semilocal if Max A is finite [5], A is quasi boolean if Spec $A = \text{Max } A$ [2] other facts and notations will be presented in context.

2   Socle of an MV-algebra

Let A be an MV-algebra and $\text{ID}(A)$ the distributive lattice of ideals of A under ideal sum "+" and ideal intersection "$\cap$". If $x \in A$, then $\text{id}(x) = \{a \in A \mid a \leq nx \text{ for some } n\}$ denotes the ideal generated by x. In general, if $\emptyset \neq X$ is a subet of A, $\text{id}(X) = \{a \in A \mid a \leq x_1 + \cdots + x_n \text{ for some } x_1, \ldots, x_n \in X\}$ is the ideal generated by X.

Definition 1. By a *minimal ideal* of A we mean a non-zero ideal $I \subseteq A$ that contains no ideals other than 0 or I. Clearly a minimal ideal is an atom in $\text{ID}(A)$.

Proposition 1. *Let I be a minimal ideal of A. Then*

(i) I is a totally ordered subset of A.

(ii) If $I \cap BA \neq \{0\}$, then there is an $e \in \mathrm{At}(BA)$ such that $I = \mathrm{id}(e)$.

(iii) If $I \cap BA = \{0\}$, then $I^2 = \{0\}$ and $I \subseteq \mathrm{Rad}\, A$.

Proof.

i) Suppose I is not linearly ordered. Hence, there are a, $b \in I$ such that $x = a\bar{b} \neq 0$ and $y = \bar{a}b \neq 0$. By [6], $x \wedge y = 0$. However, x, $y \in I$. Then $I = \mathrm{id}(x)$ and so $y \leq nx$ for some n. Therefore $y = nx \wedge y \leq n(x \wedge y) = 0$.

ii) Let I be a minimal ideal. Suppose $I \cap BA \neq \{0\}$. Then $I = \mathrm{id}(e)$ for some $e \in BA$. Suppose $0 \leq f \leq e$ for some $f \in BA$. Then $I = \mathrm{id}(f)$, so that $e \leq f$. Hence $e \in \mathrm{At}(BA)$.

iii) Suppose now that $I \cap BA = \{0\}$. I is linearly ordered, so $I^{\perp} = \{b \in A \mid a \wedge b = 0 \text{ for all } a \in I\}$ is a prime ideal. By [6], then $A/I^{\perp}$ is linearly ordered. Assume for some $a \in I$ that $na/I^{\perp} = 1 \in A/I^{\perp}$. Then $(\bar{a})^n \in I^{\perp}$ so $na \wedge (\bar{a})^n = na \wedge (\overline{na}) = 0$. But then $na \in I \cap BA$, so $a = 0$. This is not possible. Hence if $a \in I$, we must have $\mathrm{ord}(a/I^{\perp}) = \infty$. Therefore $a^2 \in I^{\perp}$. But then we have $a \wedge a^2 = a^2 = 0$. Thus for all $a \in I$, $a^2 = 0$. Now let a, $b \in I$. $a \leq b$ or $b \leq a$. In either case we obtain $ab = 0$. Thus $I^2 = \{0\}$ and $I \subseteq \mathrm{Rad}(A)$ by [2].

Definition 2. A non-zero ideal $J \subseteq A$ is called *essential* or *large* if $J \cap K \neq \{0\}$ for every ideal $K \subseteq A$, $K \neq \{0\}$.

Clearly I is essential iff $I^{\perp} = \{0\}$.

Proposition 2. *All non-zero ideals of A are essential iff A is linearly ordered.*

Proof. Clearly A linearly ordered implies all non-zero ideals are essential. Suppose then that all non-zero ideals of A are essential. Assume we have x, $y \in A$, $x \not\leq y$, $y \not\leq x$. Then $0 \neq \bar{x}y$, $0 \neq x\bar{y}$. But $x\bar{y} \wedge \bar{x}y = 0$. Let $I = \mathrm{id}(\bar{x}y)$. $z \in I$ iff $z \leq n(\bar{x}y)$ for some n. It follows that $z \wedge x\bar{y} = 0$ for all $z \in I$. Thus $I^{\perp} \neq 0$ and so I is not essential.

Proposition 3. *If A is not linearly ordered and I is minimal, then I is not essential.*

Proof. Suppose I is both minimal and essential. Since 0 is not a prime ideal, one has, for every prime ideal P, that $I \cap P \neq \{0\}$. But then $I \cap P = I$, so $I \subseteq P$ for all primes P. Hence $I = \{0\}$ and so is neither minimal nor essential.

Definition 3.

$$\mathrm{Soc}\, A = \sum \{I : I \quad \text{a minimal ideal of } A\}$$

$$\mathrm{Ess}\, A = \bigcap \{J : J \quad \text{an essential ideal of } A\}.$$

We note that $\mathrm{Soc}\, A = \{0\}$ iff A contains no minimal ideals. $\mathrm{Soc}\, A$ is called the *socle* of A.

Next we shall show that $\mathrm{Soc}\, A = \mathrm{Ess}\, A$. We adapt the proof of the similar Proposition in [10].

Definition 4. Suppose I, K are ideals of A and that $I + J = K$ for some ideal J with $I \cap J = \{0\}$. Then we call I a *direct summand* of K. We write $I \dotplus J = K$.

Proposition 4. *Suppose I is an ideal of A, $I \subseteq \mathrm{Ess}\, A$. Then I is a direct summand of* $\mathrm{Ess}\, A$.

Proof. Let $K \subseteq \mathrm{Ess}\, A$ be an ideal, maximal with respect to $K \cap I = \{0\}$. Set $H = I \dotplus K$. Let J be any ideal of A and assume that $H \cap J = \{0\}$. Let $x \in J - K$. Then $I \cap (K + \mathrm{id}(x)) \neq \{0\}$. Choose a $y \in I \cap (K + \mathrm{id}(x)), y \neq 0$. Thus $y \leq a + nx$ for some $a \in K$, $n > 0$. $y \in I$, so $y \in H$ as well. $y \leq y \wedge a + y \wedge nx = y \wedge nx \leq nx$. Hence $y \in J$, which is impossible. It follows that H is essential and we have $H = \mathrm{Ess}\, A$.

Proposition 5. *If I, J are ideals, $\{0\} \neq I \subseteq J \subseteq \mathrm{Ess}\, A$, then I is a direct summand of J.*

Proof. From Proposition 4 we know there is an ideal K such that $I \dotplus K = \mathrm{Ess}\, A$, $I \cap K = \{0\}$. Thus $J = J \cap (I + K) = I + (J \cap K)$, $I \cap (J \cap K) = \{0\}$. Therefore $J = I \dotplus (J \cap k)$.

Thus the condition on $\mathrm{Ess}\, A$ expressed in Proposition 4 is hereditary.

Proposition 6. *Suppose $\mathrm{Ess}\, A \neq \{0\}$. Then $\mathrm{Soc}\, A \neq \{0\}$.*

Proof. Let $x \in \mathrm{Ess}\, A$, $x \neq 0$. Choose an ideal $I \subseteq \mathrm{Ess}\, A$ maximal with respect to $x \notin I$. By Proposition 4, there is an ideal K with $\mathrm{Ess}\, A = I \dotplus K$. $K \neq \{0\}$ since $x \notin I$. Suppose J is an ideal, $\{0\} \neq J \subset K$. By Proposition 5, there is an ideal J' with $K = J \dotplus J'$. Then we have, since $J \cap J' = \{0\}$, $I = (I + J) \cap (I + J')$. But $x \in I + J$ and $x \in I + J'$, so $x \in I$. It follows that K is a minimal ideal and, thus, $\mathrm{Soc}\, A \neq \{0\}$.

Proposition 7. $\mathrm{Soc}\, A = \mathrm{Ess}\, A$.

Proof. If I is a minimal ideal and J an essential ideal, then $I \cap J \neq \{0\}$. Hence $I \subseteq J$ and we have $\mathrm{Soc}\, A \subseteq \mathrm{Ess}\, A$. If $\mathrm{Ess}\, A = \{0\}$, then $\mathrm{Soc}\, A = \{0\}$. Suppose then that $\mathrm{Ess}\, A \neq \{0\}$. By Proposition 6, $\mathrm{Soc}\, A \neq \{0\}$. By Proposition 4, we also know $\mathrm{Ess}\, A = \mathrm{Soc}\, A \dotplus K$ for some ideal K. Suppose there is an ideal J, $\{0\} \neq J \subset K$. Then there is an ideal J' with $K = J \dotplus J'$. Assume $x \in K$, $x \neq 0$. Let L be an ideal in K, maximal with respect to $x \notin L$. We then have $L = (L + J) \cap (L + J')$. From this we get $x \in L$, which is absurd. Hence if K is not $\{0\}$, it must be minimal. Thus $K = \{0\}$.

Let us consider which essential ideals exist in $\mathrm{Spec}\, A$. For the sake of completeness, we include the proof of the following Proposition from [4]. We have

Proposition 8. *Let P be a non-essential prime ideal. Then*

(i) $P = a^{\perp} = \{b \in A \mid b \wedge a = 0\}$ *for some* $a \in A$ *such that* $\{x \in A \mid 0 \leq x \leq a\}$ *is a linearly ordered set;*

(ii) $P \in \text{Min } A$;

(iii) *if* $P \in \text{Max } A$, *then* $P^{\perp} = \text{id}(e)$ *for some* $e \in \text{At}(BA)$ *and* $P = \text{id}(\bar{e})$.

Proof.

(i) Let $0 \neq a \in P^{\perp}$. Then $P \subseteq a^{\perp}$. Since $a \neq 0, a \notin P$. Therefore, as P is prime, $a^{\perp} \subseteq P$ and we have $P = a^{\perp}$. Now let $0 \leq x, y \leq a$ and assume $x\bar{y} \neq 0$, $\bar{x}y \neq 0$. Since P is prime, we know $x\bar{y} \in P$ or $\bar{x}y \in P$. If $x\bar{y} \in P$, then we obtain $0 \neq x\bar{y} = x\bar{y} \wedge a = 0$. Similarly if $\bar{x}y \in P$. So we must have $x \leq y$ or $y \leq x$.

(ii) Suppose Q is a prime ideal and that $Q \subseteq P = a^{\perp}$. Then $(a^{\perp})^{\perp} \subseteq Q^{\perp}$, so $a \in Q^{\perp}$. As in part i), we may infer $Q = a^{\perp}$; thus $Q = P$ and we see that $P \in \text{Min } A$.

(iii) If $P \in \text{Max } A$, then as $a \notin P$, we must have $\bar{a}^n \in P$ for some integer n [6]. That is, $\overline{na} \in P$. But $na \in P^{\perp}$ and we have $na \wedge \overline{na} = 0$; thus $na \in BA$. Let $e = na$. Clearly $a^{\perp} = e^{\perp}$, so that $P = e^{\perp}$. Now let $0 < b \leq e$, $b \in BA$. Since $\bar{e} \leq \bar{b}$, we get $P \subseteq \text{id}(\bar{b})$ which implies, since P is maximal, $P = \text{id}(\bar{b})$. Therefore, $\bar{b}e = 0$ so that $\bar{b} \leq \bar{e}$. It follows that $b = e$ and that $e \in \text{At}(BA)$.

Proposition 9. *Suppose* $M \in \text{Max } A$ *is non-essential. Then* $M^{\perp}$ *is a minimal ideal generated by an atom of* BA.

Proof. We know from Proposition 8(iii) that $M^{\perp}$ is an ideal generated by some $e \in \text{At}(BA)$. Now let $x \in M^{\perp}$, $x \neq 0$. $x \notin M$, so there is an n with $\bar{x}^n \in M$. Hence $e \wedge \bar{x}^n = 0$ and we have $e \leq nx$. Thus $e = nx$ and we infer $\text{id}(x) = M^{\perp}$. Therefore $M^{\perp} = \text{id}(e)$ is a minimal ideal.

Corollary 1. *There is a bijection between the non-essential maximal ideals of* A *and the minimal ideals of* A *that are not in* $\text{Rad } A$.

Proof. From Proposition 9, we see there is an injection from the set of non-essential maximal ideals to minimal ideals that are not in $\text{Rad } A$. Let I be a minimal ideal which is not in $\text{Rad } A$. Then $I = \text{id}(e)$ for some $e \in \text{At}(BA)$ by Proposition 1 (ii).

Then $I^{\perp} = \text{id}(\bar{e})$ and $(I^{\perp})^{\perp} = I$. Let M be a maximal ideal that contains $I^{\perp}$. If M is essential, then, as I is minimal, $I \subseteq M$. Thus $e + \bar{e} = 1 \in M$, absurd.

Therefore $\{0\} \neq M^{\perp} \subseteq (I^{\perp})^{\perp} = I$, hence $M^{\perp} = I$.

Proposition 10. *If* $\text{Min } A$ *is finite,* $\text{Spec } A - \text{Min } A$ *is the set of essential prime ideals.*

Proof. Let $m_1, \ldots, m_n$ be the minimal primes. Then $m_2 \cap \cdots \cap m_n \neq \{0\}$. But $m_1 \cap m_2 \cap \cdots \cap m_n = \{0\}$, hence m_1 is not essential. Similarly for $m_2, \ldots, m_n$.

Two natural questions arise. One, when do there exist essential minimal primes? Two, when are all maximal ideals essential?

Another question is this: we know, by Proposition 8 (ii), all primes are essential iff all minimal primes are essential. This implies $\operatorname{Soc} A = \{0\}$. Is the converse true?

Observe that from Proposition 8 (iii) that if there exists a non-essential maximal ideal, then $\operatorname{Soc} A \neq \{0\}$.

We saw above there are MV-algebras A (for instance, if $\operatorname{Min} A$ is finite) in which all maximal ideals are essential. We will now show there are MV-algebras in which no maximal ideal is essential. First,

Proposition 11. *Suppose $M \in \operatorname{Max} A$ is non-essential. Then M is an isolated point of $\operatorname{Spec} A$.*

Proof. From Proposition 8 (iii), $M = \operatorname{id}(e)$ for some $\bar{e} \in \operatorname{At}(BA)$. Thus $M \in U(\bar{e})$. If $P \in U(\bar{e})$, then $e \in P$. Hence $M \subseteq P$. It follows that $U(\bar{e}) = \{M\}$ is clopen and M is an isolated point of $\operatorname{Spec} A$.

Corollary 2. *Let $\mathcal{M}$ be the set of non-essential maximal ideals. Then $\mathcal{M}$ is open in $SpecA$.*

Corollary 3. *If no maximal ideals of A are essential, then A is semilocal and quasi-boolean, that is, A is subalgebra of $[0,1]^n$ for some positive integer n.*

Proof. Each $M \in \operatorname{Max} A$ is isolated; since $\operatorname{Max} A$ is compact, we must have $\operatorname{Max} A$ finite, so A is semilocal. From Proposition 8 (ii), we see that each $M \in \operatorname{Max} A$ must be a minimal prime, thus all primes are maximal, that is A is quasi-boolean. The remaining part of the thesis follows from [5].

Proposition 12. *Suppose $\operatorname{Soc} A$ is essential. Then the set of non-essential prime ideals of A is dense in $\operatorname{Spec} A$.*

Proof. Let $\mathcal{N}$ be the set of non-essential prime ideals. Since $\operatorname{Soc} A$ is essential, it is not zero. Hence there must be a prime ideal which is not essential. Thus $\mathcal{N} \neq \emptyset$. Let $W = \cap \mathcal{N}$. It suffices to prove that $W = \{0\}$. Suppose not. Let $E = \cap(\operatorname{Spec} A - \mathcal{N})$. As $\operatorname{Soc} A$ is essential, we must have $W \cap \operatorname{Soc} A \neq \{0\}$. But $W \cap \operatorname{Soc} A \subseteq W \cap E = \cap \operatorname{Spec} A = \{0\}$. This contradiction shows that $W = \{0\}$.

Note that $\mathcal{N} \subseteq \operatorname{Min} A$ by Proposition 8 (ii).

For some algebras, necessarily semisimple, we have a converse.

Proposition 13. *If $\mathcal{M}$ is dense in $\operatorname{Spec} A$, then $\operatorname{Soc} A$ is essential.*

Proof. Suppose $\mathcal{M}$ is dense in $\operatorname{Spec} A$. Then $\mathcal{M} \neq \emptyset$. Moreover we must have $\cap \mathcal{M} = \{0\}$ and so A is semisimple. Let I be a non-zero ideal and let $0 \neq x \in I$. The open set $U(x) \subseteq \operatorname{Spec} A$ is non-empty, thus $U(x) \cap \mathcal{M} \neq \emptyset$.

Let $M \in U(x) \cap \mathcal{M}$. We know $M = \mathrm{id}(\bar{e})$, $M^{\perp} = \mathrm{id}(e)$ for some $e \in \mathrm{At}(BA)$ by Proposition 8 (iii). Also $M^{\perp} \subseteq \mathrm{Soc}\, A$. Now $xe \in I$. If $xe = 0$, we would have $x \leq \bar{e}$ and so $x \in M$. Thus $0 \neq xe \in M^{\perp}$ and it follows that $\mathrm{Soc}\, A$ is essential.

Suppose now that A is semilocal and quasi-boolean, that is, for some $n > 0$, $A \subseteq [0,1]^n$. Assume that some $M \in \mathrm{Max}\, A$ is essential. Let $M_1, \ldots, M_n$ be all the maximal ideals different than M. Then $M_1 \cap \cdots \cap M_n \neq \{0\}$. Then we have $M \cap M_1 \cap \cdots \cap M_n \neq \{0\}$, which is impossible since A is semisimple. Thus

Proposition 14. *No maximal ideal of A is essential iff A is a subalgebra of $[0,1]^n$ for some integer $n > 0$.*

Note the above is equivalent to saying that $\mathrm{Soc}\, A = A$. Suppose $\mathrm{Soc}\, A \neq \{0\}$. Let $\mathcal{S} = \mathrm{id}(\mathrm{Soc}\, A) \in \mathrm{ID}(A)$. By Proposition 4, we know $\mathcal{S}$ is complemented, thus it is a complemented distributive lattice under "$\cap$" and "$+$". Then

Proposition 15. *If $\mathrm{Soc}\, A \neq \{0\}$, then $< \mathcal{S}, +, \cap, 0, \mathrm{Soc}\, A >$ is a Boolean algebra.*

Corollary 4. *If $\mathrm{Soc}\, A = A$, then $\mathrm{ID}(A)$ is a finite Boolean algebra.*

Conversely, suppose that $\mathrm{ID}(A)$ is a finite Boolean algebra. Let $I \in \mathrm{ID}(A)$. I has a complement J, so that $I \dotplus J = A$. Therefore, if $I \neq A$ then $I^{\perp} \neq \{0\}$. Hence the only essential ideal is A and we may infer that $\mathrm{Soc}\, A = A$. Thus,

Proposition 16. *$\mathrm{ID}(A)$ is a Boolean algebra iff $\mathrm{Soc}\, A = A$.*

From the preceding results, we see

Proposition 17. *If A is quasi-boolean and $\mathrm{Max}\, A$ is infinite, then there exist essential minimal prime ideals.*

Similar to Proposition 11 we have,

Proposition 18. *If $m \in \mathrm{Min}\, A$ is not essential, then m is an isolated point of $\mathrm{Min}\, A$.*

Proof. By Proposition 8 (i), we know $m = a^{\perp}$ for some $a \in A$. Now $U(a) \cap \mathrm{Min}\, A = \{P \in \mathrm{Min}\, A : a^{\perp} \subseteq P\}$. Thus, if $m' \in U(a) \cap \mathrm{Min}\, A$, then $m = a^{\perp} \subseteq m'$. Hence $U(a) \cap \mathrm{Min}\, A = \{m\}$ and it follows that m is an isolated point of $\mathrm{Min}\, A$.

Corollary 5. *$\mathrm{Min}\, A$ compact and infinite implies there exist essential minimal prime ideals.*

Corollary 6. *The set $\mathcal{N}$ of non-essential (minimal) prime ideals is open in $\mathrm{Min}\, A$.*

Proposition 19. *Let X be a dense subpace of $\operatorname{Spec} A$. If $P \in X$ is an isolated point of X, then P is a non-essential prime ideal.*

Proof. Assume $\{P\} = U(I) \cap X$ for some ideal $I \neq \{0\}$, where $U(I) = \{P' \in \operatorname{Spec} A : I \not\subseteq P'\}$. Therefore, $\{0\} \neq I \subseteq \bigcap\{P' \mid P' \in X, P' \neq P\}$. But then $P \cap I = \{0\}$ and so P is not essential.

Corollary 7. *If $m \in \operatorname{Min} A$ is an isolated point of $\operatorname{Min} A$, then m is not essential.*

Similarly,

Corollary 8. *Suppose $P \in \operatorname{Spec} A$ is an isolated point of $\operatorname{Spec} A$. Then P is not essential.*

We see, since $\cap \operatorname{Min} A = \{0\}$ and therefore is a dense subspace of $\operatorname{Spec} A$, that the non-essential prime ideals coincide with the isolated points of $\operatorname{Min} A$. We would like to know if there are MV-algebras A with $\operatorname{Min} A$ infinite, non-compact, and no minimal prime ideal essential? The following example answers this question in the affirmative.

To begin, note that no minimal ideal is essential iff all points of $\operatorname{Min} A$ are isolated in $\operatorname{Min} A$. That is, $\operatorname{Min} A$ has the discrete topology. The converse is true as well; if $\operatorname{Min} A$ has the discrete topology, then no minimal prime is essential.

Consider the following [4],

Example 1. Let $\mathbf{C}$ be the usual Chang algebra [6], $\mathbf{N} = \{1, 2, \ldots\}$. Let $A_1 = \mathbf{C}^{\mathbf{N}}$. Let $A_2 = \{x \in A_1 \text{ for all } n, \operatorname{ord} x_n = \infty \text{ or } \operatorname{ord} x_n < \infty\}$. A_2 is a subalgebra of A_1. Now let $I = \{x \in A_2 \mid x_n = 0 \text{ almost everywhere}\}$. I is an ideal in A_2. Let A be the subalgebra of A_2 generated by I. In [4] it is shown that if $m \in \operatorname{Min} A$, then $\{m\}$ is open, and $\operatorname{Min} A$ is infinite and non-compact. It follows that in A no minimal prime ideal is essential. We note that A is a local MV-algebra, that is has a unique maximal ideal, namely I.

We want to consider now the relation between $\operatorname{Soc} A$ and $\operatorname{Rad} A$. We first show,

Proposition 20. $\operatorname{Soc} A \subseteq \operatorname{Rad} A$ *iff* $\operatorname{Soc} A \cap \operatorname{At}(BA) = \emptyset$.

Proof. The statement is true if $\operatorname{Soc} A = \{0\}$. Suppose then that $\operatorname{Soc} A \neq \{0\}$ and $\operatorname{Soc} A \subseteq \operatorname{Rad} A$. Clearly $\operatorname{Soc} A \cap BA = \{0\}$ and therefore, $\operatorname{Soc} A \cap \operatorname{At}(BA) = \emptyset$. Suppose now that $\operatorname{Soc} A \cap \operatorname{At}(BA) = \emptyset$. If I is a minimal ideal, we must then have $I \cap BA = \{0\}$. By Proposition 1 (iii), we see that $I^2 = \{0\}$ and $I \subseteq \operatorname{Rad} A$. It follows that $\operatorname{Soc} A \subseteq \operatorname{Rad} A$.

Corollary 9. *If BA is atomless, then $\operatorname{Soc} A \subseteq \operatorname{Rad} A$.*

Corollary 10. *If A is semisimple and I a minimal ideal, then $I = \operatorname{id}(e)$ for some $e \in \operatorname{At}(BA)$. Consequently $\operatorname{Soc} A$ is generated by atoms of BA.*

Corollary 11. *If A is semisimple and BA atomless, then $\operatorname{Soc} A = \{0\}$.*

In the above corollary $\operatorname{Soc} A = \operatorname{Rad} A$. We may ask in general when does this happen or when does $\operatorname{Rad} A \subseteq \operatorname{Soc} A$? The latter clearly happens when A is semisimple. Otherwise the relation between $\operatorname{Soc} A$ and $\operatorname{Rad} A$ seems weak as the following examples show.

Example 2. Let A be the MV-algebra of sequences $< x_n >$, where $x_n \in [0,1]$ and each sequence constant almost everywhere. Let $M_i = \{< x_n >| x_i = 0\}$. Then M_i is a maximal ideal and $\bigcap_i M_i = \{0\}$. Therefore $\operatorname{Rad} A = \{0\}$. Let $F \subseteq A$ be the ideal consisting of the almost everywhere 0 sequences. Let $I_j = \{< x_n >| x_n = 0, n \neq j\}$. Then I_j is a minimal ideal of A, $I_j \subseteq F$. F is essential and $\sum_j I_j = F$. Therefore $\operatorname{Soc} A = F$.

Example 3. Let C be the Chang algebra, and let A be the algebra of sequences $< x_n >$, $x_{2n} \in C$, $x_{2n+1} \in [0,1]$, where there is a k such that for $j > 0$ we have $x_{2(k+j)} = x_{2k}$ and $x_{2(k+j)+1} = x_{2k+1}$. Let $M =< c >$ be the maximal ideal of C. Let F be the ideal of almost everywhere 0 sequences. Let $I_{2j} = \{< x_n >| x_{2j} \in M, x_n = 0, n \neq 2j\}$. Let $I_{2j+1} = \{< x_n >| x_{2j+1} \in [0,1], x_n = 0, n \neq 2j + 1\}$. The I_n are minimal ideals and every minimal ideal is some I_n. F is essential and as before, $\operatorname{Soc} A = F$. $\operatorname{Rad} A \subseteq \bigcap_i M_i$ where $M_{2i} = \{< x_n >| x_{2i} \in M\}$ and $M_{2i+1} = \{< x_n >| x_{2i+1} = 0\}$. In fact $\operatorname{Rad} A = \bigcap_i M_i$. As the alternating sequence $< x_n >\in \operatorname{Rad} A$, where $x_{2j} = c, x_{2j+1} = 0$, we see that $\operatorname{Rad} A \not\subseteq \operatorname{Soc} A$. But $\operatorname{Soc} A \not\subseteq \operatorname{Rad} A$ since $\operatorname{Soc} A \cap BA \neq \{0\}$.

3 MV-algebras of continuous functions

In the remainder of this paper we will study minimal ideals, essential ideals, and the socle in an MV-algebra, $A = \mathcal{C}(X, [0,1])$, of continuous functions from a topological space X to $[0,1]$.

A is an MV-algebra under pointwise operations [9]; $[0,1]$ is given its natural topology.

The results presented below are similar to results known in ring theory [1], [11] and [13]; we will present proofs in the MV-theoretic context. An advantage in the theory of MV-algebras consists in the fact that X can be assumed to be compact. Indeed, as pointed out in [8], X can be taken to be a completely regular Hausdorff space. Now X is dense in its Stone-Czech compactification βX [11], so $\mathcal{C}(X, [0,1])$ is isomorphic to $\mathcal{C}(\beta X, [0,1])$. Recall that a compact Hausdorff space is completely regular.

For $x \in X$, let $M_x = \{f \in A \mid f(x) = 0\}$. M_x is clearly a maximal ideal of A.

As X is assumed to be compact and Hausdorff, therefore completely regular, it's easy to see, as in in the case of rings, that the mapping $x \to M_x$ is a continuous bijection from X to $\operatorname{Max} A$. Since $\operatorname{Max} A$ is compact and Hausdorff, we have the following relation between x and M_x [11]:

Proposition 21. *The map* $x \to M_x : X \to \text{Max} \, A$ *is a homeomorphism.*

Proposition 22. x *is an isolated point of* X *iff* M_x *is a non-essential maximal ideal.*

Proof. x is isolated in X iff M_x is isolated in Max A. A is semisimple, hence Max A is dense in Spec A. By Proposition 19, M_x is non-essential. Conversely, suppose M_x is non-essential. By Proposition 11, M_x is an isolated point of Spec A, thus of Max A. It follows that x is isolated of X.

Note that in the MV-algebra $C([0,1],[0,1])$ all maximal ideals are essential.

By Corollary 1, we have

Corollary 12. *There exists a bijection between the isolated points of* X *and the minimal ideals of* A.

Another relation between the set of isolated points of X and the MV-algebra A is,

Proposition 23. *Suppose* I *is a minimal ideal of* A. *Then there is a isolated point* $x_0 \in X$ *such that* $I = \{f \in A \mid f(x) = 0 \text{ for all } x \in X, x \neq x_0\}$. *Moreover* x_0 *is unique.*

Proof. Let I be a minimal ideal. From Corollary 1, we know there is a non-essential maximal ideal M such that $M^\perp = I$. We also know that $M = \text{id}(e)$ for some idempotent e with $\bar{e} \in \text{At}(BA)$. By Proposition 22, $M = M_{x_0}$ for some unique isolated point $x_0 \in X$. Let $g : X \to [0,1]$ be defined by $g(x_0) = 0$ and $g(x) = 1$, $x \neq x_0$. Then $g \in A$, hence $g \in BA$; moreover, $\bar{g} \in M^\perp = I = \text{id}(\bar{e})$. Since $\bar{e} \in \text{At}(BA)$, we may infer that $\bar{e} = \bar{g}$. It now follows that if $f \in I$, then $f(x) = 0$ for all $x \neq x_0$.

Corollary 13. $\text{Soc} \, A = \{f \mid f = 0 \text{ except on finitely many isolated points}\}$.

Proof. If $f \in \text{Soc} \, A$, then $f \in I_{x_1} + \cdots + I_{x_n}$, where each I_{x_j} is the minimal ideal determined by the isolated point x_j. Thus $f \leq f_1 + \cdots + f_n$, $f_j \in I_{x_j}$. If $x \notin \{x_1, \ldots, x_n\}$, each $f_j(x) = 0$ and so $f(x) = 0$. Conversely, suppose $\{x_1, \ldots, x_n\}$ are isolated points and $f(x) = 0$ for $x \notin \{x_1, \ldots, x_n\}$. Let I_{x_j} be the minimal ideal determined by x_j. Let e_j be an idempotent generator for I_{x_j}, so that $e_j(x_j) = 1$. Then $(e_1 + \cdots + e_n)(x_j) = 1$. Clearly, $f \leq e_1 + \cdots + e_n \in \text{Soc} \, A$.

Corollary 14. $\text{Soc} \, A$ *is quasi-boolean, (that is, for* $f \in \text{Soc} \, A$ *there is an integer* n *such that* $nf \in BA$).

Proposition 24. *The set of isolated points of* X *is dense iff* $\text{Soc} \, A$ *is an essential ideal.*

Proof. Suppose first that the set of isolated points is a dense subset of X. Let I be a non-zero ideal and assume $\operatorname{Soc} A \cap I = \{0\}$. Let $f \in I$, $f \neq 0$. If $g \in \operatorname{Soc} A$, then $f \wedge g = 0$. Let $\mathcal{O} = f^{-1}((0,1])$. $\mathcal{O} \neq \emptyset$ and is open in X. Therefore, there is an isolated point $x \in \mathcal{O}$. x determines a minimal ideal $I_x \subseteq \operatorname{Soc} A$. $g \in I_x$ iff $g(X - \{x\}) = 0$. Thus if $g \in I_x$, $g \neq 0$ we have $(f \wedge g)(x) \neq 0$ which is absurd. Conversely, suppose $\operatorname{Soc} A$ is essential. Let $\mathcal{O}$ be a non-empty open subset of X. Let $f \in \operatorname{Soc} A$. Then, from Corollary 13 $f(x) = 0$ for all non-isolated points of X. By complete regularity, there is a non-zero function $g \in A$ such that $g(X - \mathcal{O}) = 0$. Let $I = \{h \in A \mid h(X - \mathcal{O}) = 0\}$. Then I is a non-zero ideal of A. Thus $\operatorname{Soc} A \cap I \neq \{0\}$. Let $h \in \operatorname{Soc} A \cap I$, $h \neq 0$. Then $h = 0$ on $\mathcal{O}$ and $h = 0$ on $X - \mathcal{O}$, so $h = 0$, which is absurd.

Proposition 25. *Let* $m \in \operatorname{Spec} A$, *$m$ non-essential. Then* $m \in \operatorname{Max} A \cap \operatorname{Min} A$.

Proof. By Proposition 8 (ii), we know $m = f^{\perp} \in \operatorname{Min} A$ for some $f \in A$. Thus if $m' \in \operatorname{Min} A$, $m' \neq m$, we must have $f \in m'$. If m is not maximal, then $m \subset M_{x_0}$ for some $x_0 \in X$. Let $x \in X$, $x \neq x_0$. If $m' \subseteq M_x$, $m' \in \operatorname{Min} A$, then $m' \neq m$. Hence $f \in m' \subseteq M_x$ and so $f(x) = 0$. As $f \neq 0$, we see that $f^{-1}((0, 1]) = \{x_0\}$. Since f is continuous, it follows that $\{x_0\}$ is open and so x_0 is an isolated point. But then, from Proposition 22, M_{x_0} is a non-essential maximal ideal. From Proposition 8 (ii), we infer that M_{x_0} is a minimal ideal, thus $M_{x_0} = m$, which is absurd.

Letting $Z(f)$ be the zero-set of $f \in A$, we have another characterization of essential ideals as

Proposition 26. *Let I be an ideal of A. Then I is essential iff $\bigcap_{f \in I} Z(f)$ has empty interior.*

Proof. First, assume I is essential and $U \neq \emptyset$ is contained in $\bigcap_{f \in I} Z(f)$, U open in X. Let $x \in U$. Since X is completely regular, there is a $g \in A$ with $g(x) \neq 0$, $g(X - U) = 0$. Thus, if $f \in I$, then $f \wedge g = 0$. As $g \neq 0$, we see that $I^{\perp} \neq \{0\}$, which is absurd. Conversely, suppose $I^{\perp} \neq \{0\}$. Let $0 \neq g \in I^{\perp}$. Then $g^{-1}((0,1]) \subseteq \bigcap_{f \in I} Z(f)$, so that $\bigcap_{f \in I} Z(f)$ has non-empty interior.

The next result shows that sometimes the sets $Z(f)$ are clopen.

For $x \in X$, let $\mathcal{O}_x = \{f \in A \mid Z(f) \text{ is a neighborhood of } x\}$. Let $\mathcal{I}$ be the set of isolated points of X. Then,

Proposition 27. *If $f \in \bigcap_{x \notin \mathcal{I}} \mathcal{O}_x$, then $Z(f)$ is open, and therefore clopen.*

Proof. Let $x_0 \in Z(f)$. If x_0 is an isolated point, then the open set $\{x_0\} \subseteq Z(f)$. Assume then that x_0 is not in $\mathcal{I}$. Hence, $f \in \mathcal{O}_{x_0}$. Thus there an open set U with $x_0 \in U$, and $U \subseteq Z(f)$. It follows that $Z(f)$ is open.

The sets $\mathcal{O}_x$ have other properties. Note first that the zero function of A is in $\mathcal{O}_x$. Since, evidently, $Z(f+g) = Z(f) \cap Z(g)$, we see that $\mathcal{O}_x$ is always an ideal. Moreover,

Proposition 28. *If $x \notin \mathcal{I}$, then $\mathcal{O}_x$ is an essential ideal.*

Proof. Assume $x \notin \mathcal{I}$. Let $W = \bigcap_{f \in \mathcal{O}_x} Z(f)$. Clearly $x \in W$. Suppose $y \in W$, $y \neq x$. Since X is Hausdorff and completely regular, there exists an open set U with $x \in U$, $y \notin U$ and a continuous function $g \in A$ such that $g(U) = 0$ and $g(y) \neq 0$. Since $x \in U \subseteq Z(g)$, we have $g \in \mathcal{O}_x$. As $y \in W$, we obtain $y \in Z(g)$, which is absurd. Hence $W = \{x\}$. Since x is not an isolated point, the interior of W is empty. By Proposition 26, we may conclude that $\mathcal{O}_x$ is essential.

Proposition 29. Soc $A = \bigcap_{x \notin \mathcal{I}} \mathcal{O}_x$ iff for any $h \in BA$ such that $X - \mathcal{I} \subseteq Z(h)$, we have $Z(h)$ cofinite.

Proof. Suppose first that Soc $A = \bigcap_{x \notin \mathcal{I}} \mathcal{O}_x$ and that $h \in BA$, is such that $X - \mathcal{I} \subseteq Z(h)$. As $h \in BA$, we know that $Z(h) = X - Z(1 - h)$ and so is clopen. Let $x \notin \mathcal{I}$. Then $x \in Z(h)$; since $Z(h)$ is open, we see that $h \in \mathcal{O}_x$. Hence $h \in$ Soc A. By Corollary 13, $X - Z(h)$ is finite. Assume now, the converse condition. By Proposition 26, we see Soc $A \subseteq \bigcap_{x \notin \mathcal{I}} \mathcal{O}_x$. Let $g \in \bigcap_{x \notin \mathcal{I}} \mathcal{O}_x$. Then $Z(g)$ is clopen by Proposition 27. Thus, the function h such that $h(Z(g)) = 0$, $h(X - Z(g)) = 1$ is continuous and lies in BA. Clearly, $Z(h) = Z(g)$. It is also clear that $X - \mathcal{I} \subseteq Z(g)$. Since $Z(g) = Z(h)$, we have, by the assumed condition on $Z(h)$, that $Z(g)$ is cofinite. By Corollary 13, we may infer that $g \in$ Soc A.

The next, and final, result shows that under rather mild restrictions there always exist non-prime essential ideals.

Proposition 30. *Suppose the cardinality of X is at least 3 and that X has at least two non-isolated points, x, y. Then $I = \{f \in A \mid x, y \in Z(f)\}$ is a non-prime essential ideal.*

Proof. Clearly, I is an ideal. Let $z \in X$, $z \neq x$, $z \neq y$. Thus, there is an $f_1 \in A$ with $f_1(x) = 0$, $f_1(z) \neq 0$. Similarly, there is an $f_2 \in A$, $f_2(y) = 0$, $f_2(z) \neq 0$. Hence, the function $f_1 \wedge f_2 \in I$ and $f_1 \wedge f_2 \neq \{0\}$. Therefore $I \neq \{0\}$. Now let U, V be disjoint open sets, $x \in U$, $y \in V$. There are f, $g \in A$ such that $f(U) = 0$, $f(y) \neq 0$ and $g(V) = 0$, $g(x) \neq 0$. We see, then, that $f \wedge g \in I$ but $f \notin I$, $g \notin I$, and so, I is not prime. Since x, y are not isolated points, we know from Proposition 28 that $\mathcal{O}_x$ and $\mathcal{O}_y$ are essential ideals. But clearly, $\mathcal{O}_x \cap \mathcal{O}_y \subseteq I$, and thus I is essential.

Acknowledgment 1. Thanks are indebited to the referee for useful suggestions.

References

1. F. Azarpanah, *Essential ideals in $C(X)$*, Per. Math. Hung. **31**, (2) (1995), 105–112.

2. L. P. Belluce, *Semisimple and complete MV-algebras*, Alg. Univ. **29**, (1992), 1–9.

3. L. P. Belluce, *Regular rings and MV-algebras*, preprint.

4. L. P. Belluce, A. Di Nola and S. Sessa, *The prime spectrum of an MV-algebra*, Math. Logic Quarterly, **40**, (1994), 331–346.

5. L. P. Belluce and S. Sessa, *Orthogonal decompositions of MV-spaces*, Mathware and Soft Computing, **4**, (1997), 5–22.

6. C. C. Chang, *Algebraic analysis of many valued logics*, Trans. Amer. Math. Soc., **88**, (1958), 467–490.

7. C. C. Chang, *A new proof of the completeness of the Łukasiewicz axioms*, Trans. Amer. Math. Soc., **93**, (1959), 74–80.

8. R. Cignoli and A. Torrens, *Boolean products of MV-algebras: Hypernormal MV-algebras*, J. Math. Anal. and Appl. **193**, (1996), 637–653.

9. A. Di Nola and S. Sessa, *MV-algebras of continuous functions*, in "Non-Classical Logics and Their Applications to Fuzzy Subsets: A Handbook of the Foundations of Fuzzy Set Theory", (U. Höhle and E. P. Klement, Eds.), Kluwer, Boston, (1994), 23–32.

10. Carl Faith, *Ring theory*, Springer-Verlag, New York (1976).

11. L. Gillman and M. Jerison, *Rings of continuous functions*, Springer-Verlag, New York (1976).

12. C. S. Hoo, *Maximal and essential ideals in MV-algebras*, Mathware and Soft Computing, Vol. II, No. 3, (1995), 181–96.

13. O. A. S. Karamzadeh and M. Rostami, *On the intrinsic topology and some related ideals of $C(X)$*, Proc. Amer. Math Soc. **93**, no. 1 (1985), 173–184.

14. D. Mundici, *An interpretation of AF C^*-algebras in Łukasiewicz sentential calculus*, J. Funct. Anal., **98**, (1986), 15–63.

Industrial Applications of Soft Computing

Marco Branciforte[1], Riccardo Caponetto[1], Mario Lavorgna[1], and Luigi Occhipinti[1]

Soft Computing Group, STMicroelectronics, Stradale primosole 50, 95100 Catania, Italy, e-mail: riccardo.caponetto@st.com

Abstract. In this paper two industrial applications of soft computing methodologies, developed by STMicroelectronics, are described. The main idea, in designing dedicated ICs based on soft computing paradigm, is to produce competitive devices characterized by high MIQ (Machine Intelligence Quotient).

1 Introduction

Soft computing, according to the definition given in [1], is mainly the combination of fuzzy logic FL, neural networks NNs and genetic algorithm GAs. The strength of the soft computing approach lies in its capacity to deal with uncertainty and imprecision. Two fundamental aspects distinguish the combination of the three methodologies: time complexity increases while flowing from FL to NNs and finally in GAs while a-priori knowledge decreases in the opposite direction. In the paper [1] the author outlines, also, the need to design and produce machines characterized by an higher MIQ. With this in view, the paradigm of soft computing can enlarged, adding to the original framework new methodologies that can enrich the known-how embedded into the machine. As consequence, in the scheme proposed in figure 1, two new methodologies have been added to the standard framework, indicating also some possible links with the synergic spirit that characterize soft computing.

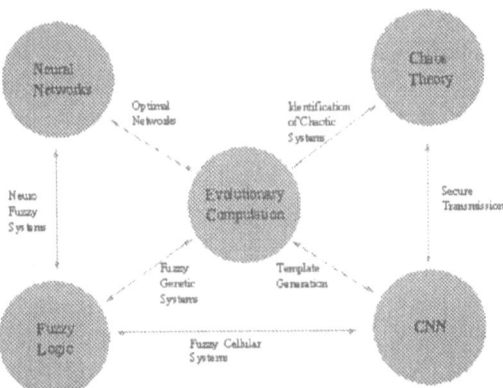

Fig. 1. Enlarged soft computing paradigm

Links between FL, NNs and GAs have been studied by many authors with thousand of variations and applications. Connections between FL, GAs, Cellular Neural Networks CNNs, and chaos theory are younger and more and more studied. In [6] a new distributed system, based on identical cells with a fuzzy core, have been introduced and successfully applied for complex phenomena modeling; in [4] CNNs have been used to generate programmable chaotic signals with application secure transmission, in [3] optimization capabilities given by GAs have been used to determine the parameters of a chaotic circuit starting from acquired data, and finally in [5] GAs have been used to automatically design CNNs template in image processing application avoiding a standard time expensive trail and error procedure. The following parts of this paper are dedicated to the description of two industrial applications, with high MIQ, of soft computing. The first, proposed in section II, concerns with a control application of the three main components of soft computing. The second, in section III, comes from an application of the CNNs paradigm for autonomous machine navigation system. Both ideas are going to be concretized for market application with the realization of a dedicated ICs.

2   Soft computing in control application

First industrial application of fuzzy logic have been based on fuzzy controller with fixed structure. It means that, off-line, the fuzzy system was defined then optimized, probably using GAs, and finally has been downloaded either on dedicated hardware or on standard micro. Even if fuzzy controllers are robust, the aging of the system and external parameter variations could decrease overall system performance. So the need to realize controllers on-line adaptable by using neural networks. In figure 2 it is reported a scheme of the internal structure of the on-line adaptable neuro fuzzy system we developed.

It is based on a DPRAM (Dual Port RAM), a MUXRAM (Multiplexer), the MICROCORE that represents the decisional part of the overall system, the FUZZY CORE dedicated only to fuzzy operations, the FUZZY COPROCESSOR dedicated to fuzzy rules calculus and finally the ALU (Arithmetic Logic Unit) used only to work with algebraic operations. The system process the information as it follows. The code, represented by fuzzy rules, learning rules and other instructions, is memorized on an external EEPROM. At the startup the code is downloaded from the EEPROM to the DPRAM and then interpreted by the MICROCORE. If the current instruction if a fuzzy operation the MICROCORE send a signal to the MUXRAM. This signal enable the FUZZYCORE to read from the DPRAM. If the current instruction is not fuzzy it will be executed by the MICROCORE or, in particular, by the ALU if algebraic. The neural optimization is achieved by upgrading the parameters of the neuro-fuzzy controller via back propagation algorithm [8]. This algorithm, is generally applied off line and uses a series of data to learn and

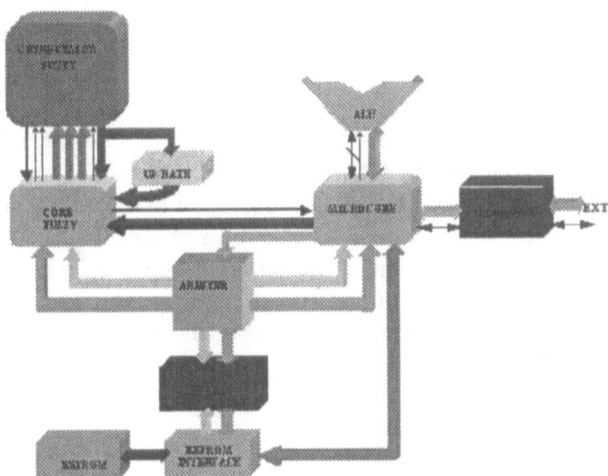

Fig. 2. Internal architecture of the developed on line adaptable fuzzy controller

modifies the weights of the net. It is easy to understand that the hardware implementation of this algorithm represents the crucial part in the realization of the on-line adaptable fuzzy controller.

3   RD-CNNS for autonomous system navigation

The problem of artificial locomotion is known to represent a difficult task when coping with multi-actuated systems controlled by one (central) or more (distributed) digital processors. Nevertheless, the studies carried out from biologists revealed that the underlying mechanism in natural locomotion can be revisited in terms of complex dynamic phenomena such as the generation and propagation of auto-waves in active media. CNNs have been found powerful architectures to model dynamics commonly met in living tissues [7]. Among the main phenomena encountered are auto-waves, and Turing patterns. Since these phenomena are all solutions of the so-called reaction-diffusion partial differential equation (RD-PDE) structure, the CNNs able to reproduce them are called RD-CNNs [7]. In particular, in [10] the paradigm of the Central Pattern Generator (CPG), commonly used in neurobiology [9], has been suitably implemented into a RD-CNN, in order to generate artificial locomotion in a ring-worm-like walking robot. Moreover, a hexapod walking robot has been introduced: the CNN hardware setup was able to generate the fast medium and slow gait locomotion type and also to implement the transition between this type of gait, by using the Turing pattern framework. In the following sections of this paper it will be presented a hexapod walking robot designed and realized in the STMicroelectronics laboratory by adopting the paradigm of RD-CNNs. In particular the system is composed by the following parts:

- The mechanical structure consisting of six legs driven by twelve servomotors and a body that contains the servos controller and ultrasonic sensors conditioning board.
- A CNN hardware board consisting in an array of 12x1 cells designed and realized to implement the RD-CNN to obtain the auto waves.
- Two CNNs hardware boards consisting in an array of 2x2 cells designed and realized to implement the RD-CNN to obtain the Turing patterns.
- A control board to implement the different types of gait (fast, medium and slow) driven by one of the two previously introduced CNN, and a switching system connecting the CNN to the servo system.
- A circuit controlled by the second CNN, generating Turing patterns, to change direction to avoid obstacles.

The present walking robot is equipped with discrete components, future version of the system will be done by using dedicated integrate circuit developed in our laboratory.

3.1   Reaction-Diffusion Cellular Neural Network model

An example of RD-PDE, discrete in space, which generates complex phenomena, is:

$$\dot{x}_{1,i,j} = -x_{1,i,j} + (1 + \mu + \epsilon)y_{1,i,j} - s_1 y_{2,i,j} + i_1 +$$
$$D_1(y_{1,i+1,j} + y_{1,i,j+1} + y_{1,i-1,j} + y_{1,i,j-1} - 4y_{1,i,j}$$
$$\dot{x}_{1,i,j} = -x_{2,i,j} + s_2 y_{2,i,j} + (1 + \mu - \epsilon)y_{2,i,j} + i_2 + \qquad (1)$$
$$D_2(y_{2,i+1,j} + y_{2,i,j+1} + y_{2,i-1,j} + y_{2,i,j-1} - 4y_{2,i,j}$$

with $y_i = 0.5(|x_i + 1| - |x_i - 1|)$, $i = 0, 1, ..., M - 1$ and $j = 0, 1, ..., N - 1$

The following RD-CNN with appropriate constant template values describes the above RD-PDE's:

$$\dot{x}_{ij} = -x_{ij} + A * y_{ij} + B * u_{ij} + I \qquad (2)$$

where

$x_{ij} = [x_{1,i,j} \ x_{2,i,j}]$; $y_{ij} = [y_{1,i,j} \ y_{2,i,j}]$; $u_{ij} = [u_{1,i,j} \ u_{2,i,j}]$.

are the state, the output and the input of the CNN respectively, while A, B and I are the feedback, control and bias templates. The cloning templates are:

$$A = \begin{pmatrix} A_{11} & A_{12} \\ A_{21} & A_{22} \end{pmatrix}$$
$$B = 0$$
$$I = [I_1 \ I_2]'$$

where

$$A_{11} = \begin{pmatrix} 0 & D_1 & 0 \\ D_1 & -4D_1 + \mu + \epsilon + 1 & D_1 \\ 0 & D_1 & 0 \end{pmatrix}$$

$$A_2 = \begin{pmatrix} 0 & D_2 & 0 \\ D_2 & -4D_2 + \mu - \epsilon + 1 & D_2 \\ 0 & D_2 & 0 \end{pmatrix}$$

$$A_{12} = \begin{pmatrix} 0 & 0 & 0 \\ 0 & -s_1 & 0 \\ 0 & 0 & 0 \end{pmatrix}$$

$$A_{21} = \begin{pmatrix} 0 & 0 & 0 \\ 0 & s_2 & 0 \\ 0 & 0 & 0 \end{pmatrix}$$

Some theorems proved in [11] and [12] show that the above system, for suitable choices of its parameters, is able to obtain pattern formation or autonomous wave propagation. We have realized three CNNs, the first one is a 12x1 CNN to obtain autonomous wave propagation to command the legs of the walking robot, the others are two 2x2 CNN have been designed to obtain pattern formation to control the type of gait and the robot direction. In figure 3 is reported the schematic of a cell able to generate both phenomena and in figure 4 is depicted an auto-wave front.

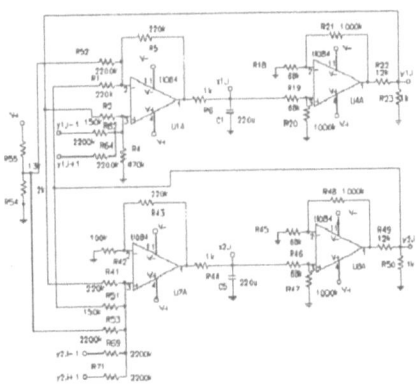

Fig. 3. Cell schematic.

3.2   The Hexapod

The prototype of walking robot built in our laboratory to study and implement the strategy of locomotion control is the hexapod reported in figure 5.

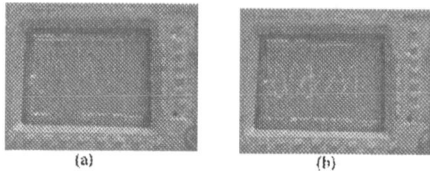

Fig. 4. Auto-waves front.

Each leg has two degrees of freedom and is moved by two servomotors: one drives the vertical position of the foot, while the other one drives the rotation of the leg to realize the locomotion steps. The cinematic has been designed so that both of the motions in each leg can be driven by the same output of one CNN cell, strictly resembling some biological cases where the same impulse triggers an ensemble of muscular fibers which lead to a coordinated motion.

Fig. 5. The hexapod walking robot.

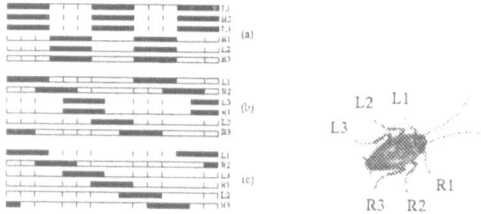

Fig. 6. The three types of gait (up fast, middle medium, down slow)

Figure 6 depicts the various patterns of motion of the insect legs with respect to the ground during the fast gait (Fig. 6a), the medium gait (Fig. 6b), and the slow one (Fig. 6c). Here each horizontal bar represents the time that each leg spends detached from ground (white part), or on ground (black part). The most common gait of insects is the fast gait (Fig. 6a). It is also called alternating tripod, since it is possible to distinguish two "tripods": L1, R2, L3 and R1, L2, R3. The legs belonging to each tripod are moved synchronously. In the fast gait configuration L1, R2, L3 and R1, L2, R3 are opposite in phase one another. If the auto-waves are considered in order to trigger the artificial insect legs to realize this gait, the state of one given cell (for instance C1), will drive the tripod L1, R2, L3, while another cell, showing an auto-wave front, in opposition of phase with C1 (Fig. 4b), will drive R1, L2, R3. Roughly speaking, a particular Turing pattern, implemented by one of the two 2x2 CNNs, imposes a particular locomotion type by sending the suitable binary commands, to connect the CNN showing auto-waves to the leg system figure 7. Indeed, the type of gait realized by insects is not only the "fast gait" depicted up to now, but also the "medium" and the "slow" gait, represented in Fig. 6b-c. From this figure it can be realized that all the types of gait can be viewed as modification of the basic fast gait, by a suitable modulation of the delay of each leg with respect to the other ones belonging to the same tripod. In fact the medium gait derives from the fast gait by introducing a delay among the legs belonging to each tripod: R2 has a phase delay with respect to L1, and the same delay exists between L3 and R2. Similar considerations hold for the opposite tripod. Moreover the diagonally opposite legs couples, L1 and R3, and L3 and R1 proceed synchronously. When the phase delay involves also these two couples the slow gait onsets.

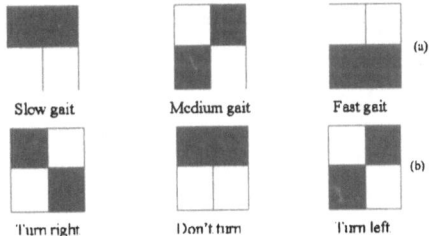

Fig. 7. Turing patterns for gait control

Our aim is to establish a suitable strategy to connect the 12 cells among themselves and to the robot legs to realize all the types of locomotion of Fig.6. It is apparent that in biology the spiking frequency of neurons cannot vary as a mere consequence of the walking speed. The hypothesis is that specific commands from other neural sites (CNN generating Turing patterns), stimulated directly from ultrasonic sensors, can activate or inhibit specific connections in the CNN that generates the auto-wave, giving rise to different pathways which change the locomotion type. The scheme for the locomotion control designed and realized in this paper is depicted in figure 8. A CNN ring consisting of 12 cells is considered. Here suitable initial conditions lead to the onset of an auto wave front (Fig. 4), which, in steady state conditions, proceeds with constant speed and unchanged amplitude periodically visiting all the cells. The connections among the state variables of the cells and the hexapod legs are also reported. For the fast gait configuration, the alternating tripod requires that the legs of each tripod proceed in phase: therefore they have to be driven by the same cell, for example C1. The other three legs must be connected together to another cell, for example C4. A further constraint to satisfy is that the two tripods must be in opposition of phase. In this sense the reaction-diffusion mechanism plays a fundamental role: the self-adaptation of the phase among the cells. In fact, in the case at hand, if the cell C6 is directly connected to the cell C1, in the sub-ring thus formed (Fig. 8a) the diffusion process induces the auto-wave oscillation to take place in the cell C1 directly after the cell C6. A kind of phase rearrangement therefore takes place, in which the cells C1 and C4 result to be in opposition of phase each other. The speed of the fast gait can be selected very easily, by choosing a particular sub-ring built of an even number of cells: of course, the more the number of cells in the sub-ring, the slower the speed of this type of gait. The medium gait (Fig. 6b) is characterized by a constant phase delay among the three legs belonging to the same tripod, but the last leg of one tripod is in phase with the first one of the other tripod. Such configuration can be easily realized with the scheme of Fig. 8b. The sub-ring thus formed realizes this type of gait. Similar discussions can be done to depict the slow gait, by comparing Fig. 6c with Fig. 8c. It should be noted that the transition among these types of gaits can be done in real time in the analog frame proposed, and the phase rearrangement among the cells is a function of the cell number.

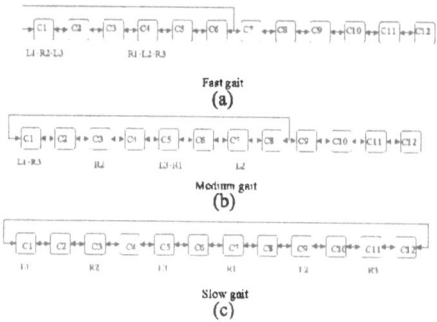

Fig. 8. Cell ring

The hardware implementation of this scheme has been realized as schematically reported in Fig. 9a. Here the three main blocks represent three sub-chain to realize the three sub-ring introduced above, while the little boxes represent digitally driven analog switches, whose task is to realize the suitable connection among cells for each gait. Table 1 reports the connections among the cells of each sub-chain that have to be joined via the switches to realize each type of gait. The last issue to be addressed involves the connections among each cell state variables and the robot legs to realize each of the gait types (Fig. 6). These are depicted in Fig. 9b, where, similar to Fig. 9a, the small boxes stand for switches, now connecting the robot legs to the corresponding cell state variables. The connections are summarized in Table 2. It has to be outlined that the analog switches used for realizing the connections among the sub-chains and the robot legs are directly driven by the 2x2 CNN generating Turing patterns (Fig. 7a). The strategy adopted to avoid the obstacles and change speed is based on the following. The hexapod is equipped with two ultrasonic sensors displaced in the front part of the robot. These sensors are used both for obstacles avoidance and changing in speed. The speed of the robot is proportional to the signal coming from the sensors. The changing of direction is decided comparing the two signals and if the smaller is less than a threshold the future direction will be changed accordingly. In particular the sensors output are directly connected to the two CNNs generating the Turing pattern responsible of the changing in speed and direction (7a-b).

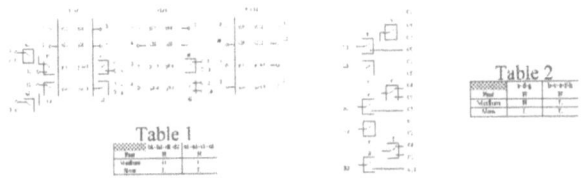

Fig. 9. Analog switches

From a mechanical point of view if the hexapod have to turn, for example, left the rotation servos of the left side of the robot receive an opposite signal with respect to the vertical servos. In figure 10 the entire control system is shown.

4 Conclusion

In this paper two industrial applications of Soft Computing techniques have been described. According to market request this activity is primary devoted to the design of dedicated architecture.

References

1. L. A. Zadeh, "Fuzzy logic, neural network and soft computing", Commun. ACM, vol 37, pp 77-84, March 1994.
2. P. Arena, L. Fortuna , M. Branciforte, "Reaction-Diffusion CNN Algorithms to Generate and Control Artificial Locomotion", IEEE Transaction on Circuits and Systems, I: Fundam. Theory and Applic, Vol. 46 N2, pp.259-266, Feb. 1999

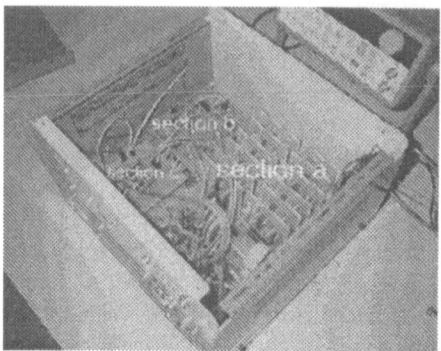

Fig. 10. The hardware structure. Section a: 12x1 CNN generating auto waves; section b: two CNN generating Turing pattern; section c: circuit to control the change of direction.

3. R. Caponetto, G. Manganaro, L.Fortuna, M.G. Xibilia, "Chaotic System Identification via Genetic Algorithms", 1st IEE/IEEE International Conference GALESIA, 12-14 September Sheffield UK, 1995.
4. R. Caponetto, M. Criscione, L. Fortuna, D. Occhipinti, L. Occhipinti, "Synthesis of a Programmable Chaos Generator, based on CNN Architectures, with Applications in Chaotic Communication", CNNA '98, London UK, 14-17 April 1998.
5. R. Caponetto, M. Lavorgna, A. Martinez, L. Occhipinti, "Cellular Neural Network Simulator for Image Processing", CNNA '98, London UK, 14-17 April 1998.
6. M. Lavorgna, L. Occhipinti, R. Caponetto, L. Fortuna, G. Di Bernardo, "Cellular Fuzzy Processor: New Architecture to explore Complexity in Locally Interconnected System", IEEE Int. Conf. On Electronics Circuits and Systems Lisbon 7-10 Sept. 1998.
7. L. O. Chua, M. Hasler, G. S. Moschytz, J. Neirynck, "Autonomous Cellular Neural Networks: A Unified Paradigm for Pattern Formation and Active Wave Propagation", IEEE Trans. on Circuits and Systems- Part I, vol. 42 no.10, pp.559-577, 1995.
8. D. E. Rumelhart, McClelland, "Parallel Distributed Processing: Exploration in the microstructure of cognition", pp.318-362, MIT Press, 1986.
9. G. M. Shepherd, "Neurobiology", Oxford Univ. Press, third edition, 1994.
10. P. Arena, L. Fortuna, M. Branciforte, " Reaction-Diffusion CNN algorithms to generate and control artificial locomotion" IEEE Trans. on Circuits and Systems in press.
11. P.Arena, L. Fortuna, and G. Manganaro, "Self organization in a two layer CNN." IEEE Trans. Circuits Syst. I, Vol. 45, pp.157-162, 1998
12. P.Arena, L. Fortuna, and G. Manganaro, "A CNN for pattern formation and active wave propagation." in Proc. European Conf. Circuit Theory Design, ECCTD '97, Budapest, Ungary, Aug. 1997

Spatial Diversity in Reaction-Diffusion Fuzzy Cellular Networks

Maide Bucolo, Maria Chiara Cutuli, Luigi Fortuna, and Alessandro Rizzo

Dip. Elettrico Elettronico e Sistemistico - Universitá degli Studi di Catania
Viale A. Doria 6, 95125 Catania Italy - Luigi.Fortuna@dees.unict.it

Abstract. This paper deals with two main points. Firstly, a new type of *Reaction-Diffusion Fuzzy Cellular Network,* suitable to describe complex phenomena is defined. The second part of the work is devoted to investigate on the role played by diversity in this type of networks. In particular, the regularizing role of non organized spatial perturbation introduced in the membership functions of the proposed architecture is analyzed.

1 Introduction

In this paper, after having introduced some main concepts regarding the role of Reaction-Diffusion (RD) equations [1], and having discussed the possibility of implementing them by using the RD-CNN paradigm, a new type of *RD Fuzzy Cellular Network* is presented. A particular type of RD fuzzy cellular network has been already presented in literature [2]. In the present work, a more general model for such a network is proposed. Moreover, the effects of spatial dissymmetry in such a type of distributed architecture are analyzed. In particular, the differences between generating the spatial diversity by means of a chaotic distribution rather than a random law are emphasized. A series of simulation experiments will be presented to this purpose.

2 Preliminaries

Linear and nonlinear Partial Differential Equations (PDEs) are commonly used to describe several phenomena in many fields, like physics, nature, economy, and engineering. They also characterize microscopic systems when fields are introduced [3]. A very popular PDE is the well known Reaction-Diffusion Equation (RD):

$$\frac{\partial \phi(x,t)}{\partial t} = D\nabla^2\phi(x,t) + f(\phi(x,t),\lambda) \tag{1}$$

where $\phi(x,t)$ can be a vector or a scalar function, x the position vector, t the time. The reactive term is given by $f(.)$, which is usually a nonlinear function of its argument. The control parameters are expressed through the λ vector. D represents the diffusion matrix. The diffusion mechanism is described by means of the well known *Laplacian* operator ∇^2. A more complicated version of (1) arises when not ordered fluctuations occur. The new equation becomes:

$$\frac{\partial \phi(x,t)}{\partial t} = D\nabla^2 \phi(x,t) + f(\phi(x,t),\lambda) + g(\phi(x,t))\eta(x,t) \qquad (2)$$

The signal $\eta(x,t)$ indicates a non ordered fluctuation signal. The equation (2) is the so-called *Stochastic Reaction-Diffusion Equation*. Both equations (1) and (2) are formal models to describe complex systems [1], usually being literally named *the equations of the complexity* [4]. The importance assumed in the last period by the theme of complexity stimulates to conceive a new architecture, suitable to implement RD equations also in real time.

Cellular Neural Networks (CNN) [5] have been widely used to solve the RD equation (1) in a spatial discretized form. The local connectivity of the CNN architecture and the dynamics of each cell often allow us to match in a complete and precise way the discretized version of RD equations. It will appear therefore that a physical distributed system can often assume the electrical model of a particular CNN structure. In this way, CNN architectures (circuits) provide a good paradigm for describing complex systems. However, some models of complex system, even if constituted by locally connected nonlinear units, are better described by *linguistic rules* rather than equations, which are often not known or difficult to set up. With this in mind, a useful alternative to the description of complex systems by means of partial differential equations is based on *Fuzzy Cellular Networks*, which will be discussed in the next section.

3   Fuzzy Cellular Networks

Fuzzy Cellular Networks (FCN) are derived from the fusion of the main concepts characterizing the so-called Cellular Neural Networks (CNN) with the peculiarity of Fuzzy Logic Theory. These networks join the features of local connectivity of the CNNs, useful to describe the interactions in complex systems, with the powerful adaptability of fuzzy systems. The literature [2,6] widely reports studies that reproduce the results derived by using such type of array in various fields, in particular in image processing and in modeling RD-based processes. The structure of a Fuzzy Cellular Network, illustrated in Fig. 1, is very similar to that of a CNN, being constituted by identical elementary units, whose behavior is described by a set of fuzzy rules.

In this study, each cell C_{ij} is characterized by its fuzzy state variable X_{ij}, that is influenced by the state of each other cell within its neighborhood of r radius. The influence of the neighboring cells is taken into account by the fuzzy variable N_{rij}. In order to set up an approach which can parallel that of CNN, in which the reactive part of the system is represented by the dynamics of the single cell, whereas the diffusive one is provided by local interconnections, the fuzzy rules describing the generic FCN cell can be

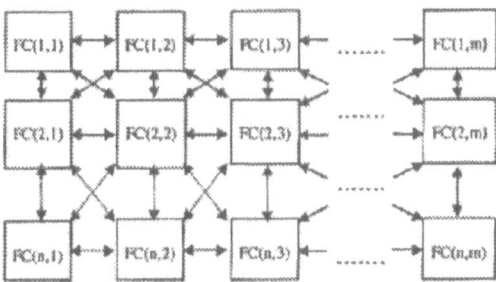

Fig. 1. Structure of a Fuzzy Cellular System.

divided into two subsets. The first one (R_1) defines the dynamics of the status of the generic cell, taking also into account the influence of the neighboring ones. The related rules assume the form:

$$\text{IF } X_{ij}(k) \text{ is } A_l \text{ AND } N_{rij}(k) \text{ is } B_l \text{ THEN } X_{ij}(k+1) \text{ is } p_l \qquad (3)$$

being:

- A_l, B_l are the membership functions of the fuzzy set for the input variables in the l*th*-rule;
- p_l is the value of the crisp output function in the l*th*-rule.

The second set of rules R_2 is constituted by the rules which take into account the local connectivity. In other words, this set of rules permits to establish the strength of the influence of the neighboring cells on a given cell, thus evaluating the variable N_{rij} on the basis of the value of the variables X_{nm}, with n and m belonging to the neighborhood of the cell ij. For the generic cell whose state is X_{ij}, rules are of the following form:

$$\text{IF } X_{(nm)_1}(k) \text{ is } C_l \text{ AND } X_{(nm)_2}(k) \text{ is } D_l \cdots \text{ AND } X_{(nm)_p}(k) \text{ is } E_l$$
$$\text{THEN } N_{rij}(k) \text{ is } q_l$$

$$(4)$$

where:

- the couples of indexes $(nm)_1$, $(nm)_2$, $\cdots$, $(nm)_p$ refer to cells belonging to the neighborhood of the cell X_{ij};
- C_l, D_l, E_l are the membership functions of the fuzzy set for the input variables in the l*th*-rule;
- q_l is the value of the crisp output function in the l*th*-rule.

The previous sets of rules are completed by taking into account some boundary conditions like the Neumann or the Dirichlet conditions, or the toroidal ones [5].

4   Reaction-Diffusion Fuzzy Cellular Networks

This kind of architecture has been widely studied in literature [2]. It has been used in image processing to generate autowaves, to form self-organizing patterns and so on. In the previously considered fuzzy cellular network [2], the realized model is a hybrid, nonlinear one, consisting of two fuzzy parts to define the cell dynamics and the diffusion rules, combined in a traditional differential equation framework. The following equations describe the behavior of the cell ij in a traditional FCN architecture with a planar structure, in which the radius of the neighborhood is 1.

$$\frac{\partial U_{ij}}{\partial t} = \gamma \left(F_1(U_{ij})U_{ij} + f_v V_{ij} \right) + D_u F_2(U_{ij}, N_r(U_{ij}))$$

$$\frac{\partial V_{ij}}{\partial t} = \gamma \left(g_u U_{ij} + g_v V_{ij} \right) + D_v F_2(V_{ij}, N_r(V_{ij})) \tag{5}$$

$$N_r(C_{ij}) = \gamma \frac{1}{4} \left(C_{(i-1,j)} + C_{(i+1,j)} + C_{(i,j+1)} + C_{(i,j-1)} \right)$$

The constant parameters of the system are $f_v, g_u, \gamma, D_u, D_v$. The terms F_1 and F_2 are respectively the output quantities derived from the fuzzy rule processing defined for each cell as follows. For the system F_1:

$$
\begin{aligned}
&\text{IF } x \text{ IS low THEN } F_1 \text{ IS negative} \\
&\text{IF } x \text{ IS medium THEN } F_1 \text{ is zero} \\
&\text{IF } x \text{ IS high THEN } F_1 \text{ is positive}
\end{aligned}
\tag{6}
$$

The introduced term is related to the nonlinearity involved into the reaction effect, while the term F_2 realizes in a fuzzy fashion the Laplace operator, which in fact includes the contribution of the neighborhood N_r, as defined in the equation set (5). It can be formalized as follows:

$$
\begin{aligned}
&\text{IF } x \text{ IS low AND } N_r \text{ IS low THEN } F_2 \text{ IS zero} \\
&\text{IF } x \text{ IS low AND } N_r \text{ IS medium THEN } F_2 \text{ IS medpos} \\
&\text{IF } x \text{ IS low AND } N_r \text{ IS high THEN } F_2 \text{ IS positive} \\
&\text{IF } x \text{ IS medium AND } N_r \text{ IS low THEN } F_2 \text{ IS medneg} \\
&\text{IF } x \text{ IS medium AND } N_r \text{ IS medium THEN } F_2 \text{ IS zero} \\
&\text{IF } x \text{ IS medium AND } N_r \text{ IS high THEN } F_2 \text{ IS medpos} \\
&\text{IF } x \text{ IS high AND } N_r \text{ IS low THEN } F_2 \text{ IS negative} \\
&\text{IF } x \text{ IS high AND } N_r \text{ IS medium THEN } F_2 \text{ IS medneg} \\
&\text{IF } x \text{ IS high AND } N_r \text{ IS high THEN } F_2 \text{ IS zero}
\end{aligned}
\tag{7}
$$

In the new version of RD Cellular Network discussed in this paper the following model will be used:

$$
\begin{aligned}
U_{i,j}(k+1) = {}& F_1(U_{i,j}(k), U_{i,j}(k-1)) + \\
{}+{}& F_2(U_{i-1,j}(k+1), U_{i+1,j}(k+1), U_{i,j-1}(k+1), U_{i,j+1}(k+1))
\end{aligned}
\tag{8}
$$

The fuzzy term F_1 describes the whole reaction effect, that in many application is commonly modeled as an autonomous slow-fast dynamics. Usually, the fuzzy function F_1 depends on the particular application, whereas the F_2 rule set describes a fuzzy Laplacian, considering a unitary neighborhood radius, with the following input fuzzy sets:

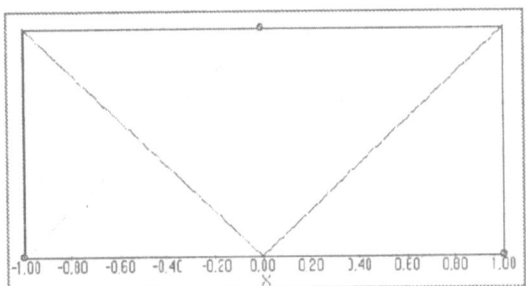

Fig. 2. Input fuzzy sets for the fuzzy Laplacian operator F_2.

The crisp Fuzzy sets of the Laplacian output will be illustrated in the following figure.

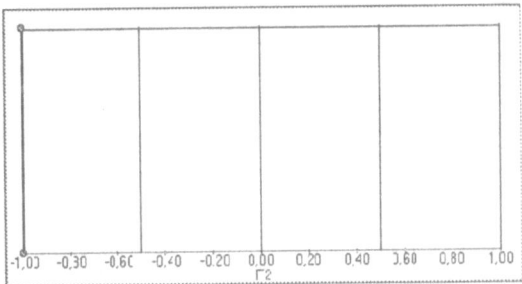

Fig. 3. Output fuzzy set for the fuzzy Laplacian operator F_2.

It is worth noticing that the diffusion coefficient D does not appear any more in the new formulation. This is due to the fact that the diffusion strength can be modulated by changing the position of the output fuzzy sets of the fuzzy Laplacian.

The new formulation (8) does not explicitly include differential operators such those contained in the previous formulation (5), thus straightly coping with the intrinsic discrete nature of the phenomena studied and the computational implementation issues. Moreover, in the previous formulation the actual reaction dynamics is still specified by partial differential relationships among variables. The fuzzyness is only exploited in defining the nonlinearity contained in the formulation through the fuzzy term F_1, and to set up an alternative fuzzy Laplacian operator. The previously adopted fuzzy Laplacian rely on the definition of an average of the neighborhood, whereas the fuzzy Laplacian rules considered into the new approach are defined on the basis of the influence of the single cells. For example, for a planar implementation with a unitary neighboring radius, it assumes the following form:

IF $x[i][j+1]$ IS medium AND $x[i][j-1]$ IS medium AND $x[i][j]$ IS low
THEN F_2 IS medpos
IF $x[i][j+1]$ IS high AND $x[i][j-1]$ IS high AND $x[i][j]$ IS low
THEN F_2 IS positive
IF $x[i][j+1]$ IS low AND $x[i][j-1]$ IS low AND $x[i][j]$ IS medium
THEN F_2 IS medneg
IF $x[i][j+1]$ IS high AND $x[i][j-1]$ IS high AND $x[i][j]$ IS medium
THEN F_2 IS medpos
IF $x[i][j+1]$ IS low AND $x[i][j-1]$ IS low AND $x[i][j]$ IS high
THEN F_2 IS negative
IF $x[i][j+1]$ IS medium AND $x[i][j-1]$ IS medium AND $x[i][j]$ IS high
THEN F_2 IS medneg
IF $x[i][j+1]$ IS low AND $x[i][j-1]$ IS low AND $x[i][j]$ IS low
THEN F_2 IS zero
IF $x[i][j+1]$ IS medium AND $x[i][j-1]$ IS medium AND $x[i][j]$ IS medium
THEN F_2 IS zero
IF $x[i][j+1]$ IS high AND $x[i][j-1]$ IS high AND $x[i][j]$ IS high
THEN F_2 IS zero

$$(9)$$

It is clear that the previous approach requires the knowledge of the partial differential equations underlying the system considered, thus diminishing in some way the power of the fuzzy logic paradigm, which is very useful in formalizing systems from a behavioral point of view. The advantage of the approach presented in this paper consists of setting up a framework in which the whole dynamics of a cell constituting a complex system can be described by behavioral linguistic rules, instead of equations.

5   Experimental Results

The aim of the experiments carried out in this work is to show that spatial non symmetric disorder could lead to some regularization in RD Fuzzy Cellular Networks. In particular, the effects of slightly changing the membership function shape, rather than the rules themselves, is emphasized. Moreover, the introduction of a deterministic disorder like chaos, instead of a random disorder can give better results in term of regularization. Two key-points must be taken into account in our study. The first one is to reach the regularization by using low diffusion effects. The second one is related to the introduction of a low level of spatial uncertainty. It will be shown that introducing a small uncertainty in the system parameters, without imposing changes into the system structure, can lead the system toward a common dominant behavior. The considered RD Fuzzy Cellular Network is generated by a one-dimensional array of chaotic autonomous fuzzy cells. The number of chaotic oscillators is 128. This number has been chosen in order to have a model comparable with other studies, where the same number of oscillators have been considered [7,8]. Cells have been coupled according to the Laplacian fuzzy operator described in (9). The chaotic fuzzy cell oscillator, constituting the "reactive" part of our experimental setup, has been established by a set of fuzzy rules which permits to generate chaotic maps with a given Lyapunov Exponent, as derived in [9]. According to this approach, two input fuzzy variables are considered: the nominal state $x(k)$, and the uncertainty with respect to this variable $d(k)$. For each of the two variables, the fuzzy sets whose membership function shape (parameters) characterizes the generated chaotic series can be established, according to the required Lyapunov exponent. In particular, five fuzzy sets have been established for $x(k)$ and for $d(k)$, whereas for the output variable two crisp fuzzy sets have been considered. The rules generating the oscillator are summarized in the following:

IF D IS very large pos AND $x[i][j]$ IS large right THEN D_s IS large neg
IF D IS very large pos AND $x[i][j]$ IS large left THEN D_s IS small pos
IF D IS very large pos OR D IS very large neg AND $x[i][j]$ IS large right
THEN F_1 IS small left
IF $x[i][j]$ IS large left OR $x[i][j]$ IS large right NOT $x[i][j]$ AND D
IS very large neg THEN D_s IS small pos
IF D IS very large pos OR D IS very large neg AND $x[i][j]$
IS large left THEN F_1 IS small right
IF D IS very large pos OR D IS very large neg AND NOT D AND
$x[i][j]$ IS large right THEN F_1 IS small right
IF D IS very large pos OR D IS very large neg AND NOT D AND
$x[i][j]$ IS large left THEN F_1 IS small left
IF $X[i][j]$ IS small left THEN F_1 IS large left
IF $X[i][j]$ IS small right THEN F_1 IS large right
IF D IS zero THEN D_s IS zero
IF D IS small pos THEN D_s IS medium neg
IF D IS large pos THEN D_s IS very large neg
IF D IS small neg THEN D_s IS medium pos
IF D IS large neg THEN D_s IS very large pos

$$(10)$$

where $D \equiv d(k)$, $D_s \equiv d(k+1)$, $x[i][j] \equiv x[i][j](k)$, and F_1 is evaluated at the time step $k+1$.

Once the reactive part of the structure has been defined, in order to describe completely the system, the rules characterizing the Laplacian must be taken into account, as described in fuzzy system (9). The diffusion coefficient is then modulated by the position of the crisp output fuzzy sets F_2.

In the following, the performed experiments will be described:

- *Experiment 1*: Let us consider independent oscillators, without diffusive coupling. The chaotic spatio-temporal behavior of the system is evident (Fig. 4).

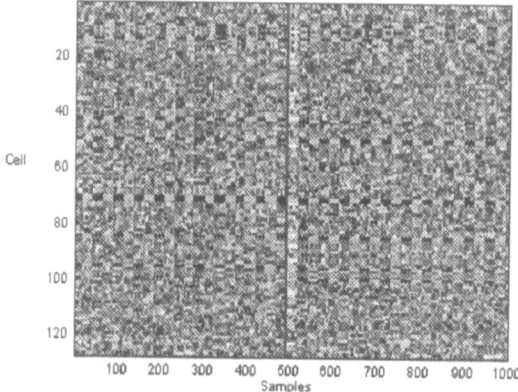

Fig. 4. Spatio-Temporal chaotic behavior of independent oscillators.

- *Experiment 2*: Let us introduce a diffusive coupling effect. Even if a regularization effect is evident, in this case some oscillators reach a zero steady state condition, indicating the fact that the diffusion effect is excessive (Fig. 5).

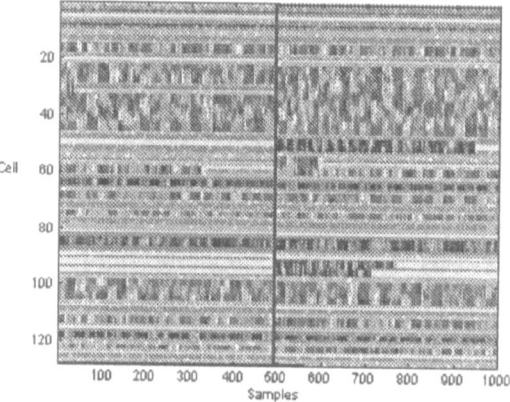

Fig. 5. Spatio-Temporal behavior of the system with large diffusion effect.

- *Experiment 3*: In order to decrease the diffusion effect, a different fuzzy output for F_2 has been chosen, as depicted in Fig. 6. The regularization can be still noticed, with no oscillator reaching the previous zero steady state (Fig. 7).

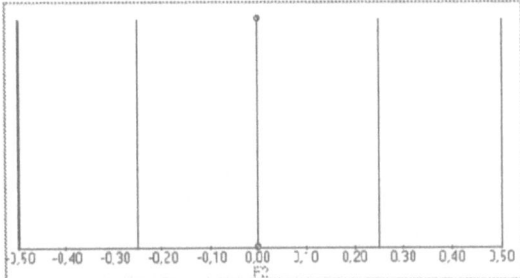

Fig. 6. Output fuzzy sets of F_2, for the setup of Experiment 3.

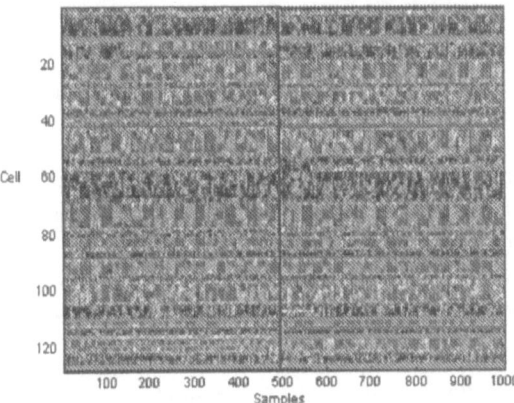

Fig. 7. Spatio-Temporal behavior of the system due to the introduction of F_2.

- *Experiment 4*: Considering the output fuzzy sets for F_2 reported in Fig. 8, the pattern of Fig. 9 are obtained. This constitutes the starting experimental setup to explore the regularization effects due to the introduction of spatial dissymmetry. Therefore, the same fuzzy sets (Fig. 8) will be ever considered for the next experiments.

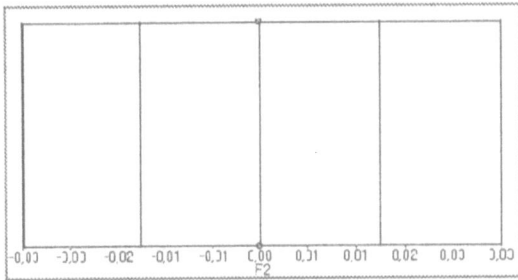

Fig. 8. Output fuzzy sets of F_2, for Experiments from 4 to 8.

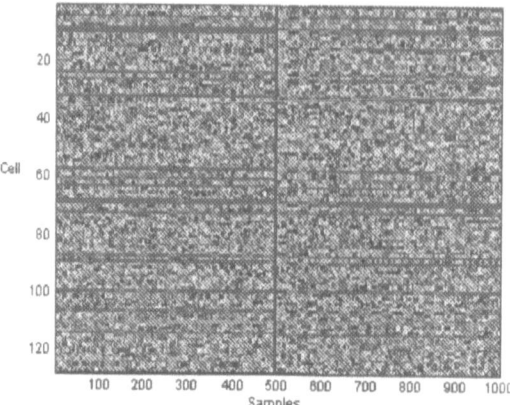

Fig. 9. Spatio-Temporal behavior of the system due to the introduction of F_2.

The next experiments show the effects of introducing a slight spatial diversity in the fuzzy cellular oscillators. In particular, the influence of slightly changing the geometric parameters defining the shape of the original membership functions are considered. The vertices of the functions are kept, while the bases of each triangular function is shrunk or enlarged of a small quantity with respect to the nominal situation.

- *Experiment 5*: The membership function shapes are slightly modified by using a white random generator. A variation of 5% around the nominal values is imposed. In Fig. 10 the obtained spatio-temporal patterns are shown. Some organized regular strips indicate a regularization effect, compared with the map in *Experiment 1*.

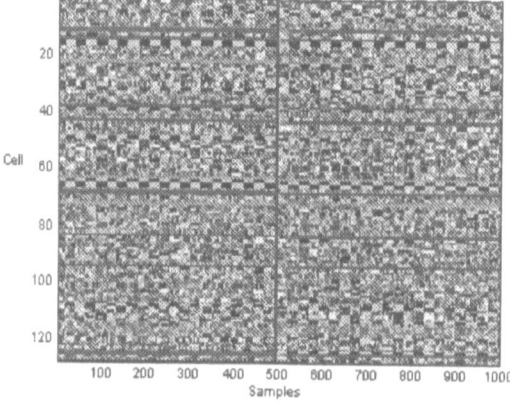

Fig. 10. Spatio-Temporal behavior with 5% random dissymmetry.

- *Experiment 6*: The experiment is formally identical to that of *Experiment 5*, but a random variation of the 10% and 20% around the nominal values is fixed. The regularization effect is more evident (Figs. 11-12). Anyway, as discussed before, our aim is to establish the minimum diversity level such as some regularization effect does still exist, as systems nature should not be strongly changed.

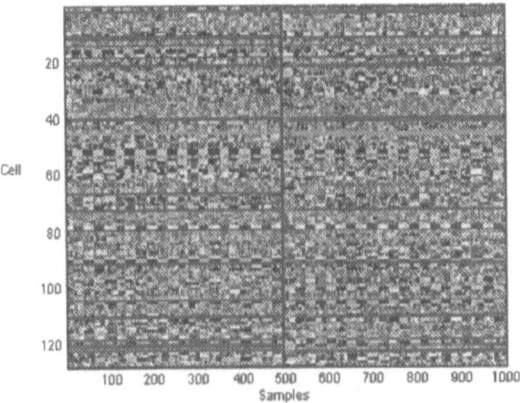

Fig. 11. Spatio-Temporal behavior with 10% random dissymetry.

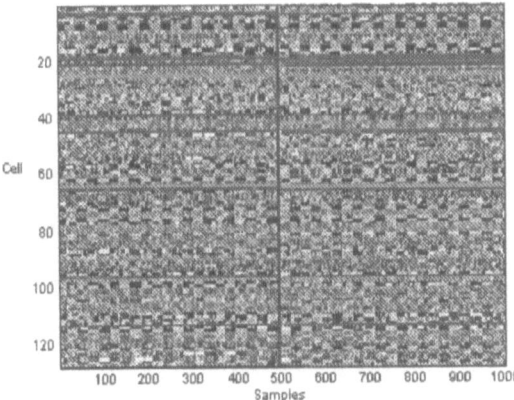

Fig. 12. Spatio-Temporal behavior with 20% random dissymmetry.

In the next experiments, a chaotic generator is used instead of a random generator. This will involve the generation of a deterministic, unpredictable diversity into the system. In particular, the chaotic generator has been obtained by using the sampling of the output of a Chua Circuit [10] with the following parameters $a = 10$, $b = 14.87$, $m_0 = -1.27$, $m_1 = -0.687$.

- *Experiment 7*: A chaotic diversity of 5% with respect to the nominal value of the shape parameters of the fuzzy membership functions is introduced. The experiment can be compared with *experiment 5*. The regularization effect is now evident, as shown in Fig. 13.

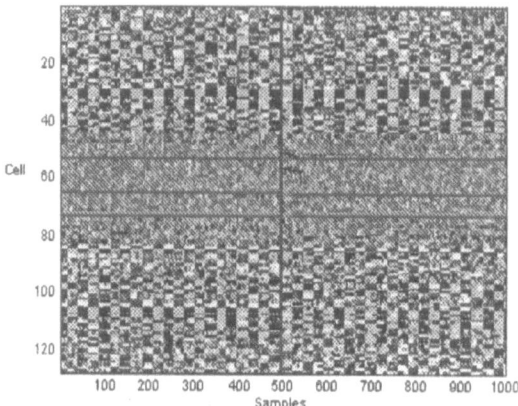

Fig. 13. Spatio-Temporal behavior with 5% chaotic dissymmetry.

- *Experiment 8*: A chaotic perturbation of 10% has been introduced. This experiment is the counterpart of *Experiment 6*, in the case of uncorrelated deterministic generator of spatial dissymmetry. The self-organization and the synchronization effect of the various oscillators is more evident than the corresponding case, as shown in Fig. 14

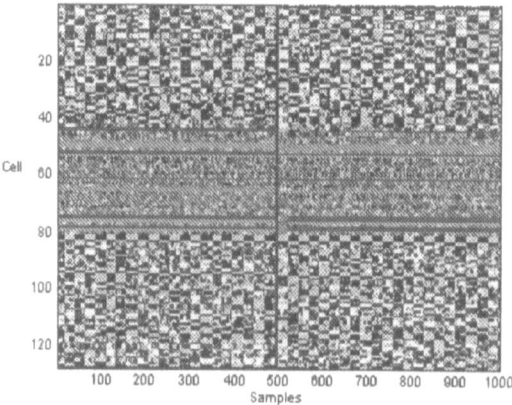

Fig. 14. Spatio-Temporal behavior with 10% chaotic dissymmetry.

6   Conclusions

Reaction-Diffusion Fuzzy Cellular Networks have been introduced with the aim of studying by simulation the effect of the introduction of spatial dissymmetry in pattern forming, regularization, and self-organization. Reaction-Diffusion Fuzzy Cellular Networks can be considered a good paradigm for studying complex systems, as shown also in former studies. In this work, a new formalization of the Reaction-Diffusion Fuzzy Cellular Networks has been introduced, allowing to take into account the intrinsic discrete nature of the phenomena considered, define a more explicit fuzzy Laplacian operator and, most important, describe the reaction processes from a behavioral point of view. The introduction of a spatial disorder has led to regularization, as often occurring in other kind of networks, and in general in complex systems. A future trend of the outlined activity is to introduce some input in each cell, in order to consider also the introduction of time-variant noise in the system, thus being able to study the *stochastic resonance* concepts in the considered case of distributed systems. In that case, a *RD Stochastic Cellular Fuzzy System* must be considered. It is opinion of the author that such a type of architecture can constitute one of the most general paradigms to explore the theme of complexity.

References

1. Y. Bar, Yam (1997) Dynamics of Complex Systems. Addison-Wesley.
2. R. Caponetto, L. Occhipinti, M. Lavorgna, L. Fortuna, G.Di Bernardo (Sept. 1998) Cellular Fuzzy Processor: a New Architecture to Explore Complexity in Local Interconnected Systems. IEEE Int. Conf. on Circuits and Systems, Lisbona.
3. J. Garcia Ojalvs, J. M. Sancho (1999) Noise in Extended Spatially Systems. Springer.
4. Y. Bar-Yam (1997) Dynamics of Complex Systems. Addison-Wesley.
5. L.O. Chua, (1998) CNN: A Paradigm for Complexity. World Scientific.
6. S. Baglio, L. Fortuna, G. Giudice, G. Manganaro (1997) Fuzzy Cellular Systems for a new Computational Paradigm. Engineering Applic. Art. Intell. Vol10, 47-52.
7. Y. Braiman, John F. Lynder, William L. Ditto (1995) Taming Spatiotemporal Chaos with Disorder. Nature Vol387, 465-467.
8. P. Arena, R. Caponetto, L. Fortuna, A. Rizzo (2000) Non Organized Deterministic Dissymmetries Induce Regularity in Spatio-Temporal Dynamics. Int. Journal Bifurcation and Chaos. World Scientific.
9. S. Baglio, L. Fortuna, G. Manganaro (1996) Design of Fuzzy Iterators to generate Chaotic Time Series with Assigned Lyapunov Exponent. Electronic Letters, Vol32, No 4.
10. R. N. Madan, (1993) Chua's Circuit: A Paradigm for Chaos. World Scientific Series on Nonlinear Science.

An Evolutionary View to the Design of Soft-Computing Agents

Stefano A. Cerri[1], Vincenzo Loia[2]

[1] LIRMM,Université Montpellier II 161 Rue Ada,34392 Montpellier Cedex 5, France, e-mail: cerri@lirmm.fr
[2] Dipartimento di Matematica ed Informatica, Soft-Computing Lab, Università di Salerno, 84081 Baronissi (SA) Italia, e-mail:loia@unisa.it

Abstract. Among the possible experiments aiming to enhance Actors (active objects) to have a behavior compatible with the requirements traditionally identified for Agents, here we discuss those integrating an evolutionary fuzzy reasoning module into Actors. The resulting framework, based on the notion of FuzzyEvoAgent, allows to realise societies of Agents evolving as a result of interactions with the environment. We propose here: 1. a formal definition of FuzzyEvoAgents; 2. an architecture in Java and 3. an application to a simple scenario in artificial life (pray and predator). The results shown in this paper confirm that the evolutionary fuzzy framework may represent an important component for ensuring the autonomy of Agents, i.e. the ability to learn from interactions with the environment.

1 Introduction

Agent technologies have recently become one of the most important growing areas in Information Technologies. Within this attractive domain, the efforts accomplished by the research and industrial communities in the development of Intelligent Agents have reported a remarkable progress in practical applications as well as in theoretical results [9] [16][14].

In order to have an idea of the amount and impact of the literature available on Agents, see at the site:
http://agents.umbc.edu/Publications_and_presentations/index.shtml.
You may find hundreds of references. Furthermore: they are daily increasing: from February 18th to March 3rd, 2000 the site included 30 new references (an average of 2 each day, possibly multiple, i.e. a Conference or a book). We may therefore be excused if our referenced papers will just consist of the ones that have influenced our research. For an industrial association interested in Agents, see http://www.fipa.org. For an European network of Excellence, see http://www.AgentLink.org.
A comprehensive synthesis of all contributions is hard to compile if not impossible. We share with [11] the view that Agents are essentially active objects (i.e. Actors) that are autonomous [6] and learn (evolve) during their life cycle. The autonomy of Agents with respect to the Actors behavior is ensured by a dynamic scheduling mechanism that selects (or generates) the next message

to serve (or the next action to perform) by evaluating a procedure at run time that accesses the available knowledge. Both [11] and [6] report about Agent Languages that operate in this way. Agents are considered, by these authors, like Operating Systems communicating with Agent Communication Languages (i.e. pragmatically tagged messages, cf. KQML [16]). [11] describes how Agents control behavior by means of Augmented Transition Networks; [6] by means of Finite State Automata and [20] by means of Production Rules: the three representations are, as known, quite similar. Furthermore, [6] introduces the notion of Cognitive Environment as a first class Abstract Data Type representing a kind of Federated DBMS that models the Knowledge of any Agent, including the mutual Knowledge. Essentially, these studies show that Agents are advanced software components that consist of:

- a suitable control with its associated model (different from that typical of servers of clients);
- a suitable memory model (in order to represent the states of multiple conversations or interactions about multiple subjects or ontologies);
- a suitable transactional protocol model (such as the one described by KQML or the STReam component of [6]).

The learning component is yet not clearly defined in these studies. In order to focus on it (the main objective of this paper) let us go back to a forthcoming paper about Agent-Oriented Software Engineering [14]. We will hopefully understand how our study specific to learning in Agents is complementary and synergist to other ones focusing more at Agent Communication, Control or Memory management.

1.1   The State of Research on Agent Languages

Jennings and Woolridge [14] characterize an Agent as *an encapsulated computer system that is situated in some environment, and that is capable of flexible, autonomous action in that environment in order to meet its design objectives.*
Further in the text, the authors explain that Agents are:

- *clearly identifiable problem solving entities with well-defined boundaries and interfaces;*
- *situated (embedded) in a particular environment: they receive inputs related to the state of that environment through their sensors and they act on the environment through their effectors;*
- *designed to fulfil a specific role (they have particular objectives to achieve, that can either be explicitly or implicitly represented within the agents;*
- *autonomous: they have control both over their internal state and over their own behavior;*
- *capable of exhibiting flexible (context-dependent) problem solving behavior: they need to be reactive (able to respond in a timely fashion to changes*

that occur in their environment in order to satisfy their design objectives) and proactive (able to opportunistically adopt new goals and take the initiative in order to satisfy their design objectives).

And further: *agent-oriented interactions generally occur through a high-level (declarative) agent communication language (typically based on speech act theory). Consequently, interactions are usually conducted at the knowledge level... when agents interact there is typically some underlying organizational context ... social interactions means existing relationships evolve ... and new relations are created ... the key abstraction models that define the agent-oriented mindset are agents, interactions and organizations ... explicit structures and mechanisms are often available for describing and managing the complex and changing web of organizational relationships that exist between agents...*

Further, Jennings and Wooldridge [14] come to the major point in agent design, i.e. the notion of Agent-Oriented Decompositions. The key to understand why this viewpoint is different from the usual one adopted in Software Engineering is highlighted by a footnote observation (page 8): *the view that decompositions based upon functions/actions/processes are more intuitive and easier to produce than those based upon data/objects is even acknowledged within the object-oriented community. In short, agent-oriented decompositions make it easier to develop complex systems.*

We have adopted this view in our research on complex systems. The function-action-process decomposition we have envisioned consists of three main directions of investigation, that hopefully will be followed by the integration of the results achieved within each of them. These directions consist of experiments:

- about the suitability of Actor architectures for designing and implementing complex software systems, such as diagnostic systems, both in closed domains and in open ones (cognitive diagnosis, as in learner-teacher interactions)[10][5] [4]. These studies use state-of-the art concurrent OO languages.
- about the overall nature and structure of the Agents architecture, including Cognitive Environments, a Dynamic scheduler of the messages available at the mailbox, and the Agent Communication Language [6]. It must be acknowledged that the architecture extends the Actor model that has showed in [2] to be extremely suitable: these studies develop new languages that enhance the autonomy of Actors to the status of autonomous Agents, in particular for modeling rational (cognitive) Agents.
- about the integration of Actors and Fuzzy reasoning in order to produce a truly adaptive, thus autonomous behavior in Agents [3], in particular reactive Agents.

Concerning the transition from Actors to Agents, a synthesis may be found in [15]. However, the authors do not include fuzziness as a mechanism for learning. In order to show how our Fuzzy Evolutionary Agents may be nicely integrated into a generic Agents architecture to offer them the learning functionality, in the following we will describe both our Actor and Agent architectures more in detail, and discuss their relations.

As a conclusion, we understand that not only autonomous interaction capacities (*reactive, proactive*) are necessary for agents but also adaptation, evolution and learning as an effect of the interactions with the environment that is supposed to change continuously. These properties can be associated not only to single Agents, but also to Societies of Agents (organizations).

1.2 Actors and Agents

The Actor model [2] is based on a software entity, named Actor, that is a computational mutable object. Mutable objects have a State, i.e. they have access to Memory where they may write intermediate values. Mutable objects, therefore, are a higher level abstraction for sequential networks. Differently from combinatorial networks, that are purely functional electronic (or logic) representations of transformations from inputs to outputs, and being purely functional, produce the same outputs at equal inputs, sequential networks at equal inputs may produce different outputs. An example is a counter, or a bank account. Typical transactional systems are intrinsically mutable objects (cf. the bank account, or a ticket reservation DBMS). The transformation of intrinsically mutable objects into state-less functions is a promising, although very risky, research area that we do not survey here.

Traditional views about languages and machines, however, do not include the meta-level (such as reflection, learning, reasoning about itself, etc.) as an essential component of the programs or processes. For instance, an Operating System (OS) is hardly to be defined in terms of state transformations of a sequential network because its behavior depends on special events such as those occurring at interrupts, or exceptions, that are essentially unpredictable and yet have an effect on its behavior. Even if we may consider the code of the OS as a program that generates processes modifing the memory accessible to the OS, the intrinsic behavior of these processes is not foreseeable because of those special events. The major task of an OS is exactly to control each of the processes it generates, as a server of multiple users, and also as a guardian of the exceptions eventually occurring. The most important control function of an OS consists in starting, suspending or resuming a process further to modifying its own accessible memory. An Actor, similarly, switches state during the computation, not only because it modifies its own memory, but also because exchanged messages may start its activities, suspend, resume and kill them when required. Basically, an Actor is quite similar to an Operating System: Actors serve messages as Operating Systems do.

Modelling software as collections of distributed, co-operative Actors is a natural evolution of object-level languages. Object-oriented programming is based on a philosophy of software development through progressive refinements, strengthening the abstraction level as a result of the use of abstract data types and information hiding. The Actors combine object-oriented and functional programming in order to make the management of concurrency easier for the user. Briefly, the Actor model can be summarized by the following scenario:

- the universe contains computational entities, called Actors;
- Actors perform computation through asynchronous, point-to-point message passing;
- each Actor is defined by its state (activity state and private memory), a mail queue to store external messages and its internal behaviour;
- an Actor's state is defined by its internal data, not sharable by other Actors. These local variables are named acquaintances;
- an Actor reacts to the external environment by executing its procedural skills, called scripts.

The Actor model is at the basis of Object-Oriented Concurrent Programming which constitutes one of the most important paradigms used to realize DAI (Distributed Artificial Intelligence) architectures (see, historically [2][12] [24]). Actors are quite well formalized: the notion of an Actor is well established in the research literature on Programming Languages. This is not the case for Agents.

Despite the considerable effort spent on Agent research, both at academic and at industrial level, there is some convergence about agents functionalities, but not any clear-cut formal theory or model that allows to distinguish when some software system performs as an agent (i.e., accepting the pragmatic attitude of Agents researchers, is an agent) and when not. For instance, Russel and Norvig in their widely accepted book on AI ([21], page 33) say explicitly *the notion of an agent is meant to be a tool for analysing systems, not an absolute characterisation that divides the world into agents and non-agents.* In [21], the authors conclude correctly that *the only concepts that yield sharp edge categories are mathematical concepts, and they succeed only because they are content free. Agents live in the real world (or some world), and real world concepts yield fuzzy categories.*
In short: agents are requirements for models, languages and applications for complex systems, where complex means semantically complex.

The same observation may be found in [16]: *From remote procedure call and remote procedure invocation (RPC and RMI) to CORBA and object request brokers, the goal has been the same. What distinguish Agent Communication Languages from such past efforts are the objects of discourse and their semantic complexity. Agent Communication Languages (ACLs) stand a level above CORBA for two reasons:*

- *ACLs handle propositions, rules and actions instead of simple objects with no semantics associated with them.*
- *An ACL message describes a desired state in a declarative language, rather than a procedure or method.*

In general, Agents own local duties. Those are associated to their ability in reasoning, in refusing orders, in negotiating commitments. To improve the intelligence of the behaviour, the agents must be equipped by autonomy that allows them to generate emerging behaviours that strengthen the performance of the agent on the fly, at run-time, without being forced to foresee all the (hardwired) possible behaviours.

1.3   Agents and Fuzziness

Soft-Computing technologies offer new tools to achieve higher functionality control even in the design of complex agent-based architecture. Recently we noticed a growing interest in combining DAI approaches with soft-computing, as a proof of an immediate benefit in sketching more efficient and manageable models of interaction, cooperation, and learning. The results of the merging are reported in different areas, with frequent intersections. Among these we underline:

- Using Fuzzy Logic as an inference engine for complex distributed, inference-based applications [22] [19][23]. Here fuzziness is usually used to represent and reason about vague knowledge with fuzzy production rules. The distributed approach is applied as an optimization boost to the construction and maintenance of fuzzy knowledge bases, an activity that is quite time consuming. Often, the deductive model inherits its properties from a variant of control in fuzzy logic.
- Using genetic computation as a powerful exploratory method to acquire knowledge, with reduced efforts in human design, in highly evolutionary environments [1]. This approach is nowadays one of the most popular ones: the simple biological metaphor at the base of genetic processing has been widely acknowledged as an efficient engine to design reactive, adaptive, and evolving Agents. The assumption that genetic algorithms are a promising approach to be used for modeling adaptive systems has been successfully verified in many experiments and in different areas such as, for instance, information retrieval/filtering [18], engineering design [13] and robot construction [20], where the genetic algorithm is chosen to engineer the action selection functionality.
- Using artificial neural networks (ANNs) to optimize Agent behavior. The impact of multi-agent strategies, often associated to evolutionary behaviors, shows to be beneficial for the transfer of knowledge across multiple functions [25] and for a successful multifunctional learning of the systems [26].

For the reasons briefly outlined above, our main goal is to extend the notion of an Actor towards that of an Agent, i.e. to inject in the Actor the competence to adapt its behaviour according to the fulfillment of its local (sub) goals. Our result is the definition and implementation of the FuzzyEvoAgent. This adaptive functionality of our Actors may be well integrated with other Agent architecture that, however, lack learning and adaptability.

After Section 2 on formal definitions, in Section 3 we will deepen the evolutionary engine that provides the mechanism to generate emerging behaviors in the agents. In Section 4 we will give a short description of our implementation and then its associated experimental applications. Finally, section 4 will present our conclusions, not just on the current results, but also with respect to our future work.

2   FuzzyEvoAgent

A FuzzyEvoAgent (FEA) is a modified version of a classical actor. Task activation for FEA's is related to the values of their acquaintances. The FEA's knowledge base includes, together with the canonical set of acquaintances, a set of fuzzy rules like the following one:

$$\text{when } acq_1 \text{ is } \alpha_{j1} \text{ and } acq_2 \text{ is } \alpha_{j2} \text{ and } \cdots \text{ and } acq_n \text{ is } \alpha_{jn} \text{ then tasks}$$
$$\text{execution is } \theta_j$$

This rule states a causality relation between the fuzzy value of acquaintances α_{ji} $(1 \leq i \leq n)$ and a fuzzy possibility distribution (θ_j) over the set of tasks to be executed.

Given the set of acquaintance values $\{acq_1, acq_2, \cdots, acq_n\}$ $\alpha_i's$ are fuzzy numbers over the domains of the corresponding $acq_i's$ and $\alpha_{ji}(acq_i)$ represents the matching degree of the actual value of the ith acquaintance with respect to the requirement of the jth rule. The fuzzy set θ_j is defined over the set of possible tasks. Let $\{task_1, \cdots, task_m\}$ be the set of tasks executable by the actor under consideration, then the above rule says that when the requirements are fully fulfilled, i.e. $\alpha_{ji}(acq_i) = 1 \ \forall i$ then $task_i$ is to be executed with possibility equal to $\theta_j(task_i)$. Since a task can be only executed or not, at all, a defuzzification is obviously required. We propose the following two schemas:

- there is only one task executed at each time; it is the one whose firing strength (see after) is maximum
- a fixed threshold is defined and all and only the tasks whose firing strength is above the threshold, are executed.

The firing strength of a task is computed via Mamdani's scheme of inference [17]. Given the set of acquaintances $acq_1, \cdots, acq_n$, the firing strength of the jth fuzzy rule is computed as $\tau_j = \alpha_{j1}(acq_1) \wedge \alpha_{j2}(acq_2) \wedge \cdots \wedge \alpha_{jn}(acq_n)$. From the fuzzy set $\bar{\theta} = \bigvee_j \tau_j \wedge \theta_j$ the firing strength of the ith task is the value $\bar{\theta}(task_i)$.

3   Evolution

As seen in the previous section, the value of the acquaintances causes the activation, for any agent, of a certain number of tasks. Task can bias the overall computation process more or less positively.

In our model, agents are assigned a gain value according to how much good was their performance with respect to the overall goals of the computation. Well performing agents (whose gain value is high) will survive, and continue carrying out their computational duties. Poor gain agents will be substituted by new agents built as a genetic mixing of the good ones.

We cluster the agents according to their duties. Agents belonging to the same class have the same acquaintances and the same set of tasks. They differ for the fuzzy rules relating acquaintances values and tasks to be performed. Genetic evolution of agents relies on the way chosen to represent their fuzzy knowledge and the procedure to produce new fuzzy rules from given ones. Recall the general fuzzy rule presented in the previous section. If the domain of the acquaintances set and the tasks set are taken to be invariant, then two rules differ for the values of the involved fuzzy sets. Furthermore, if these fuzzy sets are translated into the unit interval, to become function of $[0,1]^{[0,1]}$, then a rule can be interpreted as a fuzzy partition of the set $T = \{acq_1, \cdots, acq_n, task_1, \cdots, task_m\}$ with clusters evaluated by fuzzy numbers, i.e. function in $[0,1]^{[0,1]}$.

The fuzzy rule:

when acq_1 is α_{j1} and acq_2 is α_{j2} and $\cdots$ and acq_n is α_{jn} **then** tasks execution is θ_j

can be encoded as the fuzzy partition:

$$B = b_m^{\beta_m} b_{m-1}^{\beta_{m-1}} \cdots b_1^{\beta_1}$$

where $b_i's$ are subsets of T and $\beta_i's$ are the normalized versions of the fuzzy sets appearing in the fuzzy rule. For any i the set b_i represents the subset of T whose elements share the same normalised fuzzy number, which has been named β_i.

From this new representation of a fuzzy rule, we can use an evolutionary algorithm for generating fuzzy partitions, defined in [7], in order to produce new FEA's. In the following subsections we outline the main characteristic of this fuzzy evolutionary methodology for generating qualitative partitions of a set: we define the composition, mating and mutation operators of fuzzy strings and the genetic algorithms .

3.1   Composition Operator

Let A and B be two instances of the fuzzy string we have introduced above:

$$A = a_n^{\alpha_n} a_{n-1}^{\alpha_{n-1}} \cdots a_1^{\alpha_1} \text{ and } B = b_m^{\beta_m} b_{m-1}^{\beta_{m-1}} \cdots b_1^{\beta_1}$$

where $a_i (1 \leq i \leq n), b_j (1 \leq j \leq m)$ are subsets of T. In order to define the composition of A and B, we compute the least common refinement of the two fuzzy partitions $(\alpha_n, \alpha_{n-1}, \ldots, \alpha_1)$ and $(\beta_m, \beta_{m-1}, \ldots, \beta_1)$ obtaining the new partition $(\gamma_{m \times n}, \gamma_{m \times n-1}, \ldots, \gamma_1)$. Accordingly we compute the least common refinement of the boolean partitions $(a_n, a_{n-1}, \ldots, a_1)$ and $(b_m, b_{m-1}, \ldots, b_1)$ obtaining the new partition $(c_{n \times m}, c_{n \times m-1}, \ldots, c_1)$. So we have the new fuzzy string $C = c_{n \times m}^{\gamma_{m \times n}} c_{n \times m-1}^{\gamma_{m \times n-1}} \cdots c_1^{\gamma_1}$. The composition of A and B is then defined as the string C obtained by cutting out from C all the components $c_i^{\gamma_i}$ such that γ_i is the empty fuzzy set, i.e., $\gamma_i(x) = 0, \forall x$, and by reordering the remaining components $c_j^{\gamma_j}$, in decreasing order, according to the modal point of their fuzzy set γ_j.

For the composition operator, we have just defined, the following properties hold : closure, associativity, commutativity, idempotence, existence of a neutral element [8].

3.2   Mating

In a genetic algorithm, mating of individuals is aimed at producing new individuals sharing the characteristic of the parents, eventually leading towards optimality. Mating of two parent strings generally generates two child strings by swapping parts of the genotype between parent strings.

In fact, in the model we are going to present, mating produce one single child string. We treat inheritance from a phenotypic point of view, since we are interested in qualitative characteristic. In biological individual the exchange of genetic material produces a smooth mix of phenotypic appearance of off-spring; in our algorithm the mating of two FEA's yields a new FEA whose behaviour, in terms of fuzzy rules is obtained as a sort of the average of the fuzzy rules in the knowledge base of the parent actors. Let two actors' knowledge base contain fuzzy rules whose encodings are represented by the two sets $\{A_1, A_2, \cdots, A_n\}$ and $\{B_1, B_2, \cdots, B_m\}$ respectively; where $A_i's$ and $B_j's$ are instances of the fuzzy partition defined above. The mating of the two actors will produce a new actor with k fuzzy rules, where k is a random generated number in the set $[min(m, n), max(n, m)]$. Then the ith fuzzy rule of the new actor will be attained as $C_i = A_{li} \diamond B_{ti}$, where $1 \leq l_i \leq n$ and $1 \leq t_i \leq m$ are randomly generated. In other words, to generate the fuzzy rules of the new actor, we choose at random one rule out of each rule set of the parent actors and mate them by using the composition operator described in the previous section.

3.3   Mutation

The new actor undergoes a mutation operator. Mutation is, as usual, aimed at producing unexpected individuals, i.e. individuals which present characteristics not belonging to any of the parents. Mutation in our model is implemented by two different operations. One is meant to modify the fuzzy numbers, like a sort of noise in the transcription of the genetic code of the strings. The result is that a randomly chosen fuzzy number α is replaced with a new fuzzy number $\alpha \pm \epsilon$ where the sign of the operation and the value of the fuzzy noise ϵ are also random; the sum here is, of course, the extended operation between fuzzy numbers.

A second mutation operation is meant to operate on the structure of the partition entailed by a string. It works by swapping elements among two randomly chosen sets of the partition.

Let $A = a_n^{\alpha_n} a_{n-1}^{\alpha_{n-1}} \cdots a_1^{\alpha_1}$ and Mut_1, Mut_2 represent the two mutation operators. $Mut_1(A) = a_n^{\alpha_n} \cdots a_i^{\alpha_i + \epsilon} \cdots a_1^{\alpha_1}$, where the index i and the fuzzy number ϵ have been randomly chosen.

Being $\underline{a_i}$ and $\underline{a_j}$ two randomly generated subsets of a_i and a_j, respectively, then $a_i' = a_i / \underline{a_i} \cup \underline{a_j}$ and $a_j' = a_j / \underline{a_j} \cup \underline{a_j}$. Thus $Mut_2(A) = a_n^{\alpha_n} \cdots a_i'^{\alpha_i} \cdots a_j'^{\alpha_j} \cdots a_1^{\alpha_1}$, where i and j are the randomly chosen indexes of the sets to undergo the mutation and .

4   JAVA Implementation

This section gives some details about the implementation of a FEA in Java. The first step was to introduce a new class *Actor* in Java that provides primitives to use actor programming exploiting as super-class the existing class *Thread*. Thus an Actor inherits all the access methods of JAVA Threads; moreover it has the required characteristic according to Agha's definition of actor paradigm, particularly regarding the possibility of asynchronous message exchange. It is possible to define group of actors. This feature is accomplished by means of the class *ActorGroup*.

ActorGroup is a class useful to define operations that are to be applied concurrently on a population of actors (those belonging to the group).

RuleActor inherits the characteristic of the Actor class, but it also includes specific behavioural rules, in the light of the model described in the previous section. Since it is an abstract class, it is not possible to define objects of type RuleActor, but only objects inheriting RuleActor features. Actually, tasks and acquaintances of the actors can be determined only according to the particular application (then by the programmer).

5   Experimental Results

We tested our system within the framework of an ecosystem simulation. The testbed was a simple ecosystem including frogs and flies. The attention is focused on frogs' behaviour.

The environment is a 100x100 square matrix, with 100 frogs and 1000 flies.

Frogs are agents of the type defined above, whose behaviour is defined by fuzzy rules. At first agents' rules are defined at random. Frogs' behaviour rules define relationship between tasks activation and acquaintances values. Tasks considered are *movement toward food* and *flies eating*. Acquaintances considered are *starving time* and *flies distance*.

We expect system evolution to produce a population of frogs able to survive by selecting the optimal behavioural rules. Frogs population is evaluated every 10 sec. Frog fitness is computed by considering the capacity to achieve food within a fixed slice of time (*starving period*). As long as the amount of time from the last feeding increases, the frog's fitness is decreased. A fitness bonus is gained by a frog whenever it eats a fly. Fitness decrease is also used to punish fool behaviours (e.g. an attempt to eat a fly, when no fly is nearby). If a frog's fitness falls below a fixed value (*living energy*) the agent gets killed (i.e. the frog dies). Such an evaluation aims at selecting the best fitting frogs which undergo the evolutionary operations (mating and mutation) producing new individuals, whose behaviour is defined to be a refinement of parents' ones (mating) and also includes new strategical features (mutation), eventually leading to optimal effective behaving individuals.

In Figure 1 the average fitness of frog population and the fitness of the best frog are plotted against the generating populations.

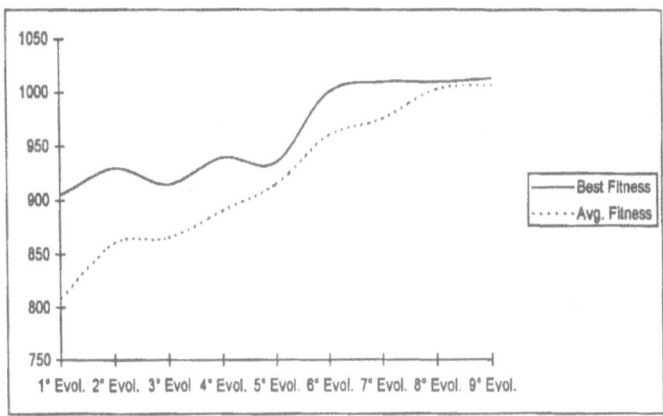

Fig. 1. Evolution of Frog FuzzyEvoAgents

6   Conclusions

We have described a general framework for evolutionary distributed, concurrent computation based on the notion of FuzzyEvoAgents. They are Active Objects capable to modify their inner capabilities according to a scheme of controlled, collective evolutionary learning. The theoretical basis as well as an experiment for an implemented prototype have been described and show quite promising results, particularly in terms of the effectiveness and efficiency of a rather simple software architecture performing a task - learning - known to be intrinsically complex.

As we have shown previously for "cognitive" or "rational" Agents, once the functionalities associated to Agents have been explored in experiments based on Actors, the integration of those functionality's into a generic Agent architecture does not present particular difficulties. In the essence, one has to recombine at the level of a programming language - the Agent language - the primitives that have been "discovered" at the level of a well performing program in another, more established language - the Actor language. Each program is a language for a restricted class of applications. Software research consists in transforming application-oriented programs into generic languages; i.e. generalizing the functionalities of software at a higher level of abstraction, yet keeping the approach simple enough for humans.

Thus, our perspective work will aim at the goal of integrating the FuzzyEvoAgent's control architecture into our current architecture for generic Agents. As the latter has been developed with a control module for the Agent's state that consists in a Production System, the integration seems straightforward. Variants - such as Augmented Transition Networks - are sets of Production Systems, where each one may be fuzzified in the same way, if required by the application.

Finally, a note on our implementations. The control of FuzzyEvoAgents has been implemented in Java by exploiting multi-threads that allow an Actor behavior. One of our current research trend consists in transforming our agents in mobile agents able to assist the user in searching useful information on the Web. The Java implementation is helping this project, due to the high level description of the agents and to the embedding of java control into web pages.

The way to coping with complexity is full of obstacles due to contradictory constraints. Each problem requires linguistic tools possessing primitives adequate to denote the problem components, but each time new primitives are developed, the human competence needed to use those primitives requires anyone to be able to master their associated properties. The Agent's recent history is not different from previous ones: tools above tools for building software systems with more and more potentialities. We have always tried to reduce at a minimum the explosive growth of tools, even if, as we know, it is impossible to cope with difficult problems with minimal resources.

Acknowledgements

This paper is essentially based on two lectures given in 1998 by Stefano Cerri and in 1999 by Vincenzo Loia. FuzzyEvoAgents are a result of a research project that has benefited of frequent discussions with many people of our Soft-Computing Lab. A particular thank to Ferdinando Cicalese, without his contribution the evolution of FuzzyEvoAgents would have been fruitless.

References

1. Ackley, D., Littman, M. (1991) Interactions between Learning and Evolution. Artificial Life II, Edited by C. Langton, C. Taylor, J. Farmer and S. Rasmussen, Addison Wesley.
2. Agha, G. (1986) A Model of Concurrent Computation in Distributed Systems. MIT Press, Cambridge, MA.
3. Cicalese, F., Di Nola, A., Loia, V. (1999) A Fuzzy Evolutionary Framework for Adaptive Agents. Proceedings of 13th International ACM Symposium of Applied Computing, 29 Feb. - 2 Mar. 1999, San Antonio, Texas, ACM Press.
4. Cerri, S.A., Gisolfi, A., Loia, V. (1999) Towards the Abstraction and Generalization of Actor-based Architectures in Diagnostic Reasoning. Collaboration between Human and Artificial Societies, Coordination and Agent-Based Distributed Computing, vol. 1624, Lecture Notes in Artificial Intelligence, J. A. Padget, Ed. Berlin Heidelberg New York: Springer-Verlag, pp. 115-131.
5. Cerri, S.A., Loia, V. (1997) A Concurrent, Distributed Architecture for Diagnostic Reasoning.Journal of User Modeling and User Adapted Interaction, vol. 7, pp. 69-105.
6. Cerri, S.A. (1999) Shifting the focus from control to communication: the STReams OBjects Environments model of communicating agents. Collaboration between Human and Artificial Societies, Coordination and Agent-Based Distributed Computing, vol. 1624, Lecture Notes in Artificial Intelligence, J. A. Padget, Ed. Berlin Heidelberg New York: Springer-Verlag, pp. 71-101.
7. Cicalese, F., Loia, V. (1988) A Fuzzy Evolutionary Approach to the Classification Problem. Int. Journal of Intelligent and Fuzzy System, Special Issue on Evolutionary Computation, (H. Takagi Ed.), Vol.6, Issue 1, 1998.
8. Cicalese, F., Gisolfi, A. (1996) Classifying through a fuzzy algebraic structure. Fuzzy Sets and System, vol. 78, pp.317-331.
9. Genesereth, M.R., Ketchpel, S.P. (1994) Software agents. Communications of the ACM, 37(7):48-53.
10. Gisolfi, A., Loia, V. (1994) Designing Complex Systems in Distributed Architectures: an ITS Perspective. Applied Artificial Intelligence, vol. 8, pp. 393-411.
11. Guessoum, Z., Briot, J-P. (1998) From Active Objects to Autonomous Agents. LIP6: Laboratoire d'Informatique de Paris VI, April 1998.
12. Hewitt, C. (1977) Viewing control structure as pattern of passing message. Artificial Intelligence, vol.8 pp.326-364.

13. Hoffmann, F. (1998) Incremental tuning of fuzzy controllers by means of evolution strategy. Proceedings of GP-98 Conference, pp. 550-556, Madison, Wisconsin.
14. Jennings, N., Wooldridge, M. (2000) Agent-Oriented Software Engineering. Handbook of Agent Technology, J. Bradshaw, Ed. Boston, MA: AAAI/MIT Press, (to appear).
15. Kafura, D., Briot, J-P.(1998) Actors and Agents. IEEE Concurrency, vol. 6, pp. 24-29.
16. Labrou, Y., Finin, T., Peng,Y. (1999) Agent Communication Languages: The Current Landscape. Intelligent Systems, vol. 14, 2, pp. 45-52.
17. Mandami, E.H., and Assilian S. (1975) An experiment in linguistic synthesis with a fuzzy logic controller. Int. J. Man Machine Studies, 7(1):1-13.
18. Moukas, A.G. (1996) Amalthaea: Information Discovery and Filtering using a Multiagent Evolving Ecosystem. Proceedings of the Conference on Practical Application of Intelligent Agents and Multi-Agent Technology, London.
19. Nebot, A., Cellier, F.E., Linkens, D.A. (1996) Synthesis of an Anaesthetic Agent Administration System Using Fuzzy Inductive Reasoning.Artificial Intelligence in Medicine, 8(3), pp.147-166.
20. Pollack, J. B., Lipson, H., Funes, P.O, Ficici, S. G., Hornby, G. (1999). Coevolutionary Robotics. The First NASA/DoD Workshop on Evolvable Hardware (EH'99). John R. Koza, Adrian Stoica, Didier Keymeulen, Jason Lohn, eds., IEEE Press
21. Russel, S. J., Norvig, P. (1995) Artificial Intelligence: A Modern Approach. Englewood Cliffs, NJ: Prentice Hall.
22. Song, H., Franklin, S., Aregahegn Negatu (1996) SUMPY: A Fuzzy Software Agent. Proceedings of the ISCA Conference on Intelligent Systems, Reno Nevada.
23. Sanz, R., Matia, F., de Antonio, A., Segarra (1998) Fuzzy Agents for ICa. Proceedings of FUZZ-IEEE 1998, pp.545-550, IEEE Press.
24. Yonezawa, A., Takoro, M. (1987) Object-Oriented Concurrent Programming. Boston, MA: MIT Press.
25. Wermter, S., Panchev, C., and Arevian, G. (1999) Hybrid Neural Plausibility Networks for News Agents. Proceedings of the Sixteenth National Conference on Artificial Intelligence, July 18-22 1999, Orlando, Florida.
26. Wang, F., Mckenzie, E. (1999) Multifunctional Learning of a Multiagent based Evolutionary Artificial Neural Network with Lifetime Learning. IEEE International Symposium on Computational Intelligence in Robotics and Automation, Monterey, California, USA, Nov., 1999, pp.332-337.

Finiteness and Duality in MV-algebras Theory

Antonio Di Nola[1] and Revaz Grigolia[2]

[1] Dipartimento di Matematica e Informatica, Universitá di Salerno,
   via S. Allende, 84081 Baronissi (SA), Italy - dinola@unina.it
[2] Institute of Cybernetics, Georgian Academy of Sciences
   Sandro Euli Str. 5, 38086 Tiblis, Georgia - grigolia@yahoo.com

Abstract. Some results about finiteness properties of MV-algebras and some dualities between categories of MV-algebras and categories of certain ordered structures are presented. Actually, finite MV-algebras are presented as algebras of words. Moreover, it is presented a duality between the category of MV-algebras which are finitely generated, having finite spectrum, and the category of finite linear dual Heyting algebras.

1 Preliminaries

A *root system* is a partially ordered set (for short, a *poset*) P such that for each $x \in P$, the final section $\{y \in P \mid y \geq x\}$ is totally ordered. Let P be a poset and $Q \subseteq P$. We then say that Q is a *down set* if, whenever $x \in Q$, $y \in P$ and $y \leq x$, it follows that $y \in Q$. Dually, $W \subseteq P$ is called an *upper set* if whenever $x \in W$, $y \in P$ and $x \leq y$, $y \in W$. The smallest down set (upper set) containing a given subset U of P we denote by $U \downarrow (U \uparrow)$; instead of $\{x\} \downarrow (\{x\} \uparrow)$ we write $x \downarrow (x \uparrow)$. A *spectral* root system is a root system $(X, \leq)$ fulfilling the following conditions:

(i) Each totally ordered subset of X has supremum and infimum in X;
(ii) If x, y are elements of X such that $x < y$, then there are s, t in X such that $x \leq s < t \leq y$, and there is no element of X between s and t.

Let X be a topological space and cl be closure operator of X. A non empty closed subset A of X is called *irreducible* if it can not be written as the union of two strictly smaller closed subsets; that is, whenever C_1 and C_2 are closed sets with $A = C_1 \cup C_2$, then either $A = C_1$ or $A = C_2$. A subset B of X is called *monogenic* iff there is an element $b \in B$ such that B is the closure of the singleton $\{b\}$. A topological space X is called *spectral* iff X is a T_0-space and every irreducible subset of X is monogenic. Let $(P, \leq)$ be a poset. We say that a spectral topology on P is *compatible* with the order $\leq$ provided that for every $x, y \in P, y \in cl(x)$ iff $x \leq y$.

We refer to [1] and [7] for background on the variety $\mathbf{D}_{01}$ of bounded distributive lattices, i.e., distributive lattices with 0 and 1. For any $L \in \mathbf{D}_{01}$ let $\mathrm{Sp}(L)$ be the set of its prime ideals. Then the *Stone space* of L, in symbols

$$\mathrm{Stone}(L)$$

is the set $Sp(L)$ equipped with the *hull-kernel* topology, i.e., the topology generated by the basis $\{\emptyset\} \cup \{\hat{x} \mid x \in L\}$, where $\hat{x} = \{I \in Sp(L) \mid x \notin I\}$. The following result is well-known (see, for istance, [19], Section 3, or Prodanov [20], Section 1):

Lemma 1. *For each finite poset X there is exactly one spectral topology compatible with the order of X*

Lemma 2. *A topological space X is homeomorphic to $Stone(L)$ for some bounded distributive lattice L iff X satisfies the following conditions:*

(a) *X is spectral;*
(b) *X is compact (in the sense that every family of open sets whose union is X contains a finite subfamily whose union is X);*
(c) *the intersection of any two compact open subsets of X is compact open;*
(d) *the compact open sets of X form a basis in X.* □

Given spectral spaces X and Y, a map $g : X \to Y$ is said to be *strongly continuous* iff $g^{-1}(A)$ is a compact open set for every compact open set $A \subseteq Y$. A familiar construction yields a contravariant functor Stone(), which turns out to be a categorical duality between $\mathbf{D}_{01}$ and the category $\mathbf{S}$ of spectral spaces satisfying conditions (b), (c) and (d) of Lemma 2, with strongly continuous maps as morphisms (see, e.g., [7]).

Let $\mathbf{S}'$ be the full subcategory of $\mathbf{S}$ whose objects are the finite members of $\mathbf{S}$. It is well-known that there is a categorical duality between $\mathbf{S}'$ and the category $\mathbf{P}'$ of finite posets with the order preserving maps as morphisms (see, e.g. [7]). It turns out that the full subcategory $\mathbf{D}'_{01}$ of the category $\mathbf{D}_{01}$, having finite lattices as objects, is a dual category of $\mathbf{P}'$. Hence, each finite distributive lattice can be represented in the form $L(X)$, where $L(X)$ is the set of all down subsets of a poset X, ordered by inclusion. We usually omit the explicit mention of the partial ordering "$\leq$" of a poset.

Let H be a bounded distributive lattice and $x, y \in H$. If there is a largest $z \in H$ such that $x \wedge z \leq y$, then such z is denoted by $x \rightharpoonup y$ and is called the *implication* or *relative pseudo-complement* of x with respect to y. A distributive lattice H with 0, in which $x \rightharpoonup y$ exists, is called by *Heyting algebra*. If there is a smallest element $z \in H$ satisfying the inequality $x \vee z \geq y$, then we denote such element z by $x \leftharpoonup y$ and say that z is the *pseudo-difference* of x and y. A distributive lattice H with 1, in which there exist both $x \rightharpoonup y$ and $x \leftharpoonup y$, is called by *bi-Heyting algebra*(or *Skolem algebra* in Esakia's terminology [14]). Having any poset X we can form a complete bounded distributive lattice $L(X)$ of all down subsets of X. For any elements $a, b \in L(X)$ there exists pseudo-difference $a \leftharpoonup b$, which is defined as $(a \smallsetminus b) \downarrow$, and the relative pseudo-complement of a with respect to b, which is defined as $X \setminus ((a \setminus b) \uparrow)$.

Let $(X, \leq)$ and $(Y, \leq')$ be posets. We say that an isotone (i.e. order-preserving) map $f : X \to Y$ is *strongly isotone* [12] if for any elements $x \in X, y \in Y f(x) \leq' y$ iff $\exists x' \in X(x \leq x'$ and $f(x') = y)$, or, in other writing, $f(x) \in y \downarrow$ iff $x \uparrow \cup f^{-1}(y) \neq \emptyset$.

It is easy to see that any strongly isotone map is isotone. Let us note that if we have finite posets X and Y and $f : X \to Y$ is a strongly isotone map, then it is easy to check that $f^{-1} : L(Y) \to L(X)$ preserves pseudo-difference.

Lemma 3. *[13]. Let $(X, \leq)$, $(Y, \leq')$ be arbitrary posets and $f : X \to Y$ be an isotone map. Then the following are equivalent:*

(1) *f is strongly isotone;*
(2) *$\forall y \in Y \quad (f^{-1}(y)) \downarrow = f^{-1}(y \downarrow)$;*
(3) *$\forall x \in X \quad (f(x)) \uparrow = f(x \uparrow)$;*
(4) *if U is an upper set of X (of Y, resp.), then $f(U)$ ($f^{-1}(U)$, resp.) is an upper set of Y (of X, resp.).*

Let X, Y be posets and $f : X \to Y$ be an onto strongly isotone map. Then f^{-1} generates a partition of X. Let E denote the corresponding equivalence relation. Let $E(W) = \cup_{x \in W} E(x)$ for any upperset W of X. We call $E(W)$ the *saturation* of W (by E). Then, according to condition (4), with the above notations, we have:

Theorem 1. *The saturation $E(W)$ of any upper set $W \subseteq X$ is an upper set.*

Proof. Let us observe that $E(W) = f^{-1}(f(x))$. Then the assertion immediately follows from item (4) of Lemma 3.

Theorem 2. *Let X be a poset and E be an equivalence relation on X such that the saturation of any upper set of X is upper set. Then E-saturated down sets of X are closed under pseudo-difference $Y_1 \leftarrow Y_2 = (Y_1 \backslash Y_2) \downarrow$, where Y_1 and Y_2 are any saturated down sets of X.*

Proof. Let $f : X \to X/E$ be the natural map such that $f(x) = x/E$ for every $x \in X$. Then the partition corresponding to the equivalence relation E is $ker(f)$, and $(f^{-1})(x/E) = E(x)$. Then the proof follows from item (2) of Lemma 3.

2 MV-algebras

It is well known that MV-algebras are algebraic models of infinite-valued Lukasiewicz logic and also an example of a special class of residuated structures.

An *MV-algebra* is a structure $A = (A, \oplus, \circ, *, 0, 1)$ such that $(A, \oplus, 0)$ is an abelian monoid, and, moreover, $x \oplus 1 = 1$, $0^* = 1$, $x^{**} = x$, $x \circ y = (x^* \oplus y^*)^*$, and $(x^* \oplus y)^* \oplus y = (y^* \oplus x)^* \oplus x$, for all $x, y \in A$. Following tradition, for every $x, y \in A$, let us write $x \vee y = (x \circ y^*) \oplus y$ and $x \wedge y = (x \oplus y^*) \circ y$. Then as proved in [2], $(A, \vee, \wedge, 0, 1)$ is a bounded distributive lattice.

When A denotes an MV-algebra , $Spec A$ will denote the set of all prime ideals of A endowed with inclusion relation between prime ideals, sometimes denoted by $\subseteq$ or $\leq$. The set of all ideals of MV-algebra A, ordered by inclusion, is the universe of an algebraic lattice, which we denote by $J(A)$. Given an MV-algebra A and $a \in A$, $< a >$ denotes the principal ideal generated by a in A, i.e. $< a >= \{b \in A : b \leq na$ for some $n \in \omega\}$. The lattice $J_p(A)$ of all principal ideals of A is a sublattice of $J(A)$. The lattice operations of $J_p(A)$ satisfy the following conditions:

$$< a > \vee < b >=< a \vee b >=< a \oplus b >,$$

$$< a > \wedge < b >=< a > \cap < b >=< a \wedge b > .$$

If A, B are MV-algebras, then an MV-homomorphism $h : A \to B$ induces a homomorphism $J_p(h) : J_p(A) \to J_p(B)$ by the prescription $J_p(< a >) =< h(a) >$ for every $a \in A$. The elements of $J_p(A)$ are compact elements of the distributive lattice $J(A)$ [15].

From above we get a covariant functor $\mathbf{J_p}$ from the category of MV-algebras to the category of bounded distributive lattices. It follows from [5], that the functor $\mathbf{J_p}$ coincides with the one defined by P.Belluce in [2], here denoted by $\mathbf{Be}$. The equivalence relation $\equiv$: $x \equiv y$ iff $\forall P \in SpecA(x \in P$ iff $y \in P)$.

The canonical mapping $\beta : x \to \beta(x)$ defines Belluce's functor $\mathbf{Be}$. This equivalence relation is a congruence with respect to $\oplus$ and $\wedge$. The quotient set $\mathbf{Be}(A) = A/ \equiv$ becames a bounded distributive lattice with respect to the following operations :

$\beta(x) \vee \beta(y) = \beta(x \oplus y)$ and $\beta(x) \wedge \beta(y) = \beta(x \wedge y)$.

For any ideal I of A $\beta(I) = \{\beta(x) : x \in I\}$ is an ideal of the lattice $\mathbf{Be}(A)$. Let A be an MV-algebra. As it is well known, $SpecA$ equipped with set-theoretic inclusion is a root system. For any $x \in A$ let $supp(x) = \{J \in SpecA : x \notin J\}$. Then $S(A) = \{supp(x) : x \in A\}$ is a basis for a topology on $SpecA$. $S(A)$ ordered by inclusion is a bounded distributive lattice.

Lemma 4. *[6]. Let A be an MV-algebra. Then $S(A)$ is isomorphic to $\mathbf{Be}(A)$.*

Proof. The map sending $supp(x)$ to $\beta(x)$ is the isomorphism from the lattice $S(A)$ to the Belluce lattice $\mathbf{Be}(A)$.

Let A, B be MV-algebras and $f : A \to B$ be an embedding. Then we can define the map $\tilde{f}$ as follows: for every $P \in SpecB$

$$\tilde{f}(P) = \{a \in A : f(a) \in P\}.$$

It is easy to check that $\tilde{f}$ is a map from the poset $SpecB$ to the poset $SpecA$.

Lemma 5. *[8]. Let A, B be MV-algebras and f be homomorphism from A to B. Then $\tilde{f}$ preserves upper sets from the poset $SpecB$ to the poset $SpecA$, i.e. $\tilde{f}$ is strongly isotone.*

Let G be an abelian ℓ-group with strong unit u. By definition,

$$\Gamma(G, u) = [0, u] = \{x \in G \; : \; 0 \leq x \leq u\}$$

is the MV-algebra B, obtained by equipping the unit interval $[0, u]$ of G with the following operations: $x \oplus y = (x + y) \wedge u$, $x \circ y = (x + y - 1) \vee 0$, $x^* = u - x$, and identifying u with the unit element 1 of B. As proved in [18], up to isomorphism, for any MV-algebra B there is a unique abelian ℓ-group G with strong unit u such that B is isomorphic to $\Gamma(G, u)$. Recall, that Mundici's functor Γ is a categorical equivalence (a full, faithful, dense functor) between abelian lattice-ordered groups (for short, ℓ-groups; we refer to [3] for background on abelian ℓ-groups) with strong unit and MV-algebras [18]. A straightforward verification shows that this equivalence preserves prime spectra. Let A be an MV-algebra. Then the intersection of all maximal ideals, the radical of A, shall be denoted by $\mathrm{Rad}\, A$ and the set $\{x \in A : x^* \in \mathrm{Rad}\, A\}$ by $(\mathrm{Rad}\, A)^*$. An important example of MV-algebra is the MV-algebra C described by Chang, (see[4]). A generalization of the MV-algebra C gives rise to the class of perfect MV-algebras by the following

Definition 1. An MV-algebra A is *perfect* iff $A = (\mathrm{Rad}\, A) \cup (\mathrm{Rad}\, A)^*$.

An MV-algebra A is called $MV(C)$-*algebra* if additionally A satisfies the condition $(x \circ x) \oplus (x \circ x) = (x \oplus x) \circ (x \oplus x)$. Remark that the variety $\mathbf{MV}(C)$ of all $MV(C)$-algebras is generated by Chang algebra C [9]. Let us note that perfect MV-algebras are categorically equivalent to abelian ℓ-groups [4]. Note that $C \cong \Gamma(Z \otimes Z, (1, 0))$, where $Z \otimes Z$ is the totally ordered group obtained by the lexicographic product of the additive group of integers by itself, with strong unit $(1, 0)$.

Lemma 6. *Let A be an MV-algebra. Then the following are equivalent:*

(i) *$\mathrm{Spec}A$ is finite*
(ii) *The Belluce lattice $\mathbf{Be}(A)$ is finite.*

Proof. Assume (i).
Then the powerset $\mathcal{P}(\mathrm{Spec}A)$ is finite. Define the map $\sigma \; : \; \mathbf{Be}(A) \to \mathcal{P}(\mathrm{Spec}A)$ by $\sigma(\beta(x)) = \{J \in \mathrm{Spec}A \mid x \notin J\} = \mathrm{supp}(x)$, for all $x \in A$. Trivially, σ is (well defined and) injective. Thus, by the finiteness of $\mathcal{P}(\mathrm{Spec}A)$, condition (ii) holds.
Conversely, assume (ii).
Let $\varphi : \mathrm{Spec}A \to \mathcal{P}(\mathbf{Be}(A))$ be defined by $\varphi(J) = \{\beta(x) \in \mathbf{Be}(A) \mid x \in J\}$, for all $J \in \mathrm{Spec}A$. It follows that φ is injective: as a matter of fact, assume $J, J' \in \mathrm{Spec}A$, with $J \neq J'$, and let $x \in J' \backslash J$. Then $\beta(x) \subseteq J'$ and $\beta(x) \nsubseteq J$, i.e., $\varphi(J') \neq \varphi(J)$. Thus, from the finiteness of $\mathcal{P}(\mathbf{Be}(A))$, condition (i) follows.

Theorem 3. *Let $\mathcal{R}$ be a finite root system. Then there is an MV-algebra A such that $\mathrm{Spec}A$ and $\mathrm{Stone}\,(L(\mathcal{R}))$ are homeomorphic.*

Proof. In [6] it is shown that each spectral root system is isomorphic to the poset of prime ideals of an MV-algebra, ordered by inclusion. Since each finite root system is trivially a spectral root system, we have that each finite root system is isomorphic to the poset of prime ideals of an MV-algebra A, and by Lemma 1 we conclude that $SpecA$ and $\text{Stone}(L(\mathcal{R}))$ are homeomorphic.

In section 3 finite MV-algebras are represented by MV-algebras of words. In section 4 finite MV-algebras are shown to be dual with *uniform symmetric models*. In section 5 it is showed that the category of finitely generated $MV(C)$-algebras, having finite spectrum of prime ideals is dual to the category of finite linear dual Heyting algebras.

3   Finite MV-algebras of words

Let $A = \{a_1, \ldots, a_n\}$ be a finite alphabet and $K = \{k_1, \ldots, k_n\}$ be a set of non negative integers, *the set of bounds*, and n be a positive integer. Let A^* be the set of all finite strings $x_1 x_2 \ldots x_t$ of elements from A, including the empty string $\perp$. Two strings $x_1 \ldots x_p$ and $y_1 \ldots y_q$ in A^* are equal iff $p = q$ and $x_i = y_i$ for every i such that $1 \leq i \leq p$. Let us call *word* any element w from A^*. If $w = x_1 \ldots x_t$ is a word in $A^* \setminus \perp$, then the natural number t is the *lenght* of w, in symbols $|w| = t$. By convention $|\perp| = 0$. $|w|_{a_i}$ denotes the number of occurrences of the symbol a_i in w. Let S be a subset of A^* such that $w \in S$ iff $|w|_{a_i} \leq k_i$ for every $1 \leq i \leq n$. We shall identify words w and w' from S iff $|w|_{a_i} = |w'|_{a_i}$ for every $a_i \in A$. Let S^* be the quotient set of S, module the above identification. We assume also that the empty word, $\perp$, is an element of S^*. Then we can safely represent elements from S^* by words of the following form: $a_{i_1} \ldots a_{i_1} a_{i_2} \ldots a_{i_2} \ldots a_{i_t} \ldots a_{i_t}$ with $i_1 \ldots i_t \in \{1, \ldots, n\}$ and $i_1 < \cdots < i_t$. The symbol $\top$ shall denote the word $a_1 \ldots a_1 a_2 \ldots a_2 \ldots a_n \ldots a_n$ with $|\top|_{a_i} = k_i$ for every i. Any element $w = a_{i_1} \ldots a_{i_1} \ldots a_{i_t} \ldots a_{i_t}$ of S^* has finite length and it can be identified by the finite sequence of integers $|w|_{a_{i_1}}, \ldots, |w|_{a_{i_t}}$. Equip S^* with the following operations $\oplus$ and $\neg$. For any $w, w' \in S^*$, set $w \oplus w' = w''$ and $\neg w = w'''$ where w'' and w''' are defined as follows

$$|w''|_{a_i} = min\{|w|_{a_i} + |w'|_{a_i}, k_i\}$$

$$|w'''|_{a_i} = k_i - |w|_{a_i}$$

for every $i \in \{1, \ldots, n\}$.

Theorem 4. *The structure $S^* = (S^*, \oplus, \neg, \perp, \top)$ is an MV algebra.*

We call S^* *an MV algebra of words of A with bounding set K*, shortly MV algebra of words.

Corollary 1. *S^* is a finite MV algebra.*

Thus any MV algebra of words is finite. We shall show that, up to isomorphism, the converse is true too. Indeed, recall that for each integer $n \geq 2$, the finite set

$$S_n = \{0, \frac{1}{(n-1)}, \ldots, \frac{(n-2)}{(n-1)}, 1\}$$

is a finite totally ordered MV algebra (for short, finite MV chain), subalgebra of $[0, 1]$, and that any MV algebra A is finite iff A is isomorphic to a finite product of finite MV chains, in symbols:

$$A \cong S_{d_1} \times \ldots \times S_{d_u}$$

for some integers $2 \leq d_1 \leq \ldots \leq d_u$.

Theorem 5. *Let $A \cong S_{d_1} \times \ldots \times S_{d_u}$ be a finite MV algebra. Then there exists an MV algebra S^* of words such that $A \cong S^*$.*

Proof. Take the alphabet $A = \{a_1, \ldots a_u\}$ and the bounding set $K = \{k_1, \ldots k_u\}$ with $k_i = d_i - 1$, for every i such that $1 \leq i \leq u$. Let S^* be the MV algebra of words of A with K as bounding set. Define a map f from A to S^* as follows

$$f(\frac{1}{d_1 - 1}, \frac{2}{d_2 - 1}, \ldots, \frac{u}{d_u - 1}) = w$$

where

$$|w|_{a_i} = h_i$$

$$0 \leq h_i \leq d_i - 1$$

$$1 \leq i \leq u.$$

We shall prove that f is an isomorphism between the MV algebras A and S^*.

Claim 1. f preserves $\oplus$

Indeed let $x, x' \in A$, such that

$$x = (\frac{h_1}{k_1}, \ldots, \frac{h_u}{k_u})$$

and

$$x' = (\frac{h'_1}{k_1}, \ldots, \frac{h'_u}{k_u})$$

then we have

$$x \oplus x' = (\frac{h_1}{k_1} \oplus \frac{h'_1}{k_1}, \ldots, \frac{h_u}{k_u} \oplus \frac{h'_u}{k_u})$$

and

$$|f(x) \oplus f(x')|_{a_i} = min\{|f(x)|_{a_i} + |f(x')|_{a_i}, k_i\} = min\{h_i + h'_i, k_i\} = |f(x \oplus x')|_{a_i}$$

for $1 \leq i \leq u$. Hence, we get

$$f(x \oplus x') = f(x) \oplus f(x').$$

Claim 2. f preserves $\neg$.

Indeed for $1 \leq i \leq u$, we have

$$|\neg f(x)|_{a_i} = k_i - |f(x)|_{a_i} = k_i - h_i = |f(\neg x)|_{a_i}.$$

Hence, $\neg f(x) = f(\neg x)$.

Claim 3. f is injective

Assume $x \neq x'$. Then for some i, $h_i \neq h_i'$ and $|f(x)|_{a_i} = h_i \neq h_i' = |f(x')|_{a_i}$. Hence $f(x) \neq f(x')$.

Claim 4. f is surjective

Indeed, let $w \in S^*$ and $w = a_{i_1} \ldots a_{i_1} \ldots a_{i_t} \ldots a_{i_t} \ldots$. Then, the element $y = (y_1, \ldots, y_n) \in A$ such that $y_j = 0$ if $j \notin \{i_1, \ldots, i_t\}$; $y_j = \frac{|w|_{a_j}}{k_p}$ if $j \in \{i_1, \ldots, i_t\}$ and $j = i_p$, satisfies the condition $f(y) = w$, as required.

4   Duality between finite MV-algebras and finite posets

As well known every bounded distributive lattice can be represented as a ring of compact open down sets of a spectral space $(X, \tau \downarrow, \leq)$ satisfying to the conditions (b), (c), (d) of Lemma 2, where $\leq$ is the specialization order: $x \leq y$ iff y belongs to the closure of the singleton $\{x\}$. The set of all compact open down subsets of X is a lattice under union and intersection. It constitutes a basis for the topology $\tau \downarrow$ of X. The sets of the form $U \backslash V$, for $U, V \in D(X)$ form a basis for a topology τ on X, called the *patch topology* on X (see [16]). Spectral spaces satisfying conditions (b), (c) and (d) equipped with the patch topology and specialization order are called *Priestley Spaces* (or compact totally order-disconnected spaces). The category $\mathbf{D}_{01}$ of bounded distributive lattices with lattice homomorphisms is dually equivalent to the category $\mathbf{P}$ of compact totally-ordered disconnected spaces (Priestley spaces) [19]. We illustrate this duality. The set $\mathcal{O}(X)$ of all clopen down subsets of an object X of $\mathbf{P}$ is closed under set union and intersection. So

$$\mathcal{O}(X) = (\mathcal{O}(X); \cup, \cap, \emptyset, X)$$

is an object of $\mathbf{D}_{01}$; the set $Sp(A)$ of all prime ideals of an object A of $\mathbf{D}_{01}$ ordered by inclusion and topologised by taking the family of sets

$$\eta(a) = \{I \in Sp(A) : a \notin I\}$$

for $a \in A$, and their complements as a base, is an object of $\mathbf{P}$. Furthermore, setting

$$Sp(h) = h^{-1} : B \to A,$$

for every morphism $h : A \to B$ of $\mathbf{D}_{01}$, and

$$\mathcal{O}(f) = f^{-1} : \mathcal{O}(Y) \to \mathcal{O}(X)$$

for every morphism $f : X \to Y$ of $\mathbf{P}$, we obtain contravariant functors $\mathbf{Pr}$: $\mathbf{D}_{01} \to \mathbf{P}$ and $\mathbf{O} : \mathbf{P} \to \mathbf{D}_{01}$.

Priestley's duality has been applied to obtain a description of dually equivalent categories to various equational subcategories of $\mathbf{D}_{01}$. To our considerations will be indispensable the description one of them, namely, the category dually equivalent to the category $\mathbf{MV}_{fin}$ of finite MV-algebras passing through the category $\mathbf{S}_{fin}$ of finite Skolem (i. e. bi-Heyting) algebras with linearity conditions.

Let (X, R_X) and (Y, R_Y) be posets, where R_X and R_Y are orders on the sets X and Y, respectively. As we know from Preliminaries an isotone map $f : X \to Y$ is strongly isotone if and only if the inverse image morphism $f^{-1} : (Y, R_Y) \downarrow \to (X, R_X) \downarrow$ preserves pseudo-difference, which is equivalent to the following condition entirely expressed in terms of elements:

(1) $\forall a \in X, \forall b \in Y (f(a) R_Y b) \Rightarrow \exists a' \in X (a R_X a'$ and $f(a') = b)$

We say that an order-preserving map $f : X \to Y$ is *dual strongly isotone* if and only if the inverse image morphism $f^{-1} : (Y, R_Y) \downarrow \to (X, R_X) \downarrow$ preserves relative pseudocomplementation (implication), which is equivalent to the following condition entirely expressed in terms of elements :

(2) $\forall a \in X, \forall b \in Y (b R_Y f(a)) \Rightarrow \exists a' \in X (a' R_X a$ and $f(a') = b)$

We say that an order preserving map $f : X \to Y$ is *bi-strongly isotone* (or interval in terminology of L.L.Esakia [14]) if and only if the inverse image morphism $f^{-1} : (Y, R_Y) \downarrow \to (X, R_X) \downarrow$ preserves both implication and pseudo-difference, which is equivalent to the following condition :

(3) $(\forall c \in X, \forall a, b \in Y)(a R_Y f(c) R_Y b \Rightarrow \exists a', b' \in X (a' R_X c R_X b'$ and $f(a') = a$ and $f(b') = b))$

On any finite distributive lattice we can define both the operation of implication ($\to$) and the operation pseudo-difference ($\leftharpoondown$). From the class of all finite bi-Heyting algebras select the subclass $\mathbf{S}_{fin}$ the elements of which satisfy to two properties of linearity :

$$(x \to y) \vee (y \to x) = 1,$$

$$(x \leftharpoondown y) \wedge (y \leftharpoondown x) = 0.$$

More precisely $\mathbf{S}_{fin}$ is the category whose objects are finite bi-Heyting algebras which satisfy linearity conditions given above and morphisms are bi-Heytng homomorphisms. From the definition of $\mathbf{S}_{fin}$ we conclude that any finite chain bi-Heyting algebra and finite product of finite chain bi-Heyting algebras are elements of $\mathbf{S}_{fin}$. The algebras from $\mathbf{S}_{fin}$ we call *S-algebras*.

Let us denote by $\mathbf{Sym}_{fin}$ the category of finite posets, the objects of which are dual to the objects of $\mathbf{S}_{fin}$, and morphisms are bi-strongly isotone maps. Observe that if A is an object of $\mathbf{S}_{fin}$, then the object X_A from $\mathbf{Sym}_{fin}$, which is dual to A, is the set of all prime ideals of A ordered by inclussion, where for every $x \in X_A$ $\quad x \downarrow$ and $x \uparrow$ are chains.

Adapting on the case of S-algebras we shall formulate the theorems concerning to bi-Heyting algebras given in [14].

Theorem 6. *[14] The category $\mathbf{S}_{fin}$ is dually equivalent to the category* $\mathbf{Sym}_{fin}$.

We say that a subset A of a poset X is *bi-cone* if A is simultaneously both upper set and down set.

Theorem 7. *[14] The lattice of congruence relations of any S-algebra A is anti-isomorphic to the lattice of bicones of corresponding to A poset $(X, R) \in$* $\mathbf{Sym}_{fin}$.

From the theorem we conclude that if we take as a bicone a component of $(X, R) \in \mathbf{Sym}_{fin}$, that is a bi-cone $P \subseteq X$ such that $P = (x \downarrow) \uparrow$ for every $x \in P$, then P is a chain with respect to the order R and the component corresponds to a simple linearly ordered bi-Heyting algebra. As a consequence of the theorem we have

Corollary 2. *If II is simple algebra from $\mathbf{S}_{fin}$, then the corresponding poset is a chain.*

We say that a partition E of $(X, R) \in Ob\mathbf{Sym}_{fin}$ is *correct* if and only if the saturation $E(P)(= \{x \in X : \exists y \in P(xEy)\})$ of any upper set (down set) P of X is an upper set (down set)

Theorem 8. *Let $(X, R_X), (Y, R_Y) \in Ob\mathbf{Sym}_{fin}$ and $f : X \to Y$ be a bi-strongly isotone map. Then the partition E of X into f-fibers is correct.*

Proof. The proof immediately follows from Lemma 3.

Theorem 9. *[14]. The lattice of subalgebras of any finite bi-Heyting algebra is anti-isomorphic to the lattice of correct partitions of its dual objects.*

Let us note, that if we have $(X, R_X) \in Ob\mathbf{Sym}_{fin}$ and E is a correct partition on X, then $(X/E, R_X/E) \in Ob\mathbf{Sym}_{fin}$, where $(R_X/E)(x/E, y/E) \Leftrightarrow (\exists x' \in x/E, \exists y' \in y/E)(R_X(x', y'))$. What is more the bi-Heyting algebra of all down sets of X/E is embeded into the bi-Heyting algebra of all down sets of X.

Since any algebra A from $\mathbf{S}_{fin}$ satisfies the Stone condition: $\neg x \vee \neg\neg x = 1$, where $\neg x = x \to 0$ (and the dual Stone condition : $\dot{\neg} x \wedge \dot{\neg}\dot{\neg} x = 0$, where $\dot{\neg} x = 1 \leftarrow x$), then the set $\{\neg x : x \in A\}(= B(A))$ forms a Boolean algebra, in other words a S-algebra where the equations $x \to y = \neg x \vee y$, $x \leftarrow y = x \wedge \dot{\neg} y$

hold. Note that there is one-to-one correspondence between the lattice of ideals of the Boolean algebra $B(A)$ and the lattice of Skolem ideals of A (which is dual to the concept of Skolem filters introduced by C. Rauszer in [21]), that is such ideals I of A that if $a \in I$ then $\dot{\neg}\neg a \in I$, namely $I \to I \cap B(A)$.

Recall that the MV-algebra $S_m = (\{0, 1/m, \ldots, m-1/m, m\}, \oplus, \cdot, *, 0, 1)$ is a Skolem algebra with respect to the operations $\vee, \wedge, \to, \leftharpoonup$, which is polynomially expressed in S_m. Let us note that $\neg x = (mx)^*$ and $\dot{\neg} x = (x^m)^*$. So Skolem ideals coincide with MV-ideals on any finite MV-algebra.

Let $X_m = \{I_0, I_1, \ldots, I_{m-1}\}$ be the set of all proper prime lattice ideals of MV-algebra S_m and $I_j = \{0, \ldots, j/m\} \subset S_m$. Then $(X_m, R_m) \in Ob\mathbf{Sym}_{fin}$, where R_m is the inclusion relation. Let E_j be a partition of X_m on m/j-element classes , where j is divisor of m :

$$\{I_0, \ldots, I_{m/j-1}\}, \{I_{m/j}, \ldots, I_{2(m/j)-1}\}, \ldots, \{I_{(j-1)m/j}, \ldots, I_{m-1}\}.$$

As we know j devides m if and only if S_j is subalgebra of S_m, and the partition E_j on X_m is correct. Indeed, the classes of E_j are convex with respect to the inclusion relation. A subcategory $\mathbf{USym}_{fin}$ of the category $\mathbf{Sym}_{fin}$ will be called the category of *uniform symmetric models* if the following conditions hold:

(1) any object of $\mathbf{USym}_{fin}$ is object of $\mathbf{Sym}_{fin}$;
(2) a morphism $f : X_1 \to X_2$ is morphism of $\mathbf{USym}_{fin}$ if and only if f is morphism of the category $\mathbf{Sym}_{fin}$ and $\forall x \in X_1$ $|f(x) \downarrow\uparrow|$ devides $|x \downarrow\uparrow|$, $\forall x, y \in |x \downarrow\uparrow|$ $|f^{-1}(f(x)) \cap x \downarrow\uparrow| = |f^{-1}(f(y)) \cap x \downarrow\uparrow|$.

The morphisms of $\mathbf{USym}_{fin}$ will be called *uniform morphisms*.

Theorem 10. *The category* $\mathbf{MV}_{fin}$ *of finite* MV-*algebras with* MV-*homomorphisms is dually equivalent to the category of uniform symmetric models* $\mathbf{USym}_{fin}$ *with uniform morphisms.*

Proof. We can define a functor $\mathbf{Us}$ from the category $\mathbf{MV}_{fin}$ to the category $\mathbf{USym}_{fin}$, mapping, by $\mathbf{Us}$, any object A from $\mathbf{MV}_{fin}$ into the object $\mathbf{Us}(A) = (X, R)$ of $\mathbf{USym}_{fin}$, where X is the set of all prime lattice ideals of A, ordered by the inclusion relation R. It is obvious that the poset (X, R) is cardinal sum of finite linearly ordered posets. Indeed, since A is a finite MV-algebra, then A is isomorphic to a finite product of finite chain MV-algebras : i.e., $A \simeq S_{n_1} \times \cdots \times S_{n_k}$. $\mathbf{Us}(A) = \mathbf{Us}(S_{n_1}) \sqcup \cdots \sqcup \mathbf{Us}(S_{n_k})$, where $\sqcup$ means cardinal sum. Any component $\mathbf{Us}(S_{n_i})$ (which is bi-cone) corresponds to a homomorphic image of the MV-algebra A by a maximal MV-ideal (or maximal Skolem ideal) of A. Let now $A_1, A_2 \in \mathbf{MV}_{fin}$ and $h : A_1 \to A_2$ be MV-homomorphism. Then h is mapped by, $\mathbf{Us}$, into $\mathbf{Us}(h) = h^{-1} : \mathbf{Us}(A_2) \to \mathbf{Us}(A_1)$. We have that $\mathbf{Us}(h)$ is a morphism of $\mathbf{USym}_{fin}$. Indeed, h is a Skolem homomorphism and h^{-1} is bi-strongly isotone. Then,

h^{-1} is uniform. This easy to check on components of $\mathbf{Us}(A_2)$. Now, let us represent the objects of $\mathbf{USym}_{fin}$ as finite cardinal sums of finite chains. Let we have the cardinal sum $C_{n_1} \sqcup \cdots \sqcup C_{n_k}$ of finite chains, where C_{n_i} have n_i elements. Then we can define a functor $\mathbf{Mf}$ from the category of $\mathbf{USym}_{fin}$ to the category of $\mathbf{MV}_{fin}$, mapping an object $C_{n_1} \sqcup \cdots \sqcup C_{n_k}$ from $\mathbf{USym}_{fin}$ into the object $\mathbf{Mf}(C_{n_1} \sqcup \cdots \sqcup C_{n_k}) = S_{n_1} \times \cdots \times S_{n_k}$, up to isomorphism, of $\mathbf{MV}_{fin}$. Moreover, we have that the lattice reduct of $S_{n_1} \times \cdots \times S_{n_k}$ is isomorphic to all down sets of $C_{n_1} \sqcup \cdots \sqcup C_{n_k}$. Then we have $\mathbf{Mf}(\mathbf{Us}(A)) \simeq A$, as well as $\mathbf{Us}(\mathbf{Mf}((X,R))) \simeq (X,R)$. Given any morphism $f : (X_1, R_1) \to (X_2, R_2)$ from $\mathbf{USym}_{fin}$, we set $\mathbf{Mf}(f) = f^{-1} : \mathbf{Mf}((X_2, R_2)) \to \mathbf{Mf}((X_1, R_1))$. It is easy to see that $\mathbf{Mf}(f)$ is an MV-morphism. What is more, we have $\mathbf{Mf}(\mathbf{Us}(h)) = h$, $\mathbf{Us}(\mathbf{Mf}(f)) = f$. Consequently $\mathbf{Mf} : \mathbf{USym}_{fin} \to \mathbf{MV}_{fin}$, $\mathbf{Us} : \mathbf{MV}_{fin} \to \mathbf{USym}_{fin}$ are contravariant functors.

5   Quasi finite $MV(C)$-algebras

Recall that a *Heyting algebra* is a lattice H with 0 in which $x \to y$ exists for each $x, y \in H$. If, in addition, H satisfies the (linearity) equation $(x \to y) \vee (y \to x) = 1$, then H is said to be a *linear Heyting algebra*, for short, an $\mathcal{L}$-*algebra*. $\mathcal{L}$-algebras were considered by Horn in [17]. They can be equivalently defined as the subvariety of Heyting algebras generated by all totally ordered Heyting algebras. A *dual Heyting algebra* is a bounded distributive lattice H such that, for each $x, y \in H$, there exists the *pseudo-difference* of y and x. Let $\mathcal{H}^{op}$ denote the category of dual Heyting algebras having as morphisms lattice homomorphisms which preserve pseudo-difference. Any Priestley space is an object dual to a dual Heyting algebra if for every open subset U of X, $U \downarrow$ is open or equivalently for all open U and V, $(U \backslash V) \downarrow$ is open. Such kind of Priestley spaces we denote by $\mathcal{PH}^{op}$. The above property says that for clopen down sets U and V there exists pseudo-difference in the lattice of all clopen down sets of X. The $\mathcal{PH}$-morphism $f : X \to Y$, for objects X and Y of $\mathcal{PH}^{op}$ is $\mathcal{PH}^{op}$-morphism if for every clopen down sets U and V of Y f^{-1} preserves the operation of pseudo-difference on the lattice of clopen down sets. According to Lemma 3, it is equivalent to the assertion that f is strongly isotone map. The specialization of the Priestley duality to the case of the category of $\mathcal{H}^{op}$ and the category $\mathcal{PH}^{op}$ with strongly isotone maps is essentially a part of folklor of Duality theory. So we have

Theorem 11. *The category $\mathcal{H}^{op}$ is dually equivalent to the category $\mathcal{PH}^{op}$.*

We let $\mathcal{L}^*$ denote the subvariety of dual Heyting algebras satisfying the equation $(x \leftarrow y) \wedge (y \leftarrow x) = 0$. Algebras in $\mathcal{L}^*$ shall be naturally called $\mathcal{L}^*$-algebras. Let H, H' be $\mathcal{L}^*$-algebras. Any D_{01}-homomorphism h from H to H' such that $h(x \leftarrow y) \leq h(x) \leftarrow h(y)$ for all $x, y \in H$, shall be called $\mathcal{L}^*$-*homomorphism*. From Theorem 5 we have that E-saturated down-sets of X form an $\mathcal{L}^*$-algebra. In other words the map $f : X \to X/E$ is strongly isotone.

By abuse of notation, we also let $\mathcal{L}^*$ denote the category whose objects are $\mathcal{L}^*$-algebras and whose morphisms are $\mathcal{L}^*$-homomorphisms. A PH^{op}-space X is said to be $\mathcal{L}^*$-space if its clopen down sets satisfy linearity condition:

$$(U \backslash V) \downarrow \cap (V \backslash U) \downarrow = \emptyset$$

Observe that the order duals of dual Heyting algebras are the Heyting algebras. Horn [17] showed that $\mathcal{L}$-algebras can be characterized among Heyting algebras in terms of the order on prime filters (co-ideals). Specifically, a Heyting algebra is $\mathcal{L}$-algebra iff its set of prime lattice filters is a root system (ordering by inclusion). The same we can say for $\mathcal{L}^*$-algebras (according to Horn's assertion): a dual Heyting algebra is $\mathcal{L}^*$-algebra iff its set of prime lattice ideals is a root system. We can restrict the Theorem 20 on the categories of $\mathcal{L}^*$-algebras and $\mathcal{L}^*$-spaces.

Theorem 12. *The category $\mathcal{L}^*$ is dually equivalent to the category $\mathcal{L}^*$-spaces.*

Let $\mathcal{FL}^*$ denote the full subcategory of $\mathcal{L}^*$ whose objects are finite $\mathcal{L}^*$-algebras, and $\mathcal{FR}$ denote the category having finite root systems as objects and strongly isotone maps as morphisms.

Theorem 13. *The categories $\mathcal{FR}$ and $\mathcal{FL}^*$ are dually equivalent.*

Thus, there are dual functors between $\mathcal{FR}$ and $\mathcal{FL}^*$. Let us denote these functors by $\Phi : \mathcal{FR} \to \mathcal{FL}^*$ and $\Lambda : \mathcal{FL}^* \to \mathcal{FR}$.

We shall be interested in finitely generated MV-algebras, having finite spectrum, from variety $\mathbb{MV}(C)$, i.e. from the variety generated by perfect MV-algebras. Indeed, let $\mathbf{qF}$ denote the class of MV-algebras obtained by the following stipulations :

1)$\mathbf{qF}$ contains every finitely generated perfect chains;

2)$\mathbf{qF}$ is the smallest class containing every finitely generated perfect chain and closed under finite products, homomorphic images and subalgebras.

Let $\mathbf{qFC}$ denote the full subcategory of the category $\mathbb{MV}$ of all MV-algebras with MV-homomorphisms, whose objects are the members of $\mathbf{qF}$. We stress that $\mathbf{qF}$ is a subclass of $\mathbb{MV}(C)$. MV-algebras from $\mathbf{qF}$ shall be called *quasi-finite $MV(C)$algebras*.

Definition 2. [11]. A class of finite algebras of a given type which is closed under the operators of forming subalgebras, homomorphic images and finite products of algebras is called by *pseudo-variety of algebras*.

Motivated by application to the theory of automata, Eilenberg and Schützenberger considered in [10],[11] classes of finite monoids closed under subalgebras, homomorphic images and finite products. These consideration led to the notion of pseudovariety [11].

Definition 3. A class of finitely generated algebras of a given type with finite congruence lattices which closed under the operators of forming subalgebras, homomorphic images and finite products of algebras is called by *quasi-pseudo-variety*.

Theorem 14. *The class of all quasi-finite $MV(C)$-algebras is a quasi-pseudo-variety.*

Proof. The proof is trivial.

The restriction of the functor **Be** to q$\mathbb{F}$C shall be denoted by **Be***

Theorem 15. **Be*** *is a functor from* q$\mathbb{F}$C *to* $\mathcal{FL}^*$.

Proof. Let A be a quasi finite $MV(C)$-algebra. We claim that $\mathbf{Be}(A) \in \mathcal{FL}^*$. Indeed, $SpecA$ is a finite root system. Then $Spec\mathbf{Be}(A)$ is a finite root system too. Hence, by Theorem 22, $\mathbf{Be}(A)$ is a finite $\mathcal{L}^*$-algebra. We shall prove that **Be*** maps MV-homomorphisms into $\mathcal{L}^*$-homomorphisms. Indeed, let $h : A \to B$ be a homomorphism, where A and B are quasi finite $MV(C)$-algebras. Then, by Lemma 7, $\tilde{h}$ is strongly isotone map from $SpecB$ to $SpecA$. Hence, $\mathbf{Be}(h)$ is strongly isotone map from $Spec\mathbf{Be}(B)$ to $Spec\mathbf{Be}(A)$. By Theorem 22, $\Phi(\mathbf{Be}(h))$ is a $\mathcal{L}^*$-homomorphism from $\mathbf{Be}(A)$ to $\mathbf{Be}(B)$. It is easy to check that **Be*** is a functor.

Theorem 16. *Let A be a quasi-finite $MV(C)$-algebra. Then $(\mathbf{Be}^*(A), \vee, \wedge, \leftarrow, 1, 0)$ is finite $\mathcal{L}^*$-algebra. Conversely, if L is finite $\mathcal{L}^*$-algebra, then there exists a unique quasi-finite $MV(C)$-algebra A such that $\mathbf{Be}^*(A) \cong L$.*

Proof. Let A be quasi-finite $MV(C)$-algebra. That is $SpecA$ is isomorphic to a finite root system. Then $SpecA \cong Spec\mathbf{Be}(A)$, where $\mathbf{Be}(A)$ is $\mathcal{L}^*$-algebra. Conversely, let L be a finite $\mathcal{L}^*$-algebra. Then $SpecL$ is isomorphic to a finite root system. According to Theorem 10 there exists an MV-algebra A such that $SpecA$ is isomorphic to $SpecL$. Since $SpecA$ is a finite root system, $SpecA$ has a finite number of maximal elements, say $P_1, \ldots, P_k$ and $SpecA$ is isomorphic to a cardinal sum $\bigsqcup_{i=1}^k (P_k \downarrow)$ of roots $P_i \downarrow$. Moreover $A \cong A_1 \times \ldots \times A_k$, where A_i is an MV-algebra such that $SpecA_i \cong P_i \downarrow$. Since A_i has one maximal ideal (which we denote also by P_i), we can form perfect MV-algebras $A_i' = P_i \cup P_i^*$, $i = 1, \ldots, k$. Therefore we have the $MV(C)$-algebra $A_i' = A_1' \times \ldots \times A_k'$, such that $SpecA \cong SpecA'$. Let us consider separately a MV-algebra A_i'. A_i' is isomorphic to a subdirect product of finitely many totally ordered perfect MV-algebras, say $H_{i1}, \ldots, H_{ip}$ such that $A_i' \hookrightarrow \prod_{j=1}^p H_{ij}$. H_{ij} has finitely many proper ideals: $J_0 \subseteq J_1 \subseteq \ldots \subseteq J_t$, where J_t is maximal ideal of H_{ij}. We pick out one element $a_i \in J_i \setminus J_{i-1}$, $i = 1, \ldots, t$. Then $\{a_1, \ldots, a_t\}$ generate the subalgebra $H_{ij}' \hookrightarrow H_{ij}$, such that $SpecH_{ij} \cong SpecH_{ij}'$ and $H_{ij}' \cong S_{(1,\ldots,0)}$, where $(1, \ldots, 0) \in Z^{k+1}$. Then $\prod_{j=1}^t H_{ij}'$ is subalgebra of $\prod_{j=1}^t H_{ij}$ and $H_i = \prod_{j=1}^t (H_{ij}' \cap A_i')$ is $MV(C)$-algebra, such that $SpecA_i' \cong SpecH_i$ and H_i is finitely generated $(i = 1, \ldots, k)$. Therefore $H = H_1 \times \ldots \times H_k$ is a quasi finite $MV(C)$-algebra, such that $SpecH \cong SpecA' \cong SpecA \cong Spec\mathbf{Be}(H)$, where $\mathbf{Be}(H) \cong L$.

Let us consider totally ordered m-generated $MV(C)$-algebra, which is isomorphic to $S_{(1,0,\ldots,0)}$, where $S_{(1,0,\ldots,0)} \cong \Gamma(Z \otimes \cdots \otimes Z, (1,0,\ldots,0))$, $Z \otimes \cdots \otimes Z$ is lexicographic product of $m+1$ copies of additive group of integers Z and $(1,0,\ldots,0) \in Z^{m+1}$ is its strong unit. The generators of $S_{(1,0,\ldots,0)}$ are $c_1 = (0,0,\ldots,0,1), c_2 = (0,0,\ldots,1,0),\ldots,c_m = (0,1,\ldots,0)$.

Lemma 7. . $\mathbf{Be}(S_{(1,0,\ldots,0)})$ is m-generated $\mathcal{L}^*$-algebra.

Proof. Indeed, $\beta(c_1),\ldots,\beta(c_m)$ generate the algebra $\mathbf{Be}(S_{(1,0,\ldots,0)})$.

Thus, we can represent $\mathbf{Be}(S_{(1,0,\ldots,0)})$ as a totally ordered lattice $0 < c_1 < \cdots < c_m < 1$, which is $\mathcal{L}^*$-algebra.

Lemma 8. *Let x_i be any element of $\beta(c_i)$, $i = 1,\ldots,m$. Then the subalgebra A of $S_{(1,0,\ldots,0)}$ generated by $x_1,\ldots,x_m$ is isomorphic to $S_{(1,0,\ldots,0)}$ and, hence, $\mathrm{Spec}A \cong \mathrm{Spec}S_{(1,0,\ldots,0)}$.*

Proof. The proof is obvious.

Lemma 9. *Let $c_{i_1},\ldots,c_{i_k}$ be some sequence of generators of*

$$S\underbrace{_{(1,0,\ldots,0)}}_{m+1}$$

where $1 \leq k \leq m$. Then the subalgebra A of

$$S\underbrace{_{(1,0,\ldots,0)}}_{m+1}$$

generated by $c_{i_1},\ldots,c_{i_k}$ is isomorphic to

$$S\underbrace{_{(1,0,\ldots,0)}}_{k+1}$$

where $(1,0,\ldots,0) \in Z^{k+1}$.

Proof. The proof is trivial.

Lemma 10. *The $m+2$-element totally ordered $\mathcal{L}^*$-algebra L is generated by m generators. The generators are all the elements of L different from 0 and 1.*

Proof. It follows from the definition of pseudo-difference.

Lemma 11. *Let L_1, L_2 be finite $\mathcal{L}^*$-algebras and $\epsilon : L_1 \to L_2$ be $\mathcal{L}^*$-embedding. Then there exist $MV(C)$-algebras A_1 and A_2 such that $\mathbf{Be}(A_i) \cong L_i$ ($i = 1,2$) and MV-embedding $\delta : A_1 \to A_2$ such that $\mathbf{Be}(\delta) = \epsilon$.*

Proof. Assume that L_2 is m-generated and $g_1, \ldots, g_m$ are its generators. As it is known, L_2 is isomorphic to a subdirect product of totally ordered $\mathcal{L}^*$-algebras $H_1, \ldots, H_n$ such that every H_i is m-generated. The generators of H_i shall be denoted by $c_1^{(i)}, \ldots, c_m^{(i)}$, where some of them may be equal to each other, 1 or 0. Then, we have $L_2 \hookrightarrow H_1 \times \cdots \times H_n$, $\pi_i(g_k) = c_k^{(i)}$, $k = 1, \ldots, m$, $i = 1, \ldots, n$, i.e. $g_1 = (c_1^{(1)}, \ldots, c_1^{(n)}), \ldots, g_m = (c_m^{(1)}, \ldots, c_m^{(n)})$, where π_i is the i-th canonical projection map. L_2 consists of a subset of the set of all n-sequences each component of which is $c_j^{(i)}$ either 1 or 0. L_1 is subset of L_2, which is closed under $\mathcal{L}^*$-algebra operations : $\vee, \wedge, \leftharpoonup, 1, 0$. Let A_2 be the $MV(C)$-algebra, corresponding to $\mathcal{L}^*$-algebra L_2, which is subalgebra direct product $B_1 \times \cdots \times B_n$, generated by $g_1, \ldots, g_m$, where $B_i \cong S_{(1,0,\ldots,0)}$, $(1, 0, \ldots, 0) \in Z^{|H_i|-1}$. In other words A_2 is generated by L_2. As $MV(C)$-algebra A_1, corresponding to $\mathcal{L}^*$-algebra L_1, we take the subalgebra of A_2 which is generated by $L_1 \subseteq L_2$.

Theorem 17. *Let L_1, L_2 be finite $\mathcal{L}^*$-algebras and $h : L_1 \to L_2$ be homomorphism of L_1 into L_2. Then there exist quasi-finite $MV(C)$-algebras A_1 and A_2 such that $\mathbf{Be}(A_i) \cong L_i$ $(i = 1, 2)$ and MV-homomorphism $f : A_1 \to A_2$ such that $\mathbf{Be}(f) = h$.*

Proof. Let L_1, L_2 be finite $\mathcal{L}^*$-algebras and $h : L_1 \to L_2$ be homomorphism of L_1 into L_2. According to Theorem 27 there exist quasi-finite $MV(C)$-algebras A_1, A_2 such that $\mathbf{Be}^*(A_i) \cong L_i$ $(i = 1, 2)$. Let $\varphi_i : \mathbf{Be}^*(A_i) \to L_i$ be the isomorphism. There exists a lattice ideal J of L_1 which is equal to $h^{-1}(0)$ and $L_1/J \cong h(L_1)$. Define ideal $\varphi_i^{-1}(J)$ on $\beta(A_i)$. Then $I = \{x \in A_1 : x \in \varphi_i^{-1}(y), y \in J\}$ is an ideal of A_1. Therefore there exists homomorphism onto $f' : A_1 \to f'(A_1)$, where $f'(A_1) \cong A_1/I$. Then $\mathbf{Be}^*(f') : L_1 \to \mathbf{Be}^*(f'(A_1))$ is $\mathcal{L}^*$-homomorphism and $\mathbf{Be}^*(f'(A_1)) \cong h(L_1) \hookrightarrow L_2 \cong \mathbf{Be}^*(A_2)$. Let $\epsilon : \mathbf{Be}^*(f'(A_1)) \to \mathbf{Be}^*(A_2)$ be embedding. Then, by Lemma 32, there exist $MV(C)$-algebra which is isomorphic to $f'(A_1)$ and $MV(C)$-algebra which is isomorphic to A_2. Identifying isomorphic elements we can conclude that there exists MV-embedding $\delta : A_1 \to A_2$ such that $\mathbf{Be}^*(\delta) = \epsilon$. Then the desired homomorphism $f : A_1 \to A_2$ is the composition $\delta f'$.

According to Theorems 27 and 33 and its proof we have the construction of a functor, denoted by λ, from the category of finite $\mathcal{L}^*$-algebras to the category of quasi-finite $MV(C)$-algebras. It is easy to prove the following

Theorem 18. *Let A_1, A_2 be quasi-finite $MV(C)$-algebras and L_1, L_2 be finite $\mathcal{L}^*$-algebras such that $\beta(A_i) \cong L_i (i = 1, 2)$. Then, if $h_1, h_2 \in Hom(L_1, L_2)$ and $h_1 \neq h_2$ then $\lambda(h_1) \neq \lambda(h_2)$.*

It shoud be noted that if we have embedding $f : S_{(1,0)} \to S_{(1,0)}$, where $f((0,1)) = (0,2)$, of $S_{(1,0)}$ into the proper subalgebra of $S_{(1,0)}$ generated by $(0,2)$, then $\mathbf{Be}^*(f)$ is an isomorphism from $\mathbf{Be}^*(S_{(1,0)})$ to $\mathbf{Be}^*(S_{(1,0)})$. Combining everything above mentioned, we can formulate

Theorem 19. *The category $\mathcal{FL}^*$ of finite $\mathcal{L}^*$-algebras is equivalent to the category of quasi-finite $MV(C)$-algebras, with one-to-one correspondence between objects of the categories.*

Theorem 20. *The category $\mathcal{FR}$ of finite root systems and strongly isotone maps is dually equivalent to the category of quasi-finite $MV(C)$-algebras and MV-homomorphisms, with one-to-one correspondence between objects of the categories.*

References

1. R. Balbes, P. Dwinger, *Distributive Lattices*, University of Missoure Press Columbia, (1974).
2. L. P. Belluce, *Semisimple Algebras of Infinite-Valued Logic and Bold Fuzzy set Theory*, Canadian J. Math. **38**, (1986), 1356–1379.
3. R. Bigard, K. Keimel, S. Wolfenstein, *Groupes et Anneaux Rticuls*, Springer Lecture Notes in Mathematics, **608**, (1977).
4. C.C. Chang, *Algebraic analysis of many valued logics*, Trans. Amer. Math. Soc. **88**, (1958), 467–490.
5. R. Cignoli, A. Di Nola, A. Lettieri, *Priestley Duality and Quotient Lattices of Many-Valued Algebra*, Rendiconti del Circolo Matematico di Palermo, Serie II **XL**, (1991), 371–384.
6. R. Cignoli, A. Torrens, *Ordered Spectra and Boolean Products of MV-algebras*, Centre de Recerca Matematica, Institut d'estudis Catalans, No.172, (1992).
7. B. A. Davey, H. A. Priestley, *Introduction to Lattices and Order*, Cambridge Mathematical Textbooks, Cambridge University Press, (1990).
8. A. Di Nola, R. Grigolia, *MV-algebras in duality with labelled root systems*, Discrete Mathematics, to appear.
9. A. Di Nola and A. Lettieri, *Perfect MV-algebras are categorically equivalent to abelian ℓ-groups*, Studia Logica **53**, No.3 (1994), 417-432.
10. S. Eilenberg, *Automata, Language and Machines*, Academic Press, N.Y.,(1974).
11. S. Eilenberg, M. P. Schützenberger, *On Pseudovarieties*, Advance in Mathematics, **19**, (1976), 413–418.
12. L. Esakia, *On Topological Kripke Models*, Dokl, AN SSSR, **214**, (1974), 2, 298–301.
13. L. Esakia, *Heyting Algebras I*, Duality Theory, Metsniereba, Tbilisi, (1985), (in Russian).
14. L. Esakia, *Semantical analysis of bi-modal (temporal) logical systems*, Logic, Semantics, Methodology; Tbilisi,"Metsiniereba", (1978), 87–99.
15. G. Grätzer, *General lattice theory*, Akademie-Verlag, Berlin, (1978).
16. G. Grätzer, *Lattice Theory, First Concepts and Distributive Lattices*, W.H. Freeman and Co., San Francisco, 1972.
17. A. Horn, *Logic with Truth Values in a Linearly Ordered Heyting Algebra*, J. Symbolic Logic **34**, (1969), 395–408.
18. D. Mundici, *Interpretation of AF C^*-Algebras in Łukasiewicz Sentential Calculus*, J. Funct. Analysis **65**, (1986), 15–63.
19. H. A. Priestley, *Ordered Sets and Duality for Distributive Lattice*, Annales of Discrete Mathematics, **23**, (1984), 39–60.

20. I. Prodanov, *An Abstract Approach to the Algebraic Notion of Spectrum*, Proceedings of the Steklov Institute of Mathematics, **4**, (1984), 215–223.

21. C. Rauszer, *Semi-Boolean algebras and their applications to intuitionistic logic withdual operations*, Fund. Math., **83**(1974),3, 219-249.

Generalized Pseudo-Effect Algebras

Anatolij Dvurečenskij and Thomas Vetterlein*

Mathematical Institute, Slovak Academy of Sciences,
Štefánikova 49, SK - 814 73 Bratislava, Slovakia;
E-mail: dvurecen@mat.savba.sk and vetterl@mat.savba.sk

Abstract. We introduce generalized pseudo-effect algebras as a non-commutative version of generalized effect algebras.

The importance of the algebras of the latter type in quantum physics is based on the fact that they reflect the inner structure of subsets of the positive cone of the group of self-adjoint operators in a Hilbert space. The new algebras are designed to model subsets of group cones as well; but now the underlying po-group may be chosen arbitrarily, in particular it does not need to be abelian.

We raise the question when a generalized pseudo-effect algebra is actually representable in the positive cone of a po-group. We are able to give an affirmative answer for two special cases. Both times a property of Riesz kind is involved, defined for our algebra in a similar manner as known for po-groups.

AMS classification: 81P10, 03G12, 06F35, 06F05, 06F15.

Key words and phrases: pseudo-effect algebras, generalized effect algebras, po-groups.

1   Introduction

In the field of quantum structures, we are concerned with the foundational problem of quantum physics. Roughly spoken, our approach to this problem is the following. The basic object under study is the Hilbert space, since this is the model of all quantum theories. Now, to gain a deeper insight into the rather complicated Hilbert space formalism, we try to extract from it its physically essential parts; that is, we work on possible reductions of the somehow sophisticated analytic structure to algebraic ones. The aim thereby is to get hopefully simpler, but still physically meaningful models of quantum phenomena.

For a long time, the closed subspaces of Hilbert space have been examined, which correspond to possible measurement results in quantum experiments. These entities possess internally the structure of an orthocomplemented lattice, which gave hope to a better understanding of quantum physics insofar,

---
* The paper has been supported by the grant 2/7193/20 SAV, Bratislava, Slovakia.

as it reminds to a system of sentential-logical propositions. But the research results, although optimal, seem to be not as useful as once expected.

In more recent times, the quantum effects have been taken into consideration instead, i.e. the positive operators of a Hilbert space lying below identity. For characterizing internally the effects, it seemed to be no longer reasonable to use lattice operations, but to be natural to use their additive relations; two effects may – or may not – add up to a third one. So the set of all effects possesses the structure of a partial additive algebra, and as a such it forms a so-called effect algebra [FoBe,?]. Equivalently, we may consider the differences of pairs of comparable effects; if we do so, we are lead to so-called difference posets [KoCh].

Thus the prototype among the effect algebras as well as difference posets is a unit interval of the group of self-adjoint Hilbert space operators. Now we may easily generalize these two structures by dropping the requirement of boundedness, so as to get axioms satified also by unbounded parts of the group cone [HePu]. The structures we arrive at, simply called generalized effect algebras or generalized difference posets, respectively, are the starting point of our discussion here.

What we will do is actually to generalize the mentioned structures even further. The prototypical examples of the algebras discussed here shall not neccessarily be so special as the Hilbert space operators or even as abelian group cones are. Namely, we drop the assumption of commutativity, so as to include a much broader family of models, in particular those arising from non-commutative po-groups. For situations in which non-commutativity actually occurs, we think in particular about any model of time-ordered sequences of whatsoever physical phenomena.

So we shall introduce what will be called generalized pseudo-effect algebras. These are still partial additive structures; they are not necessarily commutative, and by assuming commutativity they specialize to generalized effect algebras. Their prototypes arise within the positive cone of any, abelian or non-abelian, partially ordered group.

It seems then natural to ask in which cases one of our algebras is actually representable just in this way. We will prove that a group representation exists in case one among two conditions holds: The algebra is either assumed to be a lower semi-lattice and to fulfil a certain condition of Riesz type, or it is assumed to be directed and to fulfil some other, slightly different, condition again of Riesz type.

2   Generalized Pseudo-Effect Algebras

Effect algebras are of major importance in quantum physics insofar, as they reflect the internal structure of the positive operators of a Hilbert space lying below identity, i.e. of the so-called quantum effects. It was this kind of algebra that once gave motivation to introduce the generalized effect algebras; the greatest element was dropped, and a structure was obtained such that every principal order ideal was an effect algebra [HePu,?]. The connection to quantum physics has been kept insofar, as the new axioms are modelled, no longer necessarily by a unit interval of the group of self-adjoint operators but, by any subset of positive operators that is closed under the formation of differences of comparable elements.

We recall that a generalized effect algebra, or GE-algebra for short, is a structure $(E; +, 0)$, where $+$ is a partial binary operation and 0 is a constant, such that, for all $a, b, c \in E$, (GE1) $a + b$ is defined iff $b + a$ is defined, in which case these elements are equal, (GE2) $(a + b) + c$ is defined iff $a + (b + c)$ is defined, in which case these elements are equal, (GE3) from $a + b = a + c$ it follows $b = c$, (GE4) from $a + b = 0$ it follows $a = b = 0$, and (GE5) $a + 0 = a$.

We shall generalize this kind of algebra by no longer requiring the axiom of commutativity. This has in fact been the idea that lead from effect algebras to pseudo-effect algebras [DvVe1], and that we now apply to the more general structure of a generalized effect algebra.

Definition 1. A structure $(E; +, 0)$, where $+$ is a partial binary operation and 0 is a constant, is called a *generalized pseudo-effect algebra*, or GPE-algebra for short, if, for all $a, b, c \in E$, the following holds.

(GPE1) $a + b$ and $(a + b) + c$ exist if and only if $b + c$ and $a + (b + c)$ exist, and in this case $(a + b) + c = a + (b + c)$.
(GPE2) If $a + b$ exists, there are elements $d, e \in E$ such that $a + b = d + a = b + e$.
(GPE3) If $a + b$ and $a + c$ exist and are equal, then $b = c$.
    If $b + a$ and $c + a$ exist and are equal, then $b = c$.
(GPE4) If $a + b$ exists and $a + b = 0$, then $a = b = 0$.
(GPE5) $a + 0$ and $0 + a$ exist and are both equal to a.

Furthermore, we define for any $a, b \in E$
$$a \leq b \quad \text{iff} \quad a + c = b \text{ for some } c \in E.$$

Lemma 1. Let $(E; +, 0)$ be a generalized pseudo-effect algebra. Then $\leq$ is a partial order on E with 0 as smallest element.

E is naturally ordered by $\leq$, i.e., for any $a, b \in E$, $a \leq b$ if and only if $a + c = b$ for some $c \in E$ if and only if $d + a = b$ for some $d \in E$.

Proof. Let $a, b, c \in E$. We have $a \leq a$, since $a + 0 = a$ by (GPE5).

$a \leq b$ and $b \leq a$ imply $a + a_1 = b$ and $b + b_1 = a$ for some $a_1, b_1 \in E$, so $(a + a_1) + b_1 = a = a + 0$ by (GPE5); this means $a_1 + b_1 = 0$ by (GPE1) and (GPE3) and $a_1 = b_1 = 0$ by (GPE4); so $a = b$ by (GPE5).

$a \leq b$ and $b \leq c$ imply $a + a_1 = b$ and $b + b_1 = c$ for some $a_1, b_1 \in E$, so $c = (a + a_1) + b_1 = a + (a_1 + b_1)$ by (GPE1), which means $a \leq c$.

By (GPE5), 0 is the smallest element.

That $\leq$ is a natural order follows easily from (GPE2). □

Example 1. The prototypical GPE-algebras arise in the following way from positive cones of po-groups.

Let $(G; +, 0, \leq)$ be a partial ordered group with $G^+ \overset{\text{def}}{=} \{g \in G : g \geq 0\}$ as its positive cone; let G_0 be a non-empty subset of G^+ such that for all $a, b \in G_0$, where $b \leq a$, also $a - b, -b + a \in G_0$.

Then $(G_0; +, 0)$, where $+$ is the group addition restricted to those pairs of elements whose sum is again in G_0, and where 0 is the group zero, is a GPE-algebra, whose order coincides with the restriction of the group order to G_0.

Recall that a subset P_0 of a poset P is called convex, if for any two elements from P_0 all elements from P lying between these two are also contained in P_0. We note that if $(G; +, 0, \leq)$ is a po-group and G_0 is a convex subset of G^+ containing 0, then $(G_0; +, 0)$ is a GPE-algebra.

For the rest of this text, by any sentence involving partial sums to hold we mean: All these sums are defined, and it holds. Similarly, with any sentence involving infima or suprema to hold, we include tacitly the statement that these infima and suprema exist.

Furthermore, in view of (GPE1), we will denote finite sums without brackets, whenever convenient.

Lemma 2. Let $(E; +, 0)$ be a generalized pseudo-effect algebra. For $a, b, c \in E$, the following holds.

(i) If $a + b$ exists, $a_1 \leq a$, and $b_1 \leq b$, then also $a_1 + b_1$ exists.
(ii) Suppose $b + c$ exists. Then $a \leq b$ if and only if $a + c$ exists and $a + c \leq b + c$.
   Suppose $c + b$ exists. Then $a \leq b$ if and only if $c + a$ exists and $c + a \leq c + b$.
(iii) Let $(c + a) \wedge (c + b)$ exist. Then we have $c + (a \wedge b) = (c + a) \wedge (c + b)$.
   Let $(a + c) \wedge (b + c)$ exist. Then we have $(a \wedge b) + c = (a + c) \wedge (b + c)$.
(iv) Let $c + (a \vee b)$ exist. Then we have $c + (a \vee b) = (c + a) \vee (c + b)$.
   Let $(a \vee b) + c$ exist. Then we have $(a \vee b) + c = (a + c) \vee (b + c)$.

Proof. (i) If $a + b$ exists and $a_1' + a_1 = a$ and $b_1 + b_1' = b$ for some a_1' and $b_1' \in E$, it follows that $(a_1' + a_1) + (b_1 + b_1')$ exists and so, by (GPE1), that $a_1 + b_1$ exists.

(ii) Suppose $b + c$ exists. Then $a \leq b$ iff, for some d, $d + a = b$ iff, for some d, $(d + a) + c = b + c$ iff $a + c$ exists and $a + c \leq b + c$. So the first part is proved; the second is shown similarly.

(iii) As $c \leq (c + a) \wedge (c + b)$, we have for some d that $c + d = (c + a) \wedge (c + b)$. Then $c + d \leq c + a, c + b$ and, by (ii), $d \leq a, b$. Suppose $x \leq a, b$. Again by (ii), we have $c + x \leq c + a, c + b$, i.e. $c + x \leq (c + a) \wedge (c + b) = c + d$ and $x \leq d$. It follows $d = a \wedge b$.

The second part is proved similarly.

(iv) As $a, b \leq a \vee b$, we have by (ii) that $c + a$ and $c + b$ exist and $c + (a \vee b) \geq c + a, c + b$. Let $x \geq c + a, c + b$. Then for some y, z we have $x = c + a + y = c + b + z$. Since $a + y = b + z \geq a \vee b$, we get by (ii) $x \geq c + (a \vee b)$. It follows $c + (a \vee b) = (c + a) \vee (b + a)$.

The second part is proved similarly.   □

We will now see how generalized effect algebras are characterized among generalized pseudo-effect algebras.

Definition 2. Let $(E; +, 0)$ be a generalized pseudo-effect algebra. We say that two elements a and b of E *commute* if $a + b$ and $b + a$ both exist and are equal.

We say that E is *commutative* if, for $a, b \in E$, $a + b$ is defined if and only if $b + a$ is defined, in which case $a + b = b + a$.

Proposition 1. Let $(E; +, 0)$ be a generalized pseudo-effect algebra. Then E is a generalized effect algebra if and only if E is commutative.

Proof. This is evident.   □

We conclude this chapter by noting how generalized pseudo-effect algebras are related to pseudo-effect algebras. This relationship is, as to be expected, the same as between generalized effect algebras and effect algebras.

We recall [DvVe1] that a pseudo-effect algebra is a structure $(E; +, 0, 1)$, where $+$ is a partial binary operation and 0, 1 are constants, such that, for all $a, b, c \in E$, (PE1) $(a + b) + c$ is defined iff $a + (b + c)$ is defined, in which case these elements are equal, (PE2) for exactly one $d \in E$ and exactly one $e \in E$ we have $a + d = e + a = 1$, (PE3) if $a + b$ exists, there are $d, e \in E$ such that $a + b = d + a = b + e$, (PE4) if $1 + a$ or $a + 1$ exists, then $a = 0$.

Proposition 2. Let $(E; +, 0)$ be a generalized pseudo-effect algebra, and let $u \in E$. Then $(E[u]; +, 0, u)$ is a pseudo-effect algebra, where $E[u] \overset{\text{def}}{=} \{a \in E : a \leq u\}$, and where $+$ is the restriction of the equally denoted operation of E to those pairs of elements of $E[u]$ whose sum is still in $E[u]$.

In particular, if a GPE-algebra $(E; +, 0)$ has a greatest element 1, then $(E; +, 0, 1)$ is a pseudo-effect algebra.

Conversely, let $(E; +, 0, 1)$ be a pseudo-effect algebra. Then $(E; +, 0)$ is a generalized pseudo-effect algebra.

Proof. To show the first part, let $(E; +, 0)$ be a GPE-algebra and $u \in E$.

To prove that (PE1) holds in the algebra $(E[u]; +, 0, u)$, let $a, b, c \leq E[u]$ such that $a + b$ and $(a + b) + c$ exist in $E[u]$. By (GPE1), $b + c$ and $a + (b + c)$ exist in

E and $(a+b)+c = a+(b+c)$ in E. Since $b+c \leq a+(b+c) = (a+b)+c \leq u$, also $b + c$ and $a + (b + c)$ are elements of $E[u]$, and $(a + b) + c = a + (b + c)$ holds in $E[u]$. Similarly, we see that if $b + c$ and $a + (b + c)$ are defined in $E[u]$, so are $a + b$ and $(a + b) + c$. It follows that (PE1) holds in $E[u]$.

Moreover, (PE2) holds in $E[u]$ due to (GPE2) and (GPE3). (PE3) holds due to (GPE2). If $u + a$ exists in $E[u]$, we have $u + a \leq u = u + 0$ by (GPE5), and by Lemma 2(ii) $a \leq 0$, so $a = 0$. Similarly, from the existence of $a + u$ we conclude $a = 0$. So (PE4) holds.

The second part of the statement is obvious from the first one.

To show the last part, let $(E; +, 0, 1)$ be a pseudo-effect algebra. For the algebra $(E; +, 0)$, (GPE1) follows from (PE1), (GPE2) from (PE3), and (GPE3), (GPE4), (GPE5) from [DvVe1, Lemma 1.4(v), (ii), (i)].   □

3   Generalized Pseudo-Effect Algebras with Riesz Properties

For the purpose of a structure theory for generalized pseudo-effect algebras, we will introduce for these algebras four properties that may be compared to the Riesz Decomposition Property as known for po-groups.

Definition 3. Let $(E; +, 0)$ be a generalized pseudo-effect algebra.

(a) For $a, b \in E$, we write $a \operatorname{com} b$ to mean that for all $a_1 \leq a$ and $b_1 \leq b$, a_1 and b_1 commute.
(b) We say that E fulfils the *Riesz Interpolation Property*, (RIP) for short, if for any $a, b, c, d \in E$ such that $a, b, \leq c, d$ there is a $x \in E$ such that $a, b \leq x \leq c, d$.
(c) We say that E fulfils the *Weak Riesz Decomposition Property*, (RDP$_0$) for short, if for any $a, b_1, b_2 \in E$ such that $a \leq b_1 + b_2$ there are $d_1, d_2 \in E$ such that $d_1 \leq b_1$, $d_2 \leq b_2$ and $a = d_1 + d_2$.
(d) We say that E fulfils the *Commutational Riesz Decomposition Property*, (RDP$_1$) for short, if for any $a_1, a_2, b_1, b_2 \in E$ such that $a_1 + a_2 = b_1 + b_2$ there are $d_1, d_2, d_3, d_4 \in E$ such that (i) $d_1 + d_2 = a_1$, $d_3 + d_4 = a_2$, $d_1 + d_3 = b_1$, $d_2 + d_4 = b_2$, and (ii) $d_2 \operatorname{com} d_3$.
(e) We say that E fulfils the *Strong Riesz Decomposition Property*, (RDP$_2$) for short, if for any $a_1, a_2, b_1, b_2 \in E$ such that $a_1 + a_2 = b_1 + b_2$ there are $d_1, d_2, d_3, d_4 \in E$ such that (i) $d_1 + d_2 = a_1$, $d_3 + d_4 = a_2$, $d_1 + d_3 = b_1$, $d_2 + d_4 = b_2$, and (ii) $d_2 \wedge d_3 = 0$.

Among these four Riesz properties, there is no pair of equivalent ones, but we may order them by strength. Furthermore, the properties (RDP$_1$) and (RDP$_2$) possess equivalent formulations, involving sums of any finite number of elements rather than only sums of pairs.

To make collections of formulas as they will appear in the sequel easier read-able, we use the following abbreviation. Let $d_{ij}, a_i, b_j \in E$, $1 \le i \le m$, $1 \le j \le n$. By

$$
\begin{array}{ccc}
d_{11} & \dots & d_{1n} \to a_1 \\
\vdots & \vdots & \vdots \\
d_{m1} & \dots & d_{mn} \to a_m \\
\downarrow & & \downarrow \\
b_1 & \dots & b_n
\end{array}
$$

we mean $d_{i1} + \cdots + d_{in} = a_i$ for $i = 1,\dots,m$ and $d_{1j} + \cdots + d_{mj} = b_j$ for $j = 1,\dots,n$.

Lemma 3. Let $(E; +, 0)$ be a generalized pseudo-effect algebra fulfilling $(\mathrm{RDP_0})$.

(i) Let $a + b$ exist. From $a\,\mathrm{com}\,c$ and $b\,\mathrm{com}\,c$ it follows $(a + b)\,\mathrm{com}\,c$.
   ¡From $a \wedge c = 0$ and $b \wedge c = 0$ it follows $(a + b) \wedge c = 0$.
(ii) If $a \wedge b = 0$ and if a and b possess an upper bound, then $a+b = b+a = a \vee b$.

Proof. (i) Let $a\,\mathrm{com}\,c$ and $b\,\mathrm{com}\,c$ hold. Suppose $d \le a + b$, $c_1 \le c$. Due to $(\mathrm{RDP_0})$ there are elements d_1 and d_2 such that $d_1 \le a$, $d_2 \le b$ and $d = d_1 + d_2$. By assumption each of d_1 and d_2 commutes with c_1; so also $d = d_1 + d_2$ commutes with c_1. It follows $(a + b)\,\mathrm{com}\,c$.

In a similar way, we prove the second part.

(ii) Let a, b, c be given such that $a \wedge b = 0$ and $c \ge a, b$. So for some x, $c = x + b$, and by $(\mathrm{RDP_0})$ there are elements $d_1 \le x$, $d_2 \le b$ such that $a = d_1 + d_2$. Since $d_2 \le a, b$ we have $d_2 = 0$, and so $a \le x$. It follows $c = x + b \ge a + b$. In particular, $a + b$ exists, and as $a + b$ is an upper bound of a and b, it follows $a + b = a \vee b$.   □

Lemma 4. Let $(E; +, 0)$ be a generalized pseudo-effect algebra.

(i) We have the implications
$$(\mathrm{RDP_2}) \Rightarrow (\mathrm{RDP_1}) \Rightarrow (\mathrm{RDP_0}) \Rightarrow (\mathrm{RIP}).$$
   The converse of any of these implications does not hold.
(ii) Let $(\mathrm{RDP_0})$ hold and E be a lower semi-lattice. Then $(\mathrm{RDP_2})$ holds.
(iii) Let $(\mathrm{RDP_1})$ hold. Given $a_1, \dots, a_m, b_1, \dots b_n \in E$, $n, m \ge 1$, such that $a_1 + \cdots + a_m = b_1 + \cdots + b_n$, there are elements $d_{11}, \dots, d_{mn} \in E$ such that

$$
\begin{array}{ccc}
d_{11} & \dots & d_{1n} \to a_1 \\
\vdots & \vdots & \vdots \\
d_{m1} & \dots & d_{mn} \to a_m \\
\downarrow & & \downarrow \\
b_1 & \dots & b_n
\end{array}
\tag{1}
$$

and such that we have

$$(d_{i+1,j} + \cdots + d_{mj}) \ \text{com} \ (d_{i,j+1} + \cdots + d_{in}); \ \ 1 \le i < m, 1 \le j < n. \ (2)$$

In case that (RDP_2) holds, we may further assume

$$(d_{i+1,j} + \cdots + d_{mj}) \wedge (d_{i,j+1} + \cdots + d_{in}) = 0; \ \ 1 \le i < m, 1 \le j < n. \ (3)$$

Proof. (i) By Lemma 3(ii), (RDP_2) implies (RDP_1). It is clear that (RDP_1) implies (RDP_0).

Assume now (RDP_0), and let $a, b \le c, d$. Then there is an $a_1 \in E$ such that $a + a_1 = c$, and from $b \le a + a_1$ follows by (RDP_0) that there are $e \le a$ and $\bar{b} \le a_1$ such that $b = e + \bar{b}$. Moreover, for some $\bar{a} \in E$ we have $a = e + \bar{a}$, and for some $\bar{c}, \bar{d}$ also $c = e + \bar{c}$ and $d = e + \bar{d}$.

By Lemma 2(ii) we have $\bar{a}, \bar{b} \le \bar{c}, \bar{d}$, and from $e + \bar{a} + \bar{b} \le a + a_1 = c = e + \bar{c}$ it follows that $\bar{a} + \bar{b} \le \bar{c}$.

Now chose $b_1 \in E$ such that $b_1 + \bar{b} = \bar{d}$, and since $\bar{a} \le b_1 + \bar{b}$, there are $\bar{\bar{a}} \le b_1$ and $f \le \bar{b}$ such that $\bar{a} = \bar{\bar{a}} + f$. Chose $\bar{\bar{b}} \in E$ such that $\bar{b} = \bar{\bar{b}} + f$.

Let $\bar{x} = \bar{\bar{a}} + \bar{\bar{b}} + f$. Then $\bar{a}, \bar{b} \le \bar{x} \le \bar{a} + \bar{b} \le \bar{c}$ as well as $\bar{x} \le b_1 + \bar{b} = \bar{d}$. So $x = e + \bar{x}$ is the interpolant required to show that (RIP) holds.

By [DvVe1, Proposition 3.3(i)] and Proposition 2, the converse of any of the implications shown does not hold.

(ii) Let $a_1, a_2, b_1, b_2 \in E$ and suppose $a_1 + a_2 = b_1 + b_2$. Set $d_1 = a_1 \wedge b_1$ and d_2, d_3 in such a way that $a_1 = d_1 + d_2$, $b_1 = d_1 + d_3$.

We claim that $d_2 \le b_2$. Indeed, from $a_1 \le b_1 + b_2$ we get by (RDP_0) that $a_1 = e_1 + e_2$ for some $e_1 \le b_1$, $e_2 \le b_2$, and from $e_1 \le a_1 \wedge b_1 = d_1$ we have $a_1 \le d_1 + b_2$, so we conclude $d_2 \le b_2$. Choose $d_4 \in E$ such that $d_2 + d_4 = b_2$. Similarly, we may choose $d_4' \in E$ such that $d_3 + d_4' = a_2$.

Furthermore, $d_2 \wedge d_3 = 0$ since by Lemma 2(iii) we have $d_1 = a_1 \wedge b_1 = (d_1 + d_2) \wedge (d_1 + d_3) = d_1 + (d_2 \wedge d_3)$. By Lemma 3(ii) we conclude $d_1 + d_2 + d_3 + d_4' = a_1 + a_2 = b_1 + b_2 = d_1 + d_3 + d_2 + d_4 = d_1 + d_2 + d_3 + d_4$, so $d_4' = d_4$. So d_1, d_2, d_3, d_4 fulfil the requirements of Definition 3(e) of (RDP_2).

(iii) The Lemma is trivial for $m = 1$ or $n = 1$. Let $m = n = 2$. If just (RDP_1) holds, there are by definition $d_{11}, d_{12}, d_{21}, d_{22} \in E$ such that (1) and (2) and holds. If even (RDP_2) holds, $d_{11}, \ldots, d_{22}$ may be chosen such that that (1) and (3) and holds, and since d_{12} and d_{21} possess an upper bound, also (2) is true by Lemma 3(ii).

Suppose now the Lemma is true for any pair of integers such that the first is less or equal to m, where $m \ge 2$, and the second is less than n, where $n \ge 3$. From this condition we shall prove that it is true also for the pair m, n. By complete induction the Lemma then follows to be true in general.

So let $a_1 + \cdots + a_m = b_1 + \cdots + b_n$. By the induction hypothesis, there are elements $d_{11}, \ldots, d_{m,n-2}, e_1, \ldots, e_m \in E$ such that

$$
\begin{array}{cccc}
d_{11} \ \cdots \ d_{1,n-2} & e_1 & \to a_1 \\
\vdots \qquad \vdots & \vdots & \vdots \\
d_{m1} \ \cdots \ d_{m,n-2} & e_m & \to a_m \\
\downarrow \qquad \downarrow & \downarrow \\
b_1 \ \ \cdots \ \ b_{n-2} & b_{n-1} + b_n
\end{array}
$$

and for $1 \le i < m$, $1 \le j \le n - 2$ we have

$$(d_{i+1,j} + \cdots + d_{mj}) \ \text{com} \ (d_{i,j+1} + \cdots + d_{i,n-2} + e_i)$$

and in case of (RDP$_2$) also

$$(d_{i+1,j} + \cdots + d_{mj}) \wedge (d_{i,j+1} + \cdots + d_{i,n-2} + e_i) = 0.$$

Moreover, from $e_1 + \cdots + e_m = b_{n-1} + b_n$ we have by the same hypothesis that there are elements $d_{1,n-1}, d_{1n}, \ldots, d_{m,n-1}, d_{mn} \in E$ such that

$$
\begin{array}{ccc}
d_{1,n-1} \ d_{1n} & \to e_1 \\
\vdots \quad \vdots & \vdots \\
d_{m,n-1} \ d_{mn} & \to e_m \\
\downarrow \quad \downarrow \\
b_{n-1} \quad b_n
\end{array}
$$

and for $1 \le i \le m - 1$

$$(d_{i+1,n-1} + \cdots + d_{m,n-1}) \ \text{com} \ d_{in}$$

and in case of (RDP$_2$) also

$$(d_{i+1,n-1} + \cdots + d_{m,n-1}) \wedge d_{in} = 0.$$

It follows that the Lemma holds for the pair m, n. $\qquad\qquad\square$

4 Embedding of Generalized Pseudo-Effect Algebras into Positive Cones of Groups

We are concerned in this chapter with the question under which conditions generalized pseudo-effect algebra are representable as a subset of the positive cone of a po-group, just in the way shown in Example 1.

Unfortunately, the question how to characterize exactly those GPE-algebras that possess a representation by means of a po-group seems to be very hard to give an answer to. Here, we will introduce two different conditions, to be called (A) and (B), that at least turn out to be sufficient for GPE-algebras to be representable that way.

The conditions are defined in the following manner; E is supposed to be a generalized pseudo-effect algebra.

(A) E is a lower semi-lattice and fulfils (RDP_0).
(B) E is directed and fulfils (RDP_1).

We note that no one of these conditions implies the other one, as seen from the following examples.

Example 2. Set $E \stackrel{\mathrm{def}}{=} (\mathbb{N} \times \{0\}) \cup (\{0\} \times \mathbb{N})$, and define a partial addition componentwise whenever this leads to a result in E. Then $(E; +, (0,0))$ is a GPE-algebra. Moreover, it is a lower semi-lattice; e.g. we have for $a, b \in \mathbb{N}$ $(a,0) \wedge (0,b) = (0,0)$ and $(a,0) \wedge (b,0) = (\min\{a,b\},0)$. As it is easily verified, E also fulfils (RDP_0). But it is not directed; for $a, b \in \mathbb{N}$ such that $a, b \geq 1$, an upper bound of $(a,0)$ and $(0,b)$ does not exist.

Example 3. Let E be the set of rational functions $f : [0,1] \to \mathbb{R}$ without singularities, define $+$ for any pair of functions as the pointwise addition and 0 as the zero constant function (see [DvVe1, Example 3.4]). Then $(E; +, 0)$ is a generalized effect algebra, so a fortiori a GPE-algebra. Moreover, it fulfils (RDP_1) and is directed, but it is not a lower semi-lattice.

For the proof of our representation theorem, we will use a variant of the so-called word technique. This technique was introduced by Baer [Baer] and Wyler [Wyl]; furthermore, to mention just one example, it has been applied to pseudo-effect algebras fulfilling (RDP_1) [DvVe2].

We will first embed a given algebra into a semigroup (Definition 4 and Lemma 5); then the conditions that are necessary and sufficient for extending a semigroup to a po-group are checked (Lemmas 6 and 7).

Definition 4. Let $(E; +, 0)$ be a generalized pseudo-effect algebra.

(i) A sequence $A = (a_1, \ldots, a_n)$ of finite, but non-zero, length with entries from E is called a *word* in E. We denote by $\mathcal{W}(E)$ the set of all words; that is

$$\mathcal{W}(E) \stackrel{\mathrm{def}}{=} \{(a_1, \ldots, a_n) : a_1, \ldots, a_n \in E, n \geq 1\}.$$

We define an *addition* in $\mathcal{W}(E)$ as the concatenation; that is

$$+ : \mathcal{W}(E) \times \mathcal{W}(E) \to \mathcal{W}(E),$$
$$((a_1, ..., a_m), (b_1, ..., b_n)) \mapsto (a_1, ..., a_m, b_1, ..., b_n).$$

(ii) We call two words A and B of E *directly similar*, in symbols $A \sim B$, if one of the following conditions is fulfilled.
   (α) $A = (a_1, \ldots, a_n)$ and $B = (a_1, \ldots, a_p + a_{p+1}, \ldots, a_n)$; $1 \le p < n$.
   (β) $A = (a_1, \ldots, a_p + a_{p+1}, \ldots, a_n)$ and $B = (a_1, \ldots, a_n)$; $1 \le p < n$.
   (γ) $A = (a_1, \ldots, a_n)$, $B = (a_1, \ldots, a_{p+1}, a_p, \ldots, a_n)$ and $a_p \wedge a_{p+1} = 0$; $1 \le p < n$.

We call two words A and B *similar*, in symbols $A \simeq B$, if there are words $A_0, \ldots, A_k$, $k \ge 0$, such that $A = A_0 \sim A_1 \sim \cdots \sim A_k = B$. In such a case we say that A and B are *connected by a chain of length k*.

We set for $a_1, \ldots, a_n \in E$, $n \ge 1$,

$$[a_1, \ldots, a_n] \overset{\text{def}}{=} \{A \in \mathcal{W}(E) : A \simeq (a_1, \ldots, a_n)\},$$

and we put

$$\mathcal{C}(E) \overset{\text{def}}{=} \{[a_1, \ldots, a_n] : a_1, \ldots, a_n \in E, n \ge 1\}.$$

Lemma 5. Let $(E; +, 0)$ be a generalized pseudo-effect algebra fulfilling (RDP$_0$).

(i) Similarity in $\mathcal{W}(E)$ is an equivalence relation compatible with $+$. $+$ being the induced relation, $(\mathcal{C}(E); +)$ is a semigroup with the neutral element $[0]$.
(ii) For $a_1, \ldots, a_n, b \in E$, $n \ge 1$, $(a_1, \ldots, a_n) \simeq (b)$ if and only if $a_1 + \cdots + a_n$ exists and equals b.

Proof. (i) $\simeq$ is by construction an equivalence relation.

¿From $A_1 \simeq A$ and $B_1 \simeq B$ it follows $A_1 + B_1 \simeq A + B$, so $+$ is definable in $\mathcal{C}(E)$.

As $\mathcal{W}(E)$ is associative, so is $\mathcal{C}(E)$, i.e. $\mathcal{C}(E)$ is a semigroup. It has $[0]$ as a neutral element, because e.g. $[a_1, \ldots, a_n] + [0] = [a_1, \ldots, a_n, 0] = [a_1, \ldots, a_n]$.

(ii) If for a word $(x_1, \ldots, x_m)$ the sum of its elements $x_1 + \cdots + x_m$ exists, the same is true for any word directly similar to $(x_1, \ldots, x_m)$, and the sum remains the same. For this is evident in case the two words are related by condition (α) or (β) of Definition 4; and if they are related by condition (γ), the two permuted elements commute by Lemma 3(ii), because they possess their sum as an upper bound and we assumed (RDP$_0$).

So the "only if" part follows by induction on the minimal length of a chain by which (b) and $(a_1, \ldots, a_n)$ are connected.

The "if" part is obvious. $\square$

Lemma 6. Let $(E; +, 0)$ be a generalized pseudo-effect algebra fulfilling (A) or (B). Let

$$(a_1, \ldots, a_m) \simeq (b_1, \ldots, b_n),$$

where $m, n \geq 1$. Then there are elements $d_{11}, \ldots, d_{mn} \in E$ such that

$$
\begin{array}{ccc}
d_{11} & \ldots & d_{1n} \;\rightarrow a_1 \\
\vdots & & \vdots \quad \vdots \\
d_{m1} & \ldots & d_{mn} \;\rightarrow a_m \\
\downarrow & & \downarrow \\
b_1 & \ldots & b_n.
\end{array}
$$

In case that (A) holds, we may further assume that, for $1 \leq i < m,\ 1 \leq j < n$ we have

$$
(d_{i+1,j} + \cdots + d_{mj}) \wedge (d_{i,j+1} + \cdots + d_{in}) = 0. \tag{4}
$$

In case that (B) holds, we may further assume that, for $1 \leq i < m,\ 1 \leq j < n$ we have

$$
(d_{i+1,j} + \cdots + d_{mj}) \ \text{com} \ (d_{i,j+1} + \cdots + d_{in}). \tag{5}
$$

Proof. The proof is by induction on the minimal length k of a chain that connects $(a_1, \ldots, a_m)$ and $(b_1, \ldots, b_n)$.

If $k = 0$, we have $(a_1, \ldots, a_m) = (b_1, \ldots, b_n)$, and the elements in the scheme

$$
\begin{array}{ccccc}
a_1 & 0 & \ldots & 0 & \rightarrow a_1 \\
0 & a_2 & \ldots & 0 & \rightarrow a_2 \\
\vdots & \vdots & \ddots & \vdots & \vdots \\
0 & 0 & \ldots & a_m & \rightarrow a_m \\
\downarrow & \downarrow & & \downarrow & \\
a_1 & a_2 & \ldots & a_m &
\end{array}
$$

obviously fulfill the statement.

Suppose the statement is true for $k-1$, $k \geq 1$; we have to prove that it then also holds for k. So let $A = (a_1, \ldots, a_m)$ and $B = (b_1, \ldots, b_n)$ be connected by a chain of length k, say $A = A_0 \sim \cdots \sim A_k = B$.

There are according to Definition 4(ii) three possibilities how $A_k = B = (b_1, \ldots, b_n)$ is constructed from A_{k-1}.

Ad (α). Let $A_{k-1} = (b_1, \ldots, b_p^1, b_p^2, \ldots, b_n)$, $b_p = b_p^1 + b_p^2$; $1 \leq p \leq n$. Then there are by hypothesis elements in E according to the scheme

$$
\begin{array}{cccccc}
d_{11} & \ldots & d_{1p}^1 & d_{1p}^2 & \ldots & d_{1n} \;\rightarrow a_1 \\
\vdots & & \vdots & \vdots & & \vdots \\
d_{m1} & \ldots & d_{mp}^1 & d_{mp}^2 & \ldots & d_{mn} \;\rightarrow a_m \\
\downarrow & & \downarrow & \downarrow & & \downarrow \\
b_1 & \ldots & b_p^1 & b_p^2 & \ldots & b_n,
\end{array} \tag{6}
$$

such that the sums of the elements placed below and right from any element fulfil the zero-infimum relation or the **com**-relation, respectively. Consider now the scheme

$$
\begin{array}{ccccc}
d_{11} & \cdots & d^1_{1p} + d^2_{1p} & \cdots & d_{1n} \\
\vdots & & \vdots & & \vdots \\
d_{m1} & \cdots & d^1_{mp} + d^2_{mp} & \cdots & d_{mn}.
\end{array}
$$

Here the lines still sum up to $a_1, \ldots, a_m$, and the unchanged columns sum up as before to $b_1, \ldots, b_{p-1}, b_{p+1}, \ldots, b_n$. Furthermore, any d^1_{ip}, $1 < i \leq m$, commutes with any $d^2_{i'p}$, $1 \leq i' < i$. This follows in case of (A) from $d^1_{ip} \wedge d^2_{i'p} = 0$ and $d^1_{ip}, d^2_{i'p} \leq b_p$ by Lemma 3(ii), and in case of (B) from $d^1_{ip} \operatorname{\textbf{com}} d^2_{i'p}$.

So also the p-th column adds up to what it should:

$$
\begin{aligned}
b_p = b^1_p + b^2_p &= d^1_{1p} + \cdots + d^1_{mp} + d^2_{1p} + \cdots + d^2_{mp} \\
&= d^1_{1p} + d^2_{1p} + d^1_{2p} + \cdots + d^1_{mp} + d^2_{2p} + \cdots + d^2_{mp} \\
&= \cdots \\
&= d^1_{1p} + d^2_{1p} + d^1_{2p} + d^2_{2p} + \cdots + d^1_{mp} + d^2_{mp}.
\end{aligned}
$$

In case of (A), the only new relations concerning zero infima are the following, where $1 \leq j < m$:

$$
(d^1_{j+1,p} + d^2_{j+1,p} + \cdots + d^1_{mp} + d^2_{mp}) \wedge (d_{j,p+1} + \cdots + d_{jn}) = 0. \qquad (7)
$$

By the same reasoning as before, the first term in (7) is equal to $(d^1_{j+1,p} + \cdots + d^1_{mp}) + (d^2_{j+1,p} + \cdots + d^2_{mp})$. Here, we know by hypothesis that both the first term inside brackets and the second one possesses together with the second term in (7) a zero infimum . Now by Lemma 3(i), (7) follows.

In a similar manner, we argue in the case of (B).

Ad (β). Let $A_{k-1} = (b_1, \ldots, b_p + b_{p+1}, \ldots, b_n)$; $1 \leq p < n$. Then there are by hypothesis elements in E according to the scheme

$$
\begin{array}{ccccccc}
d_{11} & \cdots & & e_1 & & \cdots & d_{1n} \rightarrow a_1 \\
\vdots & & & \vdots & & \vdots & \vdots \\
d_{m1} & \cdots & & e_m & & \cdots & d_{mn} \rightarrow a_m \\
\downarrow & & & \downarrow & & & \downarrow \\
b_1 & \cdots & b_p & + & b_{p+1} & \cdots & b_n,
\end{array}
$$

such that the sums of the elements places below and right from any element fulfil the **com**-relation or the zero infimum relation, respectively. Now since

(A) implies (RDP$_2$) by Lemma 4(ii) and (B) includes (RDP$_1$), we may apply Lemma 4(iii) to the equation

$$b_p + b_{p+1} = e_1 + \cdots + e_m$$

to get elements $d_{1p}, d_{1,p+1}, \ldots, d_{mp}, d_{m,p+1} \in E$ such that

$$
\begin{array}{ccc}
d_{1p} & d_{1,p+1} & \rightarrow e_1 \\
\vdots & \vdots & \vdots \\
d_{mp} & d_{m,p+1} & \rightarrow e_m \\
\downarrow & \downarrow & \\
b_p & b_{p+1} &
\end{array}
$$

and, for $1 \leq j < m$, in case of (A)

$$(d_{j+1,p} + \cdots + d_{mp}) \wedge d_{j,p+1} = 0,$$

in case of (B)

$$(d_{j+1,p} + \cdots + d_{mp}) \text{ com } d_{j,p+1}.$$

So we have

$$
\begin{array}{ccccccc}
d_{11} & \cdots & d_{1p} & d_{1,p+1} & \cdots & d_{1n} & \rightarrow a_1 \\
\vdots & & \vdots & \vdots & & \vdots & \vdots \\
d_{m1} & \cdots & d_{mp} & d_{m,p+1} & \cdots & d_{mn} & \rightarrow a_m \\
\downarrow & & \downarrow & \downarrow & & \downarrow & \\
b_1 & \cdots & b_p & b_{p+1} & \cdots & b_n. &
\end{array}
$$

As a sum when deleting any summand from it gets smaller, and as the zero-infimum relation as well as the com-relation is additive by Lemma 3(i), we see that the conditions concerning these relations are again preserved herein.

Ad (γ). Let $A_{k-1} = (b_1, \ldots, b_{p+1}, b_p, \ldots, b_n)$, $b_p \wedge b_{p+1} = 0$; $1 \leq p < n$. There are by hypothesis elements in E according to the scheme

$$
\begin{array}{ccccccc}
d_{11} & \cdots & d_{1,p} & d_{1,p+1} & \cdots & d_{1n} & \rightarrow a_1 \\
\vdots & & \vdots & \vdots & & \vdots & \vdots \\
d_{m1} & \cdots & d_{m,p} & d_{m,p+1} & \cdots & d_{mn} & \rightarrow a_m \\
\downarrow & & \downarrow & \downarrow & & \downarrow & \\
b_1 & \cdots & b_{p+1} & b_p & \cdots & b_n,
\end{array}
\qquad (8)
$$

such that the sums of the elements placed below and right from any element fulfil the zero-infimum relation or the com-relation, respectively. Consider now the scheme that is similar to (8), but in which the p-th and the $p+1$-th column are exchanged. Then clearly the columns sum up to the elements of A_k; and by Lemma 3(ii) the rows sum up to the same values as before.

Now the infimum of any of the elements $d_{1,p}, \ldots, d_{m,p}$ and of any of $d_{1,p+1}, \ldots, d_{m,p+1}$ is zero. If (B) holds, we see by the directedness of E and by Lemma 3(ii) that all these pairs fulfil also the com-condition.

Due to the additivity of the zero-infimum relation as well as of the com-relation, we see that the conditions concerning these relations are again preserved.   □

Lemma 7. Let $(E; +, 0)$ be a generalized pseudo-effect algebra fulfilling (A) or (B).

Then $C(E)$ is a semigroup such that the following holds.

(i) $[0]$ is a neutral element.
(ii) Let $a, b \in C(E)$. From $a + b = [0]$ it follows $a = b = [0]$.
(iii) Let $a, b, c \in C(E)$. From $a + b = a + c$ it follows $b = c$;
      from $b + a = c + a$ it follows $b = c$.
(iv) For any pair $a, b \in C(E)$ there are $c, \mathfrak{d} \in C(E)$ such that $a + b = c + a = b + \mathfrak{d}$.

Proof. By Lemma 4(i) and (ii), (RDP_1) holds in E in any case.

That $C(E)$ is a semigroup fulfilling (i) has been proved in Lemma 5(i).

(ii) This follows from Lemma 5(ii) and (GPE4).

(iii) We may suppose $a = [a]$. Let $b = [b_1, \ldots, b_m]$, $c = [c_1, \ldots, c_n]$. We will show that $(a, b_1, \ldots, b_m) \simeq (a, c_1, \ldots, c_n)$ implies $(b_1, \ldots, b_m) \simeq (c_1, \ldots, c_n)$; then the first part follows, and the second one is proved analogously.

By our assumption there are, by Lemma 6, elements in E according to the following scheme:

$$
\begin{array}{ccccc}
d & d_1 & \cdots & d_n & \to a \\
e_1 & e_{11} & \cdots & e_{1n} & \to b_1 \\
\vdots & \vdots & & \vdots & \vdots \\
e_m & e_{m1} & \cdots & e_{mn} & \to b_m \\
\downarrow & \downarrow & & \downarrow & \\
a & c_1 & & c_n, &
\end{array}
$$

where any pair of elements such that one is placed further up and further right than the other one has a zero infimum or fulfils the com-relation, respectively.

$a = d + d_1 + \cdots + d_n = d + e_1 + \cdots + e_m$ implies $d_1 + \cdots + d_n = e_1 + \cdots + e_m$. Using this and the fact we may exchange successive elements in a word whenever they have a zero infimum or fulfil the com-relation, we get

$$(b_1, \ldots, b_m) \simeq (e_1, e_{11}, \ldots, e_{1n}, \ldots, e_m, e_{m1}, \ldots, e_{mn})$$
$$\simeq (e_1, \ldots, e_m, e_{11}, \ldots, e_{m1}, \ldots, e_{1n}, \ldots, e_{mn})$$
$$\simeq (d_1, \ldots, d_n, e_{11}, \ldots, e_{m1}, \ldots, e_{1n}, \ldots, e_{mn})$$
$$\simeq (d_1, e_{11}, \ldots, e_{m1}, \ldots, d_n, e_{1n}, \ldots, e_{mn})$$
$$\simeq (c_1, \ldots, c_n).$$

(iv) We will prove one half only; the second one is shown analogously. Moreover, we suppose that $\mathfrak{a} = [a]$ and $\mathfrak{b} = [b]$; the claim then follows by an easy induction argument. So what we will show is that there is a word $\mathfrak{c}$ such that $(a, b) = \mathfrak{c} + (a)$.

Suppose first that E fulfils (A).

Choose $\bar{a}, \bar{b}$ such that $\bar{a} + (a \wedge b) = a$ and $\bar{b} + (a \wedge b) = b$. Then $\bar{a} \wedge \bar{b} = 0$; indeed, from $x \leq \bar{a}, \bar{b}$ it follows by Lemma 2(ii) that $x + (a \wedge b) \leq a, b$, so $x + (a \wedge b) \leq a \wedge b$, that is $x \leq 0$, and so $x = 0$. Now choose $\bar{\bar{a}}$ such that $\bar{\bar{a}} + \bar{a} = a$. Using condition (γ) of Definition 4(ii), we get $(a, b) \simeq (\bar{\bar{a}}, \bar{a}, \bar{b}, a \wedge b) \sim (\bar{\bar{a}}, \bar{b}, \bar{a}, a \wedge b) \sim (\bar{\bar{a}}, \bar{b}) + (a)$.

Suppose now that E fulfils (B).

Let c be an upper bound of a and b, which means that $a + \bar{a} = \bar{b} + b = c$ for some $\bar{a}, \bar{b} \in E$.

By (RDP_1) there are $d_1, \ldots, d_4 \in E$ such that

$$\begin{array}{cc} d_1 & d_2 \to a \\ d_3 & d_4 \to \bar{a} \\ \downarrow & \downarrow \\ \bar{b} & b. \end{array}$$

So there are elements $d_2', d_4' \in E$ such that $(a, b) = (d_1 + d_2, d_2 + d_4) \sim (d_2' + d_1, d_2, d_4) \simeq (d_2', a, d_4) \sim (d_2', a + d_4) = (d_2', d_4' + a) \sim (d_2', d_4') + (a)$. Here we used (GPE2) and the fact that due to $d_4 \leq \bar{a}$, $a + d_4$ exists.   □

By a theorem of Birkhoff (see [Fu, Theorem II.4]), the conditions (i) to (iv) of the preceding Lemma are necessary and sufficient that $\mathcal{C}(E)$ is the positive cone of some po-group. So we may give the following definition; for details we refer to [Fu].

Definition 5. Let $(E; +, 0)$ be a generalized pseudo-effect algebra fulfilling (A) or (B).

Let $\mathcal{G}(E)$ be the po-group such that $\mathcal{C}(E) = \mathcal{G}(E)^+$ and $\mathcal{G}(E) = \mathcal{C}(E) - \mathcal{C}(E) = -\mathcal{C}(E) + \mathcal{C}(E)$.

We arrive at our main theorem.

Theorem 1. Let $(E; +, 0)$ be a generalized pseudo-effect algebra fulfilling (A) or (B).

Then $\iota_E : E \to \mathcal{G}(E)$, $a \mapsto [a]$, is an isomorphic embedding of the GPE-algebra E into the directed po-group $\mathcal{G}(E)$. Furthermore, we have for $a, b, c \in E$ that

$$a + b \text{ is defined and equals } c \text{ if and only if } \iota_E(a) + \iota_E(b) = \iota_E(c). \qquad (9)$$

Proof. (i) ι_E is injective. Indeed, for $a, b \in E$, $[a] = [b]$ implies $a = b$ by Lemma 5(ii).

Let now $a, b \in E$. If the sum exists and $a + b = c$, then $(a, b) \sim (c)$, so $[a] + [b] = [c]$, which means $\iota_E(a) + \iota_E(b) = \iota_E(c)$. Conversely, if for some $c \in E$ we have $\iota_E(a) + \iota_E(b) = \iota_E(c)$, then $[a] + [b] = [c]$, which means $(a, b) \sim (c)$, so $a + b = c$ by Lemma 5(ii).

$\mathcal{G}(E)$ is directed according to [Fu, Chapter II, Proposition 1], since $\mathcal{G}(E)^+$ generates the whole of $\mathcal{G}(E)$. □

5   Properties of the Representing Group of a GPE-algebra

We shall in this section compare the properties of a GPE-algebra on the one hand and of its representing group on the other one. We are in particular concerned with the lattice operations and with Riesz properties of different kind.

First of all, we may define the Riesz properties (RIP), (RDP$_0$), (RDP$_1$), (RDP$_2$) as well as the com-relation for po-group in exact analogy to GPE-algebras. We refer to [DvVe1] for details. Let us note the following facts concerning these properties.

Remark 1. We may formulate Lemma 4 for po-groups in the following manner. Let a po-group G be given.

(i) We have the implications (RDP$_2$) $\Rightarrow$ (RDP$_1$) $\Rightarrow$ (RDP$_0$) $\Leftrightarrow$ (RIP). For the proof, see [DvVe1, Proposition 4.2(i)].

(ii) G is lattice-ordered iff (RDP$_2$) holds. For the proof, see [DvVe1, Proposition 4.2(ii)].

(iii) Let (RDP$_1$) hold. Given $a_1, \ldots, a_m, b_1, \ldots b_n \in G^+$, $n, m \geq 1$, such that $a_1 + \cdots + a_m = b_1 + \cdots + b_n$, there are elements $d_{11}, \ldots, d_{mn} \in G^+$ such that the scheme (1) and the relations (2) hold. In case that (RDP$_2$) holds, it may be further assumed that the relations (3) hold.

This follows from Lemma 4(iii), since $(G^+; +, 0)$ is a GPE-algebra.

¿From the following lemma, we see how the lattice operations and the Riesz properties are preserved from a group to a GPE-algebra constructed from its positive cone.

Proposition 3. Let G be a directed po-group and G_0 be a convex subset of G^+ containing 0. Let $(G_0; +, 0)$ be the GPE-algebra defined according to Example 1.

(i) If the po-group G fulfils one of the properties (RIP), (RDP$_0$), (RDP$_1$), or (RDP$_2$), the GPE-algebra $(G_0; 0, +)$ has the equally denoted property, respectively.
(ii) Let G fulfil (RIP). Then the infimum of some $a, b \in G_0$ exists with respect to the GPE-algebra G_0 if and only if it exists with respect to the po-group G, and both infima are equal.
If the supremum of some $a, b \in G_0$ exists with respect to the GPE-algebra G_0, then it exists with respect to the po-group G, and both suprema are equal.

Proof. (i) This follows easily from the fact that G_0 is convex and that the sum operations of G_0 and of G coincide.

(ii) The order of G_0 coincides with the one of G. We will, on the other hand, distinguish between the lattice operations of G_0 and G by appropriate indices. Let $a, b \in G_0$.

Let $a \wedge_{G_0} b$ exist. So $a \wedge_{G_0} b \le a, b$. $x \in G$ and $x \le a, b$ implies by (RIP) that there is some $y \in G$ such that $x, 0 \le y \le a, b$; we then have $y \in G_0$, since G_0 is convex, whence $x \le y \le a \wedge_{G_0} b$. It follows $a \wedge_{G_0} b = a \wedge_G b$.

If $a \wedge_G b$ exists, it is easy to see that then $a \wedge_{G_0} b = a \wedge_G b$.

Now let $a \vee_{G_0} b$ exist. So $a, b \le a \vee_{G_0} b$. $x \in G$ and $a, b \le x$ implies by (RIP) that there is some $y \in G$ such that $a, b \le y \le x, a \vee_{G_0} b$. Since then $y \in G_0$, we have $y = a \vee_{G_0} b$, whence $a \vee_{G_0} b \le x$. It follows $a \vee_{G_0} b = a \vee_G b$.   □

It is a little bit more tedious to prove that lattice operations and the properties (RDP$_1$) and (RDP$_2$) are also preserved in the other direction, from a GPE-algebra to its representing group.

Proposition 4. Let $(E; +, 0)$ be a generalized pseudo-effect algebra fulfilling (A) or (B) and $\iota_E : E \to \mathcal{G}(E)$ its po-group representation.

(i) Let E fulfil (A). Then $\mathcal{G}(E)$ is an ℓ-group. $\iota(E)$ is a convex subset of $\mathcal{G}(E)^+$ containing 0 that generates $\mathcal{G}(E)$.
(ii) Let E fulfil (B). Then $\mathcal{G}(E)$ is a directed po-group fulfilling (RDP$_1$). $\iota(E)$ is a directed convex subset of $\mathcal{G}(E)^+$ containing 0 that generates $\mathcal{G}(E)$.
(iii) ι_E preserves infima and suprema whenever they exist.

Proof. It was already established in Theorem 1 that $\mathcal{G}(E)$ is a directed po-group.

$\iota(E)$ clearly contains 0, and it is generates $\mathcal{G}(E)$ by construction. Moreover, it is convex; indeed, from $\mathfrak{a} \in \mathcal{G}(E)$, $\mathfrak{b} \in E$ and $0 \leq \mathfrak{a} \leq [\mathfrak{b}]$ it follows $\mathfrak{a} + \mathfrak{c} = [\mathfrak{b}]$ for some $\mathfrak{c} \in \mathcal{G}(E)^+$, and so $\mathfrak{a} \in \iota(E)$ by Lemma 5(ii).

(i) We will show that $\mathcal{G}(E)$ fulfils (RDP$_2$); by [DvVe1, Proposition 4.2(ii)], it then follows that $\mathcal{G}(E)$ is lattice-ordered.

So let $\mathfrak{a}_1, \mathfrak{a}_2, \mathfrak{b}_1, \mathfrak{b}_2 \in \mathcal{C}(E) = \mathcal{G}(E)^+$ such that $\mathfrak{a}_1 + \mathfrak{a}_2 = \mathfrak{b}_1 + \mathfrak{b}_2$. Let $\mathfrak{a}_1 = [a_1^1, \ldots, a_k^1]$, $\mathfrak{a}_2 = [a_1^2, \ldots, a_l^2]$, $\mathfrak{b}_1 = [b_1^1, \ldots, b_m^1]$, $\mathfrak{b}_2 = [b_1^2, \ldots, b_n^2]$, where $a_1^1, \ldots, b_n^2 \in E$. By Lemma 6 there are $d_{11}^1, \ldots, d_{ln}^4 \in E$ such that

$$
\begin{array}{cccccc}
d_{11}^1 & \ldots & d_{1m}^1 & d_{11}^2 & \ldots & d_{1n}^2 \to a_1^1 \\
\vdots & & \vdots & \vdots & & \vdots \\
d_{k1}^1 & \ldots & d_{km}^1 & d_{k1}^2 & \ldots & d_{kn}^2 \to a_k^1 \\
d_{11}^3 & \ldots & d_{1m}^3 & d_{11}^4 & \ldots & d_{1n}^4 \to a_1^2 \\
\vdots & & \vdots & \vdots & & \vdots \\
d_{l1}^3 & \ldots & d_{lm}^3 & d_{l1}^4 & \ldots & d_{ln}^4 \to a_l^2 \\
\downarrow & & \downarrow & \downarrow & & \downarrow \\
b_1^1 & \ldots & b_m^1 & b_1^2 & \ldots & b_n^2
\end{array}
\tag{10}
$$

and such that every two elements in this diagram possess the infimum 0 if one of them is placed further up and further right than the other one.

Define now $\mathfrak{d}_1 = [d_{11}^1, \ldots, d_{1m}^1, \ldots, d_{k1}^1, \ldots, d_{km}^1]$ and in a similar way also $\mathfrak{d}_2, \mathfrak{d}_3, \mathfrak{d}_4$. Due to the commutativity condition (γ) from Definition 4(ii), we then have

$$
\begin{array}{cc}
\mathfrak{d}_1 & \mathfrak{d}_2 \to \mathfrak{a}_1 \\
\mathfrak{d}_3 & \mathfrak{d}_4 \to \mathfrak{a}_2 \\
\downarrow & \downarrow \\
\mathfrak{b}_1 & \mathfrak{b}_2.
\end{array}
\tag{11}
$$

This already shows that (RDP$_0$) holds in $\mathcal{G}(E)$. By Remark 1(i) and Proposition 3(ii), we conclude that $\mathfrak{d}_2 \wedge \mathfrak{d}_3 = 0$. So (RDP$_2$) is proved.

(ii) $\iota(E)$ is directed, since E was supposed to be directed. It remains to prove (RDP$_1$).

So let $\mathfrak{a}_1, \mathfrak{a}_2, \mathfrak{b}_1, \mathfrak{b}_2 \in \mathcal{C}(E)$ such that $\mathfrak{a}_1 + \mathfrak{a}_2 = \mathfrak{b}_1 + \mathfrak{b}_2$. Let $\mathfrak{a}_1 = [a_1^1, \ldots, a_k^1]$, $\mathfrak{a}_2 = [a_1^2, \ldots, a_l^2]$, $\mathfrak{b}_1 = [b_1^1, \ldots, b_m^1]$, $\mathfrak{b}_2 = [b_1^2, \ldots, b_n^2]$, where $a_1^1, \ldots, b_n^2 \in E$. By Lemma 6 there are $d_{11}^1, \ldots, d_{ln}^4 \in E$ such that (10) holds and such that every two elements in that diagram fulfil the com-condition if one of them is placed further up and further right than the other one.

Define now $\mathfrak{d}_1, \mathfrak{d}_2, \mathfrak{d}_3, \mathfrak{d}_4$ as in the proof of (i). Due to the com-conditions, (11) then holds.

Now, for some $\mathfrak{x}, \mathfrak{y} \in \mathcal{G}(E)$, let $0 \leq \mathfrak{x} \leq \mathfrak{d}_2$ and $0 \leq \mathfrak{y} \leq \mathfrak{d}_3$. So $\mathfrak{x} + \mathfrak{x}' = \mathfrak{d}_2$ for some $\mathfrak{x}' \in \mathcal{C}(E)$, and we may apply to this equation Lemma 6 to conclude that $\mathfrak{x}$ is representable as a word such that every element in it lies below some element in $\mathfrak{d}_2$. The same applies to $\mathfrak{y}$ and $\mathfrak{d}_3$.

Now for every $x, y \in E$ such that x lies below an element occuring in the word $\mathfrak{d}_2$ and y lies below an element occuring in $\mathfrak{d}_3$, we have $x + y = y + x$. We conclude that $\mathfrak{x} + \mathfrak{y} = \mathfrak{y} + \mathfrak{x}$. So we have proved $\mathfrak{d}_2 \operatorname{com} \mathfrak{d}_3$. So (RDP$_1$) holds in $\mathcal{G}(E)$.

(iii) Since (RIP) holds in $\mathcal{G}(E)$ by parts (i), (ii) and Remark 1, this part follows from Proposition 3. $\square$

6   Categorical Equivalences of GPE-Algebras and po-Groups

After having discussed some correlations of GPE-algebras and their represent-ing groups in the previous chapter, we now turn our attention to the question how the homomorphisms of these two types of structures are related to each other.

Definition 6. A mapping $\varphi : E \to F$ between GPE-algebras is called a homomorphism if (i) for any $a, b \in E$ such that $a+b$ is defined, also $\varphi(a)+\varphi(b)$ is defined and $\varphi(a + b) = \varphi(a) + \varphi(b)$, and (ii) $\varphi(0) = 0$.

A mapping $\varphi : E \to G$ of a GPE-algebra into a po-group is called a homo-morphism if (i) $\varphi(E) \subseteq G^+$, (ii) for any $a, b \in E$ such that $a + b$ is defined, $\varphi(a + b) = \varphi(a) + \varphi(b)$, and (iii) $\varphi(0) = 0$.

We first show that the po-group $\mathcal{G}(E)$ representing a GPE-algebra E has the universal property, i.e. any homomorphism from E into another po-group factors through $\mathcal{G}(E)$.

Theorem 2. Let $(E; +, 0)$ be a generalized pseudo-effect algebra fulfilling (A) or (B), and let $\iota_E : E \to \mathcal{G}(E)$ its po-group representation.

(i) Let E fulfil (A). Then for every infimum preserving homomorphism $h : E \to H$ from E into some ℓ-group H there is a unique ℓ-group homomorphism $h^{\mathcal{G}(E)} : \mathcal{G}(E) \to H$ such that $h = h^{\mathcal{G}(E)} \circ \iota_E$.

(ii) Let E fulfil (B). Then for every homomorphism $h : E \to H$ from E into some po-group H there is a unique po-group homomorphism $h^{\mathcal{G}(E)} : \mathcal{G}(E) \to H$ such that $h = h^{\mathcal{G}(E)} \circ \iota_E$.

Proof. (i) Let H be an ℓ-group and $h : E \to H$ a homomorphism that preserves the infima.

Since $\iota_E : E \to \mathcal{G}(E)$, $a \mapsto [a]$, is injective, we may define $h^E : \iota_E(E) \to H$, $[a] \mapsto h(a)$. Then h^E preserves $+$ whenever defined in $\iota_E(E)$, and $h^E(0) = 0$. By Proposition 3(ii), h^E preserves infima.

Now we may extend h^E to the function $h^{\mathcal{C}(E)} : \mathcal{C}(E) \to H$, $[a_1, \ldots, a_k] \mapsto h(a_1) + \cdots + h(a_k)$. Indeed, the latter sum remains the same when chosing, rather than the word $(a_1, \ldots, a_k)$, a directly similar one, or consequently even a similar one, instead. $h^{\mathcal{C}(E)}$ by construction preserves $+$.

We have to show that $h^{\mathcal{C}(E)}$ preserves infima. So let $\mathfrak{a} = [a_1^1, \ldots, a_k^1]$ and $\mathfrak{b} = [b_1^1, \ldots, b_m^1]$ from $\mathcal{C}(E)$ be given; we have to show that $h^{\mathcal{C}(E)}(\mathfrak{a} \wedge \mathfrak{b}) = h^{\mathcal{C}(E)}(\mathfrak{a}) \wedge h^{\mathcal{C}(E)}(\mathfrak{b})$. By Proposition 4(i), $\mathcal{G}(E)$ is lattice ordered, so in particular it is directed and fulfils (RDP$_2$). Let $\mathfrak{a}' = [a_1^2, \ldots, a_l^2]$ and $\mathfrak{b}' = [b_1^2, \ldots, b_n^2]$ be chosen from $\mathcal{C}(E)$ such that $\mathfrak{a} + \mathfrak{a}' = \mathfrak{b} + \mathfrak{b}'$. Then by Lemma 6 there are $d_{11}^1, \ldots, d_{ln}^4 \in E$ such that the scheme (10) holds and such that every pair of elements in it one of which is further up and further right than the other one possesses the infimum 0.

¿From the zero infimum relations we get $\mathfrak{a} \wedge \mathfrak{b} = [a_1^1, \ldots, a_k^1] \wedge [b_1^1, \ldots, b_m^1] = [d_{11}^1, \ldots, d_{1m}^1, \ldots, d_{k1}^1, \ldots, d_{km}^1]$. It follows $h^{\mathcal{C}(E)}(\mathfrak{a} \wedge \mathfrak{b}) = h^{\mathcal{C}(E)}([d_{11}^1]) + \cdots + h^{\mathcal{C}(E)}([d_{km}^1]) = h(d_{11}^1) + \cdots + h(d_{km}^1)$. On the other hand, the scheme (10) and the zero infimum relations still hold if all elements are replaced by their images under h. So we also have $h^{\mathcal{C}(E)}(\mathfrak{a}) \wedge h^{\mathcal{C}(E)}(\mathfrak{b}) = [h(a_1^1) + \cdots + h(a_k^1)] \wedge [h(b_1^1) + \cdots + h(b_m^1)] = h(d_{11}^1) + \cdots + h(d_{km}^1)$. We conclude $h^{\mathcal{C}(E)}(\mathfrak{a} \wedge \mathfrak{b}) = h^{\mathcal{C}(E)}(\mathfrak{a}) \wedge h^{\mathcal{C}(E)}(\mathfrak{b})$.

We may further extend $h^{\mathcal{C}(E)}$ to $h^{\mathcal{G}(E)} : \mathcal{G}(E) \to H$ by defining $h^{\mathcal{G}(E)}(\mathfrak{a} - \mathfrak{b}) \stackrel{\text{def}}{=} h^{\mathcal{C}(E)}(\mathfrak{a}) - h^{\mathcal{C}(E)}(\mathfrak{b})$, where $\mathfrak{a}, \mathfrak{b} \in \mathcal{G}(E)^+$. For suppose $\mathfrak{a} - \mathfrak{b} = \mathfrak{c} - \mathfrak{d}$ for some $\mathfrak{c}, \mathfrak{d} \in \mathcal{G}(E)^+$; then setting $\mathfrak{b}' = \mathfrak{a} + \mathfrak{b} - \mathfrak{a} \in \mathcal{G}(E)^+$, we have $\mathfrak{a} + \mathfrak{b} = \mathfrak{b}' + \mathfrak{a}$ and hence $h^{\mathcal{C}(E)}(\mathfrak{a}) + h^{\mathcal{C}(E)}(\mathfrak{b}) = h^{\mathcal{C}(E)}(\mathfrak{b}') + h^{\mathcal{C}(E)}(\mathfrak{a})$; and we have $\mathfrak{a} + \mathfrak{d} = \mathfrak{b}' + \mathfrak{c}$ and hence $h^{\mathcal{C}(E)}(\mathfrak{a}) + h^{\mathcal{C}(E)}(\mathfrak{d}) = h^{\mathcal{C}(E)}(\mathfrak{b}') + h^{\mathcal{C}(E)}(\mathfrak{c})$; so we conclude $h^{\mathcal{C}(E)}(\mathfrak{a}) - h^{\mathcal{C}(E)}(\mathfrak{b}) = -h^{\mathcal{C}(E)}(\mathfrak{b}') + h^{\mathcal{C}(E)}(\mathfrak{a}) = h^{\mathcal{C}(E)}(\mathfrak{c}) - h^{\mathcal{C}(E)}(\mathfrak{d})$.

$h^{\mathcal{G}(E)}$ preserves $+$ on $\mathcal{G}(E)^+$, and by a similar kind of argumentation as in the last paragraph we see that it preserves $+$ on the whole group $\mathcal{G}(E)$.

We still have to prove that $h^{\mathcal{G}(E)}$ preserves infima. So let $\mathfrak{a}, \mathfrak{b} \in \mathcal{G}(E)$. Put $\mathfrak{c} = \mathfrak{a} \wedge \mathfrak{b}$. Then we have $\mathfrak{a} - \mathfrak{c}, \mathfrak{b} - \mathfrak{c} \geq 0$, and so $h^{\mathcal{G}(E)}(\mathfrak{a} \wedge \mathfrak{b}) = h^{\mathcal{G}(E)}[(\mathfrak{a} - \mathfrak{c}) \wedge (\mathfrak{b} - \mathfrak{c})] + \mathfrak{c}) = [h^{\mathcal{G}(E)}(\mathfrak{a}) - h^{\mathcal{G}(E)}(\mathfrak{c})] \wedge [h^{\mathcal{G}(E)}(\mathfrak{b}) - h^{\mathcal{G}(E)}(\mathfrak{c})] + h^{\mathcal{G}(E)}(\mathfrak{c}) = h^{\mathcal{G}(E)}(\mathfrak{a}) \wedge h^{\mathcal{G}(E)}(\mathfrak{b})$.

So $h^{\mathcal{G}(E)}$ is proved to be an ℓ-group homomorphism such that $h = h^{\mathcal{G}(E)} \circ \iota_E$, and it is unique since $\iota_E(E)$ generates $\mathcal{G}(E)$.

(ii) Let H be a po-group and $h : E \to H$ a homomorphism. Similarly as in part (i), we define h^E and extend it to $h^{\mathcal{C}(E)}$ and further to $h^{\mathcal{G}(E)}$, to see that this is the unique po-group homomorphism from $\mathcal{G}(E)$ to H such that $h = h^{\mathcal{G}(E)} \circ \iota_E$.   □

We now establish two categorical equivalences, where our objects are GPE-algebras on the one side and po-groups on the other side.

Definition 7. Let $\mathcal{GPEA}$ be the category whose objects are the generalized pseudo-effect algebras fulfilling condition (A) and whose morphisms are the infimum preserving GPE-algebra homomorphisms.

Let $\mathcal{GPEB}$ be the category whose objects are the generalized pseudo-effect algebras fulfilling condition (B) and whose morphisms are the GPE-algebra homomorphisms.

Let $\mathcal{LG}$ be the category whose objects are the pairs (G, G_0) of an ℓ-group G and of a convex subset G_0 of G^+ containing 0 that generates G, and whose morphisms $\varphi : (G, G_0) \to (H, H_0)$ are the ℓ-group homomorphisms $\varphi : G \to H$ such that $\varphi(G_0) \subseteq H_0$.

Let $\mathcal{POGR}1$ be the category whose objects are the pairs (G, G_0) of a directed po-group G that fulfils (RDP_1) and of a directed convex subset G_0 of G^+ containing 0 that generates G, and whose morphisms $\varphi : (G, G_0) \to (H, H_0)$ are the po-group homomorphisms $\varphi : G \to H$ such that $\varphi(G_0) \subseteq H_0$.

Finally, let $\Delta_A : \mathcal{GPEA} \to \mathcal{LG}$ and $\Delta_B : \mathcal{GPEB} \to \mathcal{POGR}1$ be the functors that assign to every object E from their domain the pair $(\mathcal{G}(E), E)$ and to every morphism $\varphi : E \to F$ from their domain the group homomorphism $\psi : \mathcal{G}(E) \to \mathcal{G}(F)$ subject to the condition $\psi \circ \iota_E = \iota_F \circ \varphi$.

It is clear from Proposition 4 and Theorem 2, that Δ_A and Δ_B are well-defined as functors between categories.

Theorem 3. The functor Δ_A establishes a categorical equivalence between $\mathcal{GPEA}$ and $\mathcal{LG}$.

The functor Δ_B establishes a categorical equivalence between $\mathcal{GPEB}$ and $\mathcal{POGR}1$.

Proof. It follows from the uniqueness part of Theorem 2, that Δ_A and Δ_B are full and faithful functors.

We prove next that Δ_A is isomorphism-dense. Let a pair (G, G_0) from $\mathcal{LG}$ be given; we have to show that the image of some object from $\mathcal{GPEA}$ is isomorphic to (G, G_0). Now by Proposition 3, $(G_0; +, 0)$ is a GPE-algebra fulfilling (A); so let us show that $(\mathcal{G}(G_0), \iota_{G_0}(G_0))$ is isomorphic to (G, G_0). By Theorem 2 the map $\kappa : \iota_{G_0}(G_0) \to G$, $[a] \mapsto a$, is extendible to an ℓ-group homomorphism $\kappa^{\mathcal{G}(G_0)} : \mathcal{G}(G_0) \to G$. The latter is surjective since G_0 was assumed to generate G; and we have $\kappa^{\mathcal{G}(G_0)}(\iota_{G_0}(G_0)) = G_0$. It is also injective; indeed, $\kappa^{\mathcal{G}(G_0)}([a_1, \ldots a_m] - [b_1, \ldots b_n]) = 0$ for some $a_1, \ldots a_m, b_1, \ldots b_n \in G_0$ means $a_1 + \cdots + a_m = b_1 + \cdots + b_n$; and this implies by Remark 1(iii) that $(a_1, \ldots, a_m) \simeq (b_1, \ldots, b_n)$, hence $[a_1, \ldots, a_m] - [b_1, \ldots, b_n] = 0$.

We see similarly that also Δ_B is isomorphism-dense. $\square$

References

[Baer]   R. Baer, Free sums of groups and their generalizations. An analysis of the associative law. Amer. J. Math. **71** (1949), 706-742.

[Bir]   G. Birkhoff, "Lattice Theory," Colloquium Publications 25, American Mathematical Society, Providence, 3rd ed. 1995.

[DvPu]   A. Dvurečenskij, S. Pulmannová, "New Trends in Quantum Structures," Kluwer Acadamic Publ., Dordrecht, Ister Science, Bratislava, 2000, to appear.

[DvVe1] A. Dvurečenskij, T. Vetterlein, Pseudo-Effect Algebras. I. Basic Properties. Preprint 6/2000 of the Math. Inst. of the Slovak Academy of Sciences; submitted.

[DvVe2] A. Dvurečenskij, T. Vetterlein, Pseudo-Effect Algebras. II. Group representations. Preprint 7/2000 of the Math. Inst. of the Slovak Academy of Sciences; submitted.

[FoBe]   D. J. Foulis, M. K. Bennett, Effect algebras and unsharp quantum logics, Found. Phys. **24** (1994), 1325–1346.

[Fu]   L. Fuchs, "Partially Ordered Algebraic Systems," Pergamon Press, Oxford, 1963.

[GiGr]   R. Giuntini, H. Greuling, Towards a formal language for unsharp properties, Found. Phys. **19** (1989), 931-945.

[HePu]   J. Hedlíkov, S. Pulmannov, Generalized difference posets and orthoalgebras, Acta Math. Univ. Comenianae **45** (1996), 247–279.

[KoCh]   F. Kôpka, F. Chovanec, D-posets, Math. Slovaca 44 (1994), 21-34.

[Wyl]   O. Wyler, Clans, Compos. Math. **17** (1966), 172-189.

Extension Principle and Probabilistic Inferential Process

Giangiacomo Gerla, Domenico Calabró, and Luciana Scarpati

Dip. Matematica e Informatica - Universitá di Salerno
via S. Allende, Baronissi 84081 (SA), Italy - gerla@unisa.it

Abstract. In this work we sketch out a method to design expert systems, probabilistic in nature. The inferential engine we propose is a data-base storing information about a set of "past cases ".

1 Introduction

F. Bacchus in [1] and J. Y. Halpern in [6] compare the propositions:

(i) *"a bird will fly";*
(ii) *"Tweety (a particular bird) is able to fly"*

and they underline their complete difference when we want to assign them a probability valuation. In fact, if we say that the probability of (i) is, for example, 0.9, this valuation arises from our past experience on the birds. Indeed it coincides with a statistical information about the proportion of fliers among the whole set of birds. The problem is to assign a probability to the second proposition. Indeed a probabilistic valuation seems possible only if we have a collection of elements while Tweety is a particular bird. As a matter of fact one expects only two possibilities:

- Tweety is able to fly and so (ii) is true
- Tweety is not able to fly and so (ii) is false.

G. Gerla in [3] suggests to interpret (ii) as:

"A bird with the same observable properties of Tweety is able to fly".

In accordance, when we assign to (ii) the probability 0.9 we mean that the ninety percent of all birds with the same observable properties of Tweety is able to fly. This idea was born by the conviction that our "degrees of belief" derives from the experience about a class of past cases (the birds) we consider similar to the actual case a_c (Tweety) under consideration. So, given a property α, if we want to know the probability that a_c verifies α, we have:

1. to consider the observable (relevant) properties satisfied by a_c;

2. to consider the set of past cases *similar* to a_c, i.e., the cases satisfying the same properties of a_c (Boolean valuation);

3. to determine the probability of α as the percentage of past cases similar to a_c satisfying α (numerical valuation).

Notice that two notions are on the basis of such an idea. The *"similarity"* relation which is intended as "satisfying the same observable relevant properties" and the notion of *"Boolean valuation"* that is necessary because, as it is well known, the probability valuations are not truth functional.

In this work we start from this idea to sketch out a method to design expert systems, probabilistic in nature. The inferential engine we propose is a database storing information about a set of "past cases". The inferential process consists in a querying strategy to investigate about the main observable properties of the "actual case" a_c. The resulting information enables us to isolate the past cases which are similar to a_c to give a probabilistic valuation of a "non-observable" property of a_c.

We also consider the possibility that the information about the past cases is incomplete. To this purpose we use a simple logic, Boolean in nature, we obtain by the extension principle proposed in [2] and [5]. Due to the incompleteness of the information, the resulting valuations are not probabilities but super-additive measures, in general.

Finally, in spite of the theoretical nature of the paper, in order to taste the potentialities of the proposed notions, a prototype shell for expert systems was built up. The relational data-base *Access* is the used languages. Also, *Visual Basic* is used for the interface.

2   Probability and Boolean valuations.

Recall some elementary notions that are on the basis of any approach to probability logic. In the following we denote by F the set of formulas of a zero-order language.

Definition 1. A *probability valuation* of F is any map $\mu : F \to [0,1]$ such that:

a) $\mu(\alpha) = 1$ for every tautology α;

b) $\mu(\alpha \vee \beta) = \mu(\alpha) + \mu(\beta)$ if $\alpha \wedge \beta$ is a contradiction;

c) $\mu(\alpha) = \mu(\beta)$ if α is logically equivalent to β.

Observe that if μ is a probability valuation, then $\mu(\alpha) = 0$ for every contradiction α. Indeed, in such a case, since α is logically equivalent to $\alpha \vee \alpha$ and $\alpha \wedge \alpha$ is a contradiction, by b) and c), we have that

$$\mu(\alpha) = \mu(\alpha \vee \alpha) = \mu(\alpha) + \mu(\alpha).$$

This entails that $\mu(\alpha) = 0$.

As it is well known, the probability valuations are not truth-functional. In fact, the knowledge of the probability of two formulas α and β doesn't allow to determine the probability of the composed formula $\alpha \wedge \beta$, in general. This is a strong obstacle for a probability logic. Nevertheless, the truth-functionality can be obtained by the notion of Boolean truth-functional valuation, in a sense.

Definition 2. Let B be a Boolean algebra. We say that a map $v : F \to B$ is a *truth-functional B-valuation* if the following properties hold for every $x, y \in F$,

a) $v(x \wedge y) = v(x) \wedge v(y)$;

b) $v(x \vee y) = v(x) \vee v(y)$;

c) $v(-x) = -v(x)$.

Observe

that a), b) and c) entail that $v(1) = 1$ and $v(0) = 0$. Indeed

$$v(1) = v(-x \vee x) = v(x) \vee -v(x) = 1$$

and

$$v(0) = v(x \wedge -x) = v(x) \wedge -v(x) = 0.$$

If $B = \{0, 1\}$, the truth-functional B-valuations coincide with the usual classical interpretations of F.

The more interesting case is when B coincides with the class $\mathcal{P}(S)$ of all the subsets of a set S. Indeed, in such a case we can interpret S as the set of past cases stored in a data-base or as the set of *"possible worlds"* in a Kripke semantics. In accordance, given any sentence α, we can interpret $v(\alpha)$ as the set of past cases (possible worlds) in which α is true.

Definition 3. We call *B-probability valuation* a structure (B, v, p) where B is a Boolean algebra, $v : F \to B$ is a truth-functional B-valuation and $p : B \to [0, 1]$ is a finitely additive probability.

The following proposition shows that the notion of B-probability valuation is strictly related with the notion of probability valuation.

Proposition 1. *Let (B, v, p) be a B-probability valuation and define $\mu : F \to [0,1]$ by setting $\mu(\alpha) = p(v(\alpha))$ for every $\alpha \in F$. Then μ is a probability valuation that we call associated with (B, v, p). Conversely, let $\mu : F \to [0,1]$ be any probability valuation in F. Then a Boolean algebra B and a B-probability valuation (B, v, p) exist such that $\mu(\alpha) = p(v(\alpha))$.*

Proof. The first part of the proposition is obvious. Let $\mu : F \to [0,1]$ be a probability valuation in F and denote by B the Lindenbaum algebra associated with F. This means that we set, for any formula α,

$$[\alpha] = \{\alpha' \in F : \alpha' \text{ is logically equivalent to } \alpha\}.$$

and

$$B = \{[\alpha] : \alpha \in F\}.$$

Moreover, for any $[\alpha]$ and $[\beta]$ in B,

$$[\alpha] \wedge [\beta] = [\alpha \wedge \beta], [\alpha] \vee [\beta] = [\alpha \vee \beta], -[\alpha] = [-\alpha].$$

Define the function $v : F \to B$ by setting $v(\alpha) = [\alpha]$ for every $\alpha \in F$ and define $p : B \to [0,1]$ by setting $p([\alpha]) = \mu(\alpha)$. Then it is immediate that (B, v, p) is a B-probability valuation whose associated probability valuation is μ. $\qquad\square$

3   Knowledge representation system

The starting point of the inferential process we will define is a data base storing the information about a series of past cases we consider related with the actual case. To represent this, we propose a formalism very near to the formalism proposed by Pawlak in [7]. The first distinction we have to do is between the observable properties and the not observable ones. We call *observable* the properties for which it is possible to detect in a direct way whether they are satisfied or not by the actual case. A property that will be materialized in the future is a typical example of non observable property. Now we can give the following definition.

Definition 4. A *(complete) knowledge representation system* is a structure $\Sigma = (PC, AT, OBS, tr)$ where:

- PC is a finite set whose elements we call *past cases*;
- AT is a finite set whose elements we call *attributes*;
- OBS is a subset of AT, called the set of the *observable* attributes;

- $tr : PC \times AT \to \{0,1\}$ is a function, called *information function*.

We denote by F (by F_{obs}) the set of formulas of the propositional calculus whose set of propositional variables is AT (is OBS, respectively). Obviously, tr can be extended to the whole set F of formulas by setting

$$tr(c, \alpha \wedge \beta) = min\{tr(c, \alpha), tr(c, \beta)\},$$
$$tr(c, \alpha \vee \beta) = max\{tr(c, \alpha), tr(c, \beta)\},$$
$$tr(c, -\alpha) = 1 - tr(c, \alpha).$$

Given a formula α and a case c, the equation $tr(c, \alpha) = 1$ means that α is true in c, while $tr(c, \alpha) = 0$ means that α is false in c. In other words, tr associates any past case with a classical valuation of the formulas in F. Given a set T of formulas, we say that a past case c is a *model* of T, and we write $c \models T$, if $tr(c, \alpha) = 1$ for every $\alpha \in T$.

Now, PC, as the result of the whole past experience, is a too big basis of our inferential process, in general. A more workable tool can be obtained by observing that from the point of view of the inferential apparatus we will define, it is not useful to distinguish two past cases satisfying the same properties. Then, we define an equivalence relation on the set of past cases in the following way.

Definition 5. Let $\Sigma = (PC, AT, OBS, tr)$ be a knowledge representation system and A a set of formulas. Then we define a binary relation $\cong_A$ in PC by setting

$$c_1 \cong_A c_2 \text{ if and only if } tr(c_1, \alpha) = tr(c_2, \alpha) \text{ for every } \alpha \in A.$$

If $c_1 \cong_A c_2$, we say that c_1 and c_2 are *A-indiscernible*. Then two cases are A-indiscernible if they satisfy the same properties in A. The proof of the following proposition is obvious.

Proposition 2. *Given any set A of formulas, $\cong_A$ is an equivalence relation. Moreover, if A and B are set of formulas,*

$$A \subseteq B \Rightarrow \cong_A \supseteq \cong_B .$$

In accordance with Proposition 2, given a set A of formulas and a case c, we can consider the relative equivalence class

$$[c]_A = \{c' \in PC : c' \cong_A c\}.$$

Also, we can define the quotient

$$PC_A = \{[c]_A : c \in PC\}.$$

In the following proposition, given a set A of formulas, we denote by $\mathcal{L}(A)$ the language generated by A, i.e. the set of formulas obtained from A by an iterated application of the disjunction, conjunction and negation operations.

Proposition 3. *The relation* $\cong \mathcal{L}(A)$ *coincides with* $\cong_A$. *In particular,* $\cong_F$ *coincides with* $\cong_{AT}$. *Moreover,* $|PC_A| \leq 2^{|A|}$ *and, in particular,*

$$|PC_F| = |PC_{AT}| \leq 2^{|AT|}.$$

Proof. Obvious.

We write $c_1 \cong c_2$ instead of $c_1 \cong_F c_2$ and in this case we say that c_1 and c_2 are *indiscernible*. Moreover, we write $[c]$ to denote $[c]_F$.

At this point it is natural to consider in spite of Σ^* the related quotient $\Sigma^* = (PC_F, AT, OBS, tr^*)$ where tr^* is defined by setting, for every $c \in PC_F$,

$$tr^*([c], \alpha) = tr(c, \alpha).$$

Obviously, since $|PC_F| = |PC_{AT}| \leq 2^{|AT|}$ the cardinality of Σ^* is not too big. Also, in order to preserve the whole information of the initial knowledge representation system we have to store, for each type-case $[c]$, the number of elements contained in $[c]$. This leads to the following definition.

Definition 6. *A (complete) statistical inferential basis is a structure* $\Sigma = (TC, AT, OBS, tr, w)$ *such that:*

- (TC, AT, OBS, tr) *is a (complete) knowledge representation system such that two elements of TC are always discernible;*
- $w : TC \to N$ *is a function called* weight function.

The elements of TC are called *type-cases*.

In a sense, a statistical inferential basis is obtained by an abstraction process from our past experience. For every type-case $t, w(t)$ can be seen as the number of concrete past cases represented by t. It is immediate that $|TC| \leq 2^{|AT|}$. We call *total weight* of Σ the number of the past cases represented globally by Σ, that is

$$w(\Sigma) = \Sigma\{w(c) : c \in TC\}.$$

Proposition 4. *Every statistical inferential basis Σ defines a B-probability valuation (B, v, p) in F such that:*

- B *is the Boolean algebra $\Pi(TC)$;*

- $v(\alpha) = \{c \in TC : tr(c,\alpha) = 1\}$ *is the set of type-cases satisfying* α;
- $p : B \to [0,1]$ *is the probability defined by setting, for any set* X *of cases,*

$$p(X) = \frac{\Sigma\{w(c) : c \in X\}}{w(\Sigma)}.$$

We call entropy of Σ *the entropy of the probability* p, *i.e. the number*

$$E(\Sigma) = -\Sigma\{p(c)lg(p(c)) : c \in \Sigma\}.$$

In accordance with Proposition 1, we can associate any statistical inferential basis Σ with a probability valuation μ of the formulas. It is evident that, for every formula α,

$$\mu(\alpha) = \frac{\Sigma\{w(c) : tr(c,\alpha) = 1\}}{w(\Sigma)}.$$

In other words, $\mu(\alpha)$ represents the percentage of past cases in which α is true according to the initial dates.

4 Inferential process

Imagine that we have to evaluate the probability that an actual case a_c satisfies a formula α in F. Here we call *actual case* any model of the language F and we denote by $T(a_c)$ the set of observable formulas satisfied by a_c. Obviously, we are interested to the case that α is not observable. Then, we can imagine a step-by-step inferential process resulting from a queering strategy. At step i we obtain a formula α_i as an answer to a "query" about the observable properties of the actual case a_c. After n steps, the information about our actual case is collected in a set $T = \{\alpha_1, ..., \alpha_n\} \subseteq T(a_c)$ of formulas. Given a statistical inferential basis Σ (basic information), we say that T is *satisfiable* in Σ if a typical case exists which satisfies T. The theory T enables us to obtain a new statistical inferential basis $\Sigma(T)$ from Σ. Namely, we set

$$TC(T) = \{c \in TC : c \models T\},$$

i.e., $TC(T)$ is the set of cases c satisfying T. Equivalently, $TC(T)$ is the set of past cases T-indiscernible from the actual case. Then, we can propose the following definition:

Definition 7. Let $\Sigma = (TC, AT, OBS, tr, w)$ be a statistical inferential basis and $T \subseteq F_{obs}$ a theory satisfiable in Σ. We call *statistical inferential basis defined by* T *in* S the structure:

$$\Sigma(T) = (TC(T), AT, OBS, tr, w).$$

In accordance with Proposition 4, $\Sigma(T)$ defines a B-probability valuation $(\Pi(TC(T)), v_T, p_T)$ where:

$$v_T(\alpha) = \{c \in TC : c \models T \cup \{\alpha\}\} = TC(T) \cap v(\alpha)$$

and, for any subset X of $TC(T)$,

$$p_T(X) = \frac{\Sigma\{w(c) : c \in X\}}{w(\Sigma(T))}.$$

We can extend p_T to the whole algebra $\mathcal{P}(TC)$ by setting $p_T(\{c\}) = 0$ for any $c \notin TC(T)$. Notice that in such a way we obtain the conditional probability $p(\_/TC(T))$. In fact, for any $X \subseteq TC$,

$$p_T(X) = \frac{\Sigma\{w(c) : c \in X \cap TC(T)\}}{w(\Sigma(T))} = \frac{\Sigma\{w(c) : c \in X \cap TC(T)\}}{\Sigma\{w(c) : c \in TC\}}.$$

$$\cdot \frac{\Sigma\{w(c) : c \in TC\}}{w(\Sigma(T))} = \frac{p(X \cap TC(T))}{p(TC(T))}.$$

Also, a probability valuation μ_T of the formulas is defined in such a way that

$$\mu_T(\alpha) = \frac{\Sigma\{w(c) : c \text{ satisfies } \alpha \text{ and } T\}}{\Sigma\{w(c) : c \text{ satisfies } T\}},$$

i.e, $\mu_T(\alpha)$ is the percentage of the past cases verifying α among the cases verifying T.

Now, we are able to give the main definition in this paper.

Definition 8. Let α be a formula in F, and $T \subseteq T(a_c)$ the available information on the actual case a_c. Then we call *probability* that a_c satisfies α given T, the probability of α in the statistical inferential basis associated to T.

In conclusion, we imagine an expert system whose inferential engine contains a statistical inferential basis Σ obtained by an abstraction process from a knowledge representation system. Given an actual case a_c and a formula α in the language F, the expert system furnishes a probabilistic valuation $\mu(\alpha)$ of α by a step-by-step process as follows:

1. Set $T_0 = \emptyset$ and $\Sigma_0 = \Sigma(\emptyset) = \Sigma$.

2. Given T_i and Σ_i, put $T_{i+1} = T_i \cup \{\alpha_{i+1}\}$ and $\Sigma_{i+1} = \Sigma(T_{i+1})$, where α_{i+1} as the answer to a query β about the actual case a_c, i.e. $\alpha_{i+1} = \beta$ if the answer is positive, and $\alpha_{i+1} =_\neg \beta$ if the answer is negative.

3. If the information is sufficient, goto 4, otherwise goto 2.

4. Set $\mu(\alpha) = \mu_{T_i}(\alpha)$.

In the prototype, β is selected in order to minimize the expected value of the entropy. This is achieved by minimizing the value $|\mu(\beta) - \mu(\neg\beta)|$ where μ is the valuation related to Σ_i.

Notice that in such a way we gives a precise meaning to the claim:

"*The probability that the actual case a_c satisfies a property α is given by the percentage of the cases indiscernible from a_c that in the past verified α*".

5   An extension principle for incomplete information

The inferential process in the previous section is related to the case of complete information about the past cases stored in the memory. This means that for every case c and α propositional variable, either $tr(c, \alpha)$ is equal to 0 or $tr(c, \alpha)$ is equal to 0. Assume that available information about the truth of the formulas is not complete for some past cases. This means that at least a past case c and an attribute α exist such that we are not able to say whether α is true in c or not. The question arises whether we can propose valuations probabilistic in nature in such a case, too. To deal with such a incomplete information, we need to recall some basic definitions of fuzzy set theory and an extension principle for closure operators proposed in [2] and [5].

Let $L = (L, \wedge, \vee, 0, 1)$ be a complete lattice and S a set. We call *L-fuzzy subset*, or *L-subset* of S any map s from S into L and we denote by L^S the class of L-subsets of S. Usually, L coincides with the lattice $[0, 1]$ but in this paper we are interested to the case in which L is a complete Boolean algebra. Given x in S, the value $s(x)$ is called *degree of membership* of x in s. It is immediate that L^S is a complete lattice whose operations are pointwise defined. Then, by using the set-theoretical notations, we have that the *union* of two fuzzy subsets s and s' is defined by setting, for any x in S,

$$(s \cup s')(x) = s(x) \vee s'(x).$$

The *intersection* is defined by setting, for any x in S,

$$(s \cap s')(x) = s(x) \vee s'(x).$$

Moreover, we define the inclusion relation by setting

$$s \subseteq s' \Leftrightarrow s(x) \leq (s'(x) \text{ for every } x \in S.$$

Given an L-subset s of S and $X \subseteq S$, we define $Incl(X, s)$ by setting

$$Incl(X, s) = Inf\{s(x) : x \in X\}.$$

In particular, we have that $Incl(\emptyset, X) = 1$. The number $Incl(X, s)$ is a multivalued valuation of the statement

"for every $x \in X, x$ is an element of s",

i.e. a measure of the degree of inclusion of X in s. Observe that, by identifying any subset X of S with the related characteristic function, we can consider $\mathcal{P}(S)$ as a sublattice of L^S. Then any notion in fuzzy set theory have to be an extension of the corresponding notion in classical set theory. For example, notice that if s is the characteristic function of a subset Y of S then

$$Incl(X, s) = 1 \Leftrightarrow \text{ if } X \subseteq Y.$$

We introduce the notion of fuzzy logic in an abstract way by following Tarski point of view. Recall that, given a set F whose elements we call *formulas*, we can define an *abstract crisp logic* as any compact closure operator D in the lattice $\mathcal{P}(F)$. The intended meaning of D is that, given any set X of formulas (the set of axioms), $D(X)$ is the set $\{\alpha \in F : X \models \alpha\}$ of logical consequences of X.

Recall that a *closure operator* in a set S is any map $J : \mathcal{P}(S) \to \mathcal{P}(S)$ such that, for any X and Y in $\mathcal{P}(S)$,

i) $X \leq Y \Rightarrow J(X) \leq J(Y)$,

ii) $X \leq J(X)$,

iii) $J(J(X)) = J(X)$.

J is called *compact* provided that

$$J(X) = \cap\{J(X_f) : X_f \text{ is a finite subset of } X\}$$

for any subset X of S.

Now, to obtain a suitable definition of abstract fuzzy logic, we extend the definition of closure operator to any ordered set. Indeed, given an ordered set $(G, \leq)$, we can call *closure operator* any map $J : G \to G$ such that, for x and y in G,

j) $x \leq y \Rightarrow J(x) \leq J(y)$,

jj) $x \leq J(x)$,

jjj) $J(J(x)) = J(x)$.

In place of the notion of compactness, we consider the notion of continuity. We call *directed* any family $(x_i)_{i \in I}$ of elements in G such that for every i and $j \in I$, there is $h \in I$ such that $x_i \leq x_h$ and $x_j \leq x_h$. If $(G, \leq)$ is complete, we say that J is *continuous* if

$$J(\vee_{i \in I} x_i) = \vee_{i \in I} J(x_i)$$

for every directed family $(x_i)_{i \in I}$ of elements in G. In the case $G = \mathcal{P}(S)$, the continuous closure operators coincides with the compact closure operators. The notion of an abstract fuzzy logic is obtained by substituting $\mathcal{P}(F)$ with the lattice of all the L-subsets of F.

Definition 9. An *abstract L-logic* as any continuous closure operator $D : L^F \to L^F$ is the lattice of all L-subsets of F.

If $v : F \to L$ is any L-subset of formulas (the L-subset of axioms), then we say that $D(v)$ is the *theory generated by v*.

A simple fuzzy logic can be obtained by the following extension principle enabling to extend any operator $J : \mathcal{P}(S) \to \mathcal{P}(S)$ into an operator $D^* : L^S \to L^S$.

Definition 10. Let $J : \mathcal{P}(S) \to \mathcal{P}(S)$ be an operator. Then we call *canonical extension* of J the operator $J^* : L^S \to L^S$ defined by setting:

$$J^*(s)(x) = Sup\{Incl(X_f, s) : X_f \text{ is a finite subset of } S \text{ and } x \in J(X_f)\}.$$

In particular, we can apply such a principle to the abstract crisp logics, by obtaining an abstract L-logic.

Proposition 5. *Let $D : \mathcal{P}(F) \to \mathcal{P}(F)$ be an abstract crisp logic and $D^* : L^F \to L^F$ the related canonical extension. Then D^* is an abstract L-logic (see [2] and [5]).*

Trivially, if $v : F \to L$ is any L-subset of formulas, then the L-subset $D^*(v)$ of consequences of v is defined by setting

$$D^*(v)(\alpha) = Sup\{Incl(X_f, v) : X_f \text{ is a finite set of formulas such that}$$
$$X_f \models \alpha\}$$

where we write $X_f \models \alpha$ to denote that $\alpha \in D(X_f)$.

In this paper we are interested to the case in which D is the deduction operator of the classical propositional calculus. We say that v is *consistent* if $D^*(v)(\alpha \wedge \neg\alpha) = 0$ for every formula α.

Proposition 6. *Let D be the deduction operator of the classical propositional calculus and let v be a consistent L-subset of formulas. Then, for every $\alpha, \beta \in F$:*

i)   if α is a tautology, then $D^*(v)(\alpha) = 1$,

ii)   if α is a contradiction, then $D^*(v)(\alpha) = 0$,

iii) if α entails β, then $D^*(v)(\alpha) \leq D^*(v)(\beta)$,

iv) if α is logically equivalent to β, then $D^*(v)(\alpha) = D^*(v)(\beta)$,

v) $D^*(v)(\beta) \geq D^*(v)(\alpha) \wedge D^*(v)(\alpha \to \beta)$,

vi) $D^*(v)(\alpha \wedge \beta) = D^*(v)(\alpha) \wedge D^*(v)(\beta)$,

vii) $D^*(v)(\alpha \vee \beta) \geq D^*(v)(\alpha) \vee D^*(v)(\beta)$.

Moreover,

$$D^*(v)(\alpha \vee \beta) \neq D^*(v)(\alpha) \vee D^*(v)(\beta),$$

in general.

Proof. Propositions i), ii), iii), iv), v) and vii) are evident. To prove vi) observe that, since $\alpha \wedge \beta \to \alpha$ and $\alpha \wedge \beta \to \beta$ are tautologies, $D^*(v)(\alpha \wedge \beta) \leq D^*(v)(\alpha)$ and $D^*(v)(\alpha \wedge \beta) \leq D^*(v)(\beta)$. Then,

$$D^*(v)(\alpha \wedge \beta) \leq D^*(v)(\alpha) \wedge D^*(v)(\beta).$$

Moreover, by observing that from $\alpha_1, ..., \alpha_n \models \alpha$ and $\beta_1, ..., \beta_m \models \beta$ it follows that $\alpha_1, ..., \alpha_n, \beta_1, ..., \beta_m \models \alpha$ and $\alpha_1, ..., \alpha_n, \beta_1, ..., \beta_m \models \beta$, we have that

$D^*(v)(\alpha) \wedge D^*(v)(\beta)$
$\quad = (Sup\{v(\alpha_1) \wedge ... \wedge v(\alpha_n) : \alpha_1, ..., \alpha_n \models \alpha\}) \wedge (Sup\{v(\beta_1) \wedge ... \wedge v(\beta_m) :$
$\quad\quad : \beta_1, ..., \beta_m \models \beta\})$
$\quad = Sup\{v(\alpha_1) \wedge ... \wedge v(\alpha_n) \wedge v(\beta_1) \wedge ... \wedge v(\beta_m) : \alpha_1, ..., \alpha_n \models \alpha$ and
$\quad\quad$ and $\beta_1, ..., \beta_m \models \beta\}$
$\quad \leq Sup\{v(\gamma_1) \wedge ... \wedge v(\gamma_t) : \gamma_1, ..., \gamma_t \models \alpha$ and $\gamma_1, ..., \gamma_t \models \beta\}$
$\quad = Sup\{v(\gamma_1) \wedge ... \wedge v(\gamma_t) : \gamma_1, ..., \gamma_t \models \alpha \wedge \beta\}$
$\quad = D^*(v)(\alpha \wedge \beta).$

Assume that α is undecidable in the support of v. Then, since

$$D^*(v)(\alpha) = D^*(v)(-\alpha) = 0,$$

we have that $D^*(v)(\alpha) \vee D^*(v)(-\alpha) = 0$ while $D^*(v)(\alpha \vee -\alpha) = 1$.
$\square$

6   Incomplete information

Assume that the lattice L coincides with a complete Boolean algebra, as an example, with the Boolean algebra $B = \mathcal{P}(PC)$. Then, we interpret a B-subset $v : F \to B$, by assuming that $v(\alpha)$ is the set of past cases in which we know that α is true. Moreover, it is easy to prove that $D^*(v)(\alpha) = PC$ if α is a tautology and, otherwise,

$$D^*(v)(\alpha) = \cup\{v(\alpha_1) \cap ... \cap v(\alpha_n) : \alpha_1, ..., \alpha_n \models \alpha\}.$$

This means that $D^*(v)(\alpha)$ is the set of past cases in which we can prove that α is true. Moreover, v is consistent if and only if, given any contradiction $\alpha, D^*(v)(\alpha) = \emptyset$, i.e. no case c exists such that a contradiction can be proved.

Definition 11. Let B be a Boolean algebra, $p : B \to [0,1]$ a probability and $v : F \to B$ a B-subset of F. Then we call *belief associated with v* the map $Bel(v) : F \to [0,1]$ defined by setting, for every $\alpha \in F$.

$$Bel(v)(\alpha) = p(D^*(v)(\alpha)).$$

In accordance with Proposition 6, $D^*(v)$ is not a truth-functional B-valuation. In fact, it is possible that $Bel(v)(\alpha) = Bel(v)(\neg\alpha) = 0$. This entails that, differently of the case of complete information, $Bel(v)$ is not a probability valuation, in general. The following proposition shows some basic properties of $Bel(v)$.

Proposition 7. *Let v be a consistent B-subset. Then the map $Bel(v) : F \to [0,1]$ satisfies the following properties:*

i) if α is a tautology, then $Bel(v)(\alpha) = 1$

ii) if α is a contradiction, then $Bel(v)(\alpha) = 0$

iii) if α implies β, then $Bel(v)(\alpha) \leq Bel(v)(\beta)$

iv) if α is equivalent to β, then $Bel(v)(\alpha) = Bel(v)(\beta)$

v) $Bel(v)(\alpha \vee \beta) \geq Bel(v)(\alpha) + Bel(v)(\beta) - Bel(v)(\alpha \wedge \beta)$.

Proof. We confine ourselves to prove v). Indeed,

$$\begin{aligned}
Bel(v)(\alpha \vee \beta) &= p(D^*(v)(\alpha \vee \beta)) \\
&\geq p(D^*(v)(\alpha) \cup D^*(v)(\beta)) \\
&= p(D^*(v)(\alpha)) + p(D^*(v)(\beta)) - p(D^*(v)(\alpha) \cap D^*(v)(\beta)) \\
&= Bel(v)(\alpha) + Bel(v)(\beta) - p(D^*(v)(\alpha \vee \beta)) \\
&= Bel(v)(\alpha) + Bel(v)(\beta) - Bel(v)(\alpha \wedge \beta).
\end{aligned}$$

$\square$

In the case of incomplete information, we propose the following obvious extension of Definition 4.

Definition 12. We call (partial) *knowledge representation system* any structure $\Sigma = (PC, AT, OBS, tr)$ such that $tr : PC \times F \to \{i, 1\}$ is a map from $PC \times F$ into $\{i, 1\}$.

Notice that tr is defined in the whole set F of formulas and not only in the set of propositional variables. In the case that $tr(c, \alpha) = i$, we say that α is *undetermined* in c. The information "α false in c" is stored by setting $tr(\alpha, -\alpha) = 1$. We say that two past cases c_1 and c_2 are *indiscernible* if $tr(c_1, \alpha) = tr(c_2, \alpha)$ for every formula α. As in the case of complete information, we denote by $\Sigma^* = (PC^*, AT, OBS, tr^*)$ the quotient of Σ modulo the indiscernibility relation. Also, for every equivalence class $[c]$, we consider the number of elements contained in $[c]$. It is also immediate how define the notion of (incomplete) statistical inferential basis.

Definition 13. We call *incomplete statistical inferential basis*, a structure $\Sigma = (TC, AT, OBS, tr, w)$ such that

i) (TC, AT, OBS, tr) is a partial knowledge representation system,

ii) two elements of TC are always discernible,

iii) $w : TC \to N$ is a function we call *the weight function*.

The elements of TC are called *type-cases*.

Now, given an incomplete statistical inferential basis Σ, we can set

$$v(\alpha) = \{c \in CT : tr(c, \alpha) = 1\}.$$

Then, the map $v : F \to \Pi(CT)$ is the function that associates any formula α with the set $v(\alpha)$ of cases in which we know that α was verified. Differently from the case of complete information, the map v is not truth-functional, in general. As a matter of fact, v is any map from F to $\Pi(TC)$. We interpret the B-subset v as a system of axioms (the available information). A natural semantics is obtained if we call *model* any truth-functional B-valuation m. We say that m is a model of v if $m(\alpha) \supseteq v(\alpha)$. Consider the deduction operator $D : \mathcal{P}(F) \to \mathcal{P}(F)$ of the classical propositional calculus and the related extension $D^* : B^F \to B^F$ when $B = \mathcal{P}(TC)$. To face the problem of the incomplete information, the idea is to consider D^* as a tool to complete in part our information. We can interpret the set $D^*(v)(\alpha)$ as the set of the cases in which the available information is sufficient to prove α. Moreover, we can consider the belief $Bel(v) : F \to [0, 1]$ associated with v. It is easy to prove that, given any formula α, the number $Bel(v)(\alpha)$ denotes the frequency of cases in which we have information sufficient to prove α, i.e.,

$$Bel(v)(\alpha) = \frac{\Sigma\{w(c) : c \in D^*(v(\alpha))\}}{w(\Sigma)}.$$

The inferential process from an incomplete statistical inferential basis Σ runs as in the complete statistical inferential basis. Indeed, given an actual case a_c and a formula α in the language F, we obtain a probabilistic valuation of the claim that a_c satisfies α as follows:

- we obtain a set T of formulas in F_{obs} satisfied by a_c as a result of a sequence of queries (tests)
- we define a new representation system $\Sigma(T)$ by considering only the cases satisfying T
- we consider the quantity $Bel(v)(\alpha)$.

We conclude by observing that from i) and v) in Proposition 6.1, it follows that

$$1 = Bel(v)(\alpha \vee -\alpha) \geq Bel(v)(\alpha) + Bel(v)(-\alpha).$$

By setting $Bel(v)^*(\alpha) = 1 - Bel(v)(-\alpha)$, we have that

$$Bel(v)(\alpha) \leq Bel^*(v)(\alpha).$$

Then, our inferential apparatus furnishes, for every formula α, an interval valuation $[Bel(v)(\alpha), Bel^*(v)(\alpha)]$ of α, probabilistic in nature. The intended interpretation is that the "actual" probability of α is a number in such an interval. In the case of complete information $Bel(v) = Bel^*(v)$ is a probability valuation and the interval valuation coincides with it.

References

1. F. Bacchus, *Lp*, a logic for representing and reasoning with statistical Knowledge, *Comput. Intell.*, 6 (1990) 209-231.
2. L. Biacino, G. Gerla, An extension principle for closure operators, *J. of Math. Anal. Appl.*, 198 (1996) 1-24.
3. G. Gerla, The probability that Tweety is able to fly, *International Journal of Intelligent Systems*, 9 (1994) 403-409.
4. G. Gerla, Fuzzy Logic: Mathematical Tools for Approximate Reasoning, *Kluwer Academic Publishers* (2000) Dordrecht.
5. G. Gerla, L. Scarpati, Extension principles for fuzzy set theory, *Journal of Information Sciences*, 106 (1998) 49-69.
6. J. Y. Halpern, An analysis of first-order logic of probability, *Artificial Intelligence*, 46 (1990) 331-350.
7. Z. Pawlak, Rough sets, *International Journal of Information and Computer Science*, 11 (1982) 344-356.

An Algebraic Tool for Classification in Fuzzy Environments

Antonio Gisolfi, Luigi Di Lascio, and Enrico Fischetti

Dip. Matematica e Informatica, Universitá di Salerno via S. Allende, 84081 Baronisi (SA) Italy - gisolfi@unisa.it

Abstract. This paper illustrates the behavior of a commutative l-monoid, endowed with a suitable operation of composition, as regards the problem of classification with fuzzy attributes. The concepts of relevance and similarity are introduced, then a mechanism for weighing the attributes is shown. Finally, a case study concerning graphology is illustrated in

1   Introduction

This paper shows the peculiar features of an algebraic structure useful to get fuzzy clustering and, more generally, a qualitative classification of a set of elements.

The starting point is the following: suppose that some objects are characterized by the fuzzy attribute A, i.e. each object satisfies A to some extent that can be expressed by a linguistic value. Then objects can be classified according to the degree A is satisfied with. For example, if a set of patients is ill with flu, $P=\{p_1, p_2, p_3, p_4, p_5\}$, then the attribute A="body temperature" can induce a classification on P according to the temperature of each individual, expressed by a label ranging in the set E= (low, average, high, very high). In this sense the classification is of the type:

$\{ p_2, p_3 \}$ *average temperature*, $\{ p_1, p_5 \}$ *very high temperature*

$\{ p_4 \}$ *high temperature*, $\{\emptyset\}$ *low temperature*.

It is worth noting that such classification, strictly speaking, cannot be viewed as fuzzy since the attribute string is endowed with linguistic labels but it still partitions the set U in the classical way. Thus it is more appropriate to speak of *classification by means of fuzzy attributes* since each class of the partition gets associated with a linguistic variable that is a fuzzy number [28].

The paper is organized as follows. Section 2 illustrates the basic features of the structure and the operation of composition is introduced. Next Section deals with the technique of linguistic approximation that allows to manage the results of the composition. Section 4 is devoted to deepen on the foundational aspects and the properties of the resulting commutative l-monoid are discussed. Tools for data analysis are illustrated in Section 5, their importance for applications is clear in the next Section. A case study concerning problems arising in graphology is presented in detail in section 7.

2   The fuzzy algebraic structure

The main features of an algebraic structure that is suitable for developing classification strategies are described in this section. Thanks to such structure we can apply an algebraic approximation to the problem of generating appropriate clusters of objects characterized by fuzzy attributes. The values of the attributes are expressed in terms of linguistic labels, and thus are handled as fuzzy numbers. Now we state formally the properties of the structure.

Let $U = \{u_1, u_2, \ldots, u_t\}$ be a universe of discourse and X a set of fuzzy triangular numbers. A fuzzy set of type 2 [22] is a function $s_2: U \rightarrow X$.

For each element u_i, $s_2(u_i)$ represents the "degree of compatibility" of the element u with respect to the given fuzzy attribute. Any $s_2(u_i)$ stands for some term of a suitable linguistic variable. We claim that s_2 represents a classification of U with respect to the attribute A. We also say that s_2 is the attribute A. Each term of the linguistic variable suitable for the attribute chosen is represented by a fuzzy triangular number in [0, 1].

Given an attribute A whose values range in the set of linguistic labels $\alpha = \{\alpha_1, \alpha_2, \ldots, \alpha_n\}$ then A generates a classification of the elements of U and the latter is called *attribute string*.

Definition 1

An attribute string on a set U is a string of the form

$a_n^{\alpha n} \, a_{n-1}^{\alpha n-1} \ldots a_2^{\alpha 2} \, a_1^{\alpha 1}$ *along with an n+1-uple (k, d_n, d_{n-1}, ..., d_2, d_1) called historical string such that:*

1) $a_i \subseteq U$ $\forall$ i=1..n

2) $\{a_n, a_{n-1}, \ldots, a_2, a_1\}$ is a partition of U

3) α_j, called linguistic label, is a fuzzy triangular number belonging to the interval [0,1], $\forall$ j=1..n.

4) $\alpha_n > \alpha_{n-1} > \ldots > \alpha_2 > \alpha_1$

5) k, d_n, d_{n-1},..., d_2, d_1 are integer positive numbers.

In the following we refer to a_n, a_{n-1}, ..., a_2, a_1 as the *first parts* and to α_n, α_{n-1}, ..., α_2, α_1 as the *second parts* of the string.

The historical string is useful to record the evolution of the string and this will be shown later. Attribute strings whose historical strings contain only 1 are said *native strings*.

Let S(U) be the set of attribute strings on U. In particular, in S(U), the following strings are present: U^{NI} (1,1) and U^{NC} (1,1), where NI denotes the "no information" used when no information is available to decide to what extent the elements of U satisfy the attribute A, whereas NC denotes the label "not compatible" used when the elements of U are not compatible with the attribute A (for example, the set of plants is not compatible with the attribute "social behavior").

Example 1

Let the universe of discourse be U={white, red, light-blue, violet, yellow, brown}={b, r, c, v, g, m} and suppose that we aim to classify these colors with respect to the attribute "brightness". Assume that the attribute can take values in the range {high, average, low}, then the following string is induced on U:

brightness = $\{w, lb, y\}^{high}$ $\{r\}^{average}$ $\{v, b\}^{low}$,

where we express that white, light-blue and yellow have high levels of brightness, red is average, whereas violet and brown are not bright.

Given several attribute strings, each representing a classification in U, we look for an operation that can give a finer classification of the information contained in the original strings.

The basic idea that leads to a suitable operation can be better understood if we consider the two-fold meaning of a digit in a number, namely the roles of *value* and *position*. In the case of our strings, the *value* of the generic element a_i is the set of the elements of U which a_i represents, whereas the value corresponding to the *position* is given by the label α_i. Thus we can introduce a variant of the classical multiplication of natural numbers. However, it is necessary to distinguish the operations related to the two parts of the strings, namely one operation on the *first parts* (subsets of U), and a different one on the *second parts* (fuzzy sets).

Given two strings A and B, their composition is an ordered string in which the first part is obtained by applying the operation * to the first parts of A and B, and the second part is given by applying to the corresponding second parts of A and B.

Definition 2

The operation of binary composition of attribute strings $\Diamond:S(U)\times S(U)\rightarrow S(U)$ is defined as follows:

If $A = a_n^{\alpha n}\ a_{n-1}^{\alpha n-1}\ldots a_2^{\alpha 2}\ a_1^{\alpha 1}\ (k_A,\ d_{A,n},\ldots,d_{A,2},d_{A,1})$ and

$B = b_m^{\beta m}\ b_{m-1}^{\beta m-1}\ldots b_2^{\beta 2}\ b_1^{\beta 1}\ (k_B,\ d_{B,m},\ldots,d_{B,2},d_{B,1})$ *are strings*

then:

$C = A\Diamond\ B = c_{m+n-1}^{\gamma m+n-1}\ldots c_2^{\gamma 2}\ c_1^{\gamma 1}(k_A+k_B,d_{C,m+n-1},\ldots,d_{C,2},d_{C,1})$

where we have:

- *for $n\geq m$*

for the first parts:

$$c_i=\begin{cases} \cup_{j=1,\ldots,i}\ a_{i-j+1}\cap b_j & 1\leq i\leq m-1 \\ \cup_{j=1,\ldots,m}\ a_{i-j+1}\cap b_j & m\leq i\leq n-1 \\ \cup_{j=i-n+1,\ldots,m}\ a_{i-j+1}\cap b_j & n\leq i\leq m+n-1 \end{cases}$$

for the second parts:

$$\gamma i=\begin{cases} \frac{1}{(k_A+k_B)*dc,i}\sum_{j=1,\ldots,i}\ d_{B,j}*d_{A,i-j+1}*(k_A*\alpha_{i-j+1}+k_B*\beta_j) & 1\leq i\leq m-1 \\ \frac{1}{(k_A+k_B)*dc,i}\sum_{j=1,\ldots,m}\ d_{B,j}*d_{A;i-j+1}*(k_A*\alpha_{i-j+1}+k_B*\beta_j) & m\leq i\leq n-1 \\ \frac{1}{(k_A+k_B)*dc,i}\sum_{j=i-n+1,\ldots,m}\ d_{B,j}*d_{A,i-j+1}*(k_A*\alpha_{i-j+1}+k_B*\beta_j) & n\leq i\leq m+n \end{cases}$$

where:

$$d_{C,i}=\begin{cases} \sum_{j=1,\ldots,i}\ (d_{B,j}*d_{A,i+1-j}) & 1\leq i\leq m-1 \\ \sum_{j=1,\ldots,m}\ (d_{B,j}*d_{A,i+1-j}) & m\leq i\leq n-1 \\ \sum_{j=i-n+1,\ldots,m}\ (d_{B,j}*d_{A,i+1-j}) & n\leq i\leq m+n-1 \end{cases}$$

- *for $n\leq m$*

for the first parts:

$$c_i=\begin{cases} \cup_{j=1,\ldots,i}\ a_{i-j+1}\cap b_j & 1\leq i\leq n-1 \\ \cup_{j=i-n+1,\ldots,i}\ a_{i-j+1}\cap b_j & n\leq i\leq m-1 \\ \cup_{j=i-n+1,\ldots,m}\ a_{i-j+1}\cap b_j & m\leq i\leq m+n-1 \end{cases}$$

for the second parts:

$$\gamma_i = \begin{cases} \frac{1}{(k_A+k_B)*dc,i} \sum_{j=1,\dots,i} d_{B,j} *d_{A,i-j+1} *(k_A *\alpha_{i-j+1} + k_B *\beta_j) \\ \hspace{8cm} 1 \leq i \leq n\text{-}1 \\ \frac{1}{(k_A+k_B)*dc,i} \sum_{j=i-n+1,\dots,i} d_{B,j} *d_{A;i-j+1} *(k_A *\alpha_{i-j+1} + k_B *\beta_j) \\ \hspace{8cm} n \leq i \leq m\text{-}1 \\ \frac{1}{(k_A+k_B)*dc,i} \sum_{j=i-n+1,\dots,m} d_{B,j} *d_{A,i-j+1} *(k_A *\alpha_{i-j+1} + k_B *\beta_j) \\ \hspace{8cm} m \leq i \leq m+n\text{-}1 \end{cases}$$

where:

$$d_{C,i} = \begin{cases} \sum_{j=1,\dots,i} (d_{B,j} * d_{A,i+1-j}) \; 1 \leq i \leq n\text{-}1 \\ \sum_{j=i-n+1,\dots,i} (d_{B,j} * d_{A,i+1-j}) \; n \leq i \leq m\text{-}1 \\ \sum_{j=i-n+1,\dots,m} (d_{B,j} * d_{A,i+1-j}) \; m \leq i \leq m+n\text{-}1 \end{cases}$$

Moreover, for each attribute string A one has $A \lozenge U^{NI} = U^{NI} \lozenge A = U^{NI}$ and $A \lozenge U^{NC} = U^{NC} \lozenge A = A$.

We note that if $\alpha = [a, b, c]$ and $\beta = [d,e,f]$ are number fuzzy, then the operation on the second parts, above defined, gives: $\gamma = [(a+d)/2, (b+e)/2, +(c+f)/2]$. [11, 17]

This operation, at first look, appears very complex, anyway it is just the application of the algorithm for the multiplication of integers to the first parts, to the second parts and to the attribute strings. By composing two classifications, a third one arises where the information contained in the original classifications is synthetized. The initial strings, not generated via composition, are characterized by a sequence of 1, and thus they are called *native* strings.

Example 2

Consider the string of Example 1:
brightness = $\{w,lb,y\}^{high}$ $\{r\}^{average}$ $\{v,b\}^{low}$
and suppose that the same linguistic labels are used also for the attribute "warmth". We have: warmth = $\{r,y\}^{high}$ $\{b,v\}^{average}$ $\{lb,w\}^{low}$, whose interpretation is immediate. Applying the operator $\lozenge$ to the first parts, we get:

| | $\{w,lb,y\}$ | $\{r\}$ | $\{v,b\}$ * |
	$\{r,y\}$	$\{b,v\}$	$\{lb,w\}$ =
	$\{lb,w\}\cap\{w,lb,y\}$	$\{lb,w\}\cap\{r\}$	$\{lb,w\}\cap\{v,b\}$
	U	U	
$\{b,v\}\cap\{w,lb,y\}$	$\{b,v\}\cap\{r\}$	$\{b,v\}\cap\{v,b\}$	
U		U	
$\{r,y\}\cap\{w,lb,y\}$	$\{r,y\}\cap\{r\}$	$\{r,y\}\cap\{v,b\}$	-
c_5	c_4	c_3	c_2	c_1

and consequently:

| | $\{w,lb,y\}$ | $\{r\}$ | $\{v,b\}$ * |
	$\{r,y\}$	$\{b,v\}$	$\{lb,w\}$
	$\{lb,w\}$	$\emptyset$	$\emptyset$
	U	U	
	$\emptyset$	$\emptyset$	$\{v,b\}$	-
	U	U	
$\{y\}\{r\}$	$\emptyset$	-
$\{y\}\{r\}$	$\{lb,w\}\{v,b\}$	$\emptyset$

Thus the first parts of the composite string are:

$(\{w,lb,y\}, \{r\}, \{v,b\}) * (\{r,y\}, \{b,v\}, \{lb,w\}) = (\{y\}, \{r\}, \{lb,w\}, \{v,b\}, \emptyset)$

Suppose that the linguistic labels are represented by the following fuzzy numbers: high = [0.7,1,1], average = [0.4,0.5,0.8], low = [0,0.3,0.5].

If we suppose that "brightness" and "warmth" are native strings, then we have:

brightness = $\{w, lb, y\}^{[0.7,1,1]}$ $\{r\}^{[0.4,0.5,0.8]}$ $\{v, b\}^{[0,0.3,0.5]}$ (1,1,1,1)

warmth = $\{r, y\}^{[0.7,1,1]}$ $\{b, v\}^{[0.4,0.5,0.8]}$ $\{lb, w\}^{[0,0.3,0.5]}$ (1,1,1,1),

namely $k_B = 1$, $k_W = 1$ and $d_{B,i}=1$ $\forall i$ and $d_{W,j}=1$ $\forall j$.

In this case the historical string is obtained as follows:

| | | $d_{B,3}$ | $d_{B,2}$ | $d_{B,1}$ $\otimes$ |
		$d_{W,3}$	$d_{W,2}$	$d_{W,1}$
		$d_{3,1}$	$d_{2,1}$	$d_{1,1}$
		$\oplus$	$\oplus$	$\oplus$
	$d_{3,2}$	$d_{2,2}$	$d_{1,2}$	-
	$\oplus$	$\oplus$	$\oplus$	
$d_{3,3}$	$d_{2,3}$	$d_{1,3}$		
$d_{L\Diamond C,5}$	$d_{L\Diamond C,4}$	$d_{L\Diamond C,3}$	$d_{L\Diamond C,2}$	$d_{L\Diamond C,1}$

where any $d_{i,j} = d_{B,i} * d_{W,i}$. The symbol $\otimes$ denotes the no carry product and the symbol $\oplus$ the no carry sum. Thus we have:

$$
\begin{array}{cccc}
1 & 1 & 1 & \otimes \\
1 & 1 & 1 & = \\
\hline
1 & 1 & 1 & \\
\oplus & \oplus & \oplus & \\
1 & 1 & 1 & - \\
\oplus & \oplus & \oplus & \\
1 & 1 & 1 & - \\
\hline
1 \; 2 & 3 & 2 & 1 \\
\end{array}
$$

and thus: $d_{B\Diamond W,1}=1$ $d_{B\Diamond W,2}=2$ $d_{B\Diamond W,3}=3$, $d_{B\Diamond W,4}=2$ $d_{B\Diamond W,5}=1$.
Moreover: $K_{B\Diamond W}=k_B+k_W=1+1=2$.

Then, the operation on the second parts gives:

		$[0.7,1,1]$	$[0.4,0.5,0.8]$	$[0,0.3,0.5]$ $\circ$
		$[0.7,1,1]$	$[0.4,0.5,0.8]$	$[0,0.3,0.5]$ =
		$a_{3,1}$	$a_{2,1}$	$a_{1,1}$
	$a_{3,2}$	$a_{2,2}$	$a_{1,2}$	-
$a_{3,3}$	$a_{2,3}$	$a_{1,3}$	-	
γ_5	γ_4	γ_3	γ_2	γ_1

where the element $a_{i,j}$ is given by: $a_{i,j} = d_{W,j} * d_{B,i-j+1} * (k_B*\alpha_{i-j+1} + k_W*\beta_j)$
and thus:

$a_{1,1}=1*1*(1*[0,0.3,0.5]+1*[0,0.3,0.5]) = [0,0.6,1]$

$a_{2,1}=1*1*(1*[0.4,0.5,0.8]+1*[0,0.3,0.5]) = [0.4,0.8,1.3]$

$a_{3,1}=1*1*(1*[0.7,1,1]+1*[0,0.3,0.5]) = [0.7,1.3,1.5]$

$a_{1,2}=1*1*(1*[0,0.3,0.5]+1*[0.4,0.5,0.8]) = [0.4,0.8,1.3]$

$a_{2,2}=1*1*(1*[0.4,0.5,0.8]+1*[0.4,0.5,0.8]) = [0.8,1,1.6]$

$a_{3,2}=1*1*(1*[0.7,1,1]+1*[0.4,0.5,0.8]) = [1.1,1.5,1.8]$

$a_{1,3}=1*1*(1*[0,0.3,0.5]+1*[0.7,1,1]) = [0.7,1.3,1.5]$

$a_{2,3}=1*1*(1*[0.4,0.5,0.8]+1*[0.7,1,1]) = [1.1,1.5,1.8]$

$a_{3,3}=1*1*(1*[0.7,1,1]+1*[0.7,1,1]) = [1.4,2,2]$

and consequently: $\gamma_1=1/2*a_{1,1}=1/2*[0,0.6,1]=[0,0.3,0.5]$

$\gamma_2 = 1/4^*(a_{2,1}+a_{1,2}) = 1/4^*([0.4,0.8,1.3] + [0.4,0.8,1.3]) = [0.2,0.4,0.65]$

$\gamma_3 = 1/6^*(a_{3,1}+a_{2,2}+a_{1,3}) =$

$= 1/6^*([0.7,1.3,1.5]+[0.8,1,1.6]+[0.7,1.3,1.5]) = [0.36,0.6,0.76]$

$\gamma_4 = 1/4^*(a_{3,2}+a_{2,3}) = 1/4^*([1.1,1.5,1.8]+[1.1,1.5,1.8]) = [0.55,0.75,0.9]$

$\gamma_5 = 1/2^*a_{3,3} = 1/2^*[1.4,2,2] = [0.7,1,1]$.

By combining the result for the first parts with that for the second ones, we get:

brightness$\lozenge$warmth=

$\{y\}^{[0.7,1,1]}\{r\}^{[0.55,0,75,0.9]}\{lb,w\}^{[0.36,0.6,0.76]}\{v,b\}^{[0.2,0.4,0.65]}\emptyset^{[0,0.3,0.5]}$

We need just another step. In fact, in the final string fuzzy numbers are to be converted in the corresponding linguistic expressions. In brightness $\lozenge$ warmth, for example, $\gamma_5 = [0.7,1,1]$ corresponds to the label "high". It is worth noting that the operation $\lozenge$ generates values γ_i not corresponding to any of the original labels. This holds true for γ_2, γ_3, γ_4. In order to tackle this problem we have to deal with linguistic approximation.

3   The linguistic approximation

Given a set of linguistic labels E=$\{e_1, e_2,, e_k\}$, represented by triangular numbers $\alpha = \{\alpha_1, \alpha_2, \ldots, \alpha_k\}$, where $\alpha_i = [l_i, m_i, r_i]$, and given a fuzzy number $\beta = [l, m, r]$, the problem of approximating β in E is to formulate a suitable linguistic expression for β whose elements belong to E. In other words we aim to single out the linguistic expression of β within E. The problem can be tackled in the following way.

Suppose that the mean value of β, denoted by m, lies in the interval $[m_i, m_{i+1}]$, whose extremes are the mean values of the fuzzy numbers α_i and α_{i+1}, and let d be the quantity $m_{i+1}-m_i$. Then we apply the following approximation:

(i) if m $\in \{m_i, m_i + d/10\}$ then we approximate β with e_i ;

(i) if m $\in \{m_i + (d/10), m_i + (3/10)^*d\}$ then we say that β "next to" α_i and we write nt[e_i];

(ii) if m $\in \{m_i + (3/10)^*d, m_i + (7/10)^*d\}$ then β is "included between" α_i and α_{i+1} we write ib[e_i,e_{i+1}];

(iii) if m $\in \{m_i + (7/10)^*d, m_i + (9/10)^*d\}$ then we say that β is "before " α_{i+1} and we write b[e_{i+1}];

(iv) se m $\in \{m_i + (9/10)^*d, m_{i+1}\}$ then we approximate β with e_{i+1}.

It is worth noting that our approximation technique provides an upper bound to the number of the obtainable labels. Their number cannot exceed the value 4n-3, where n denotes the original number of linguistic labels.

Example 3

In Example 2 we have obtained:
brightness$\Diamond$warmth=
$\{y\}^{[0.7,1,1]}$ $\{r\}^{[0.55,0.75,0.9]}$ $\{$lb, $\qquad\qquad$ w$\}^{[0.36,0.6,0.76]}$ $\{$v,
b$\}^{[0.2,0.4,0.65]}$ $\emptyset^{[0,0.3,0.5]}$.

In this case E={low, average, high} and $\alpha=\{[0,0.3,0.5]$, $[0.4,0.5,0.8]$, $[0.7,1,1]\}$. We have to translate each fuzzy number of the composition into a linguistic expression:

$\gamma_1=[0, 0.3, 0.5]$	$\gamma_2=[0.2, 0.4, 0.65]$	$\gamma_3=[0.36, 0.6, 0.76]$
$\gamma_4=[0.55, 0.75, 0.9]$	$\gamma_5=[0.7, 1, 1]$	

For γ_1, m=0.3, thus m$\in[0.3,0.5]=[m_1,m_2]$ and d=0.2, then m satisfies

$0.3 \leq m \leq 0.3+d/10$ and according to (i) we get the following approximation: $\gamma_1=[0, 0.3, 0.5] \approx$ low.

In a similar way we obtain:

$\gamma_2= [0.2,0.4,0.65] \approx$ ib[low, average]

$\gamma_3= [0.36,0.6,0.76] \approx$ nt[average]

$\gamma_4 = [0.55,0.75,0.9] \approx$ ib[average,high]

$\gamma_5 = [0.7,1,1] \approx$ high

Thus the final result is:

brightness$\Diamond$warmth=
$\{y\}^{high}$ $\{r\}^{ib[average,high]}$ $\{c,b\}^{nt[average]}$ $\{v,m\}^{ib[low,average]}$ $\emptyset^{low}$

A slight change in notation allows to write:

brightness$\Diamond$warmth=
$[y]^{high}$ $[r]^{ib[average,high]}$ $[lb,w]^{nt[average]}$ $[v,b]^{ib[low,average]}$ $[-]^{low}$

where the brackets "{ }" are replaced by "[ ]" and the empty set is denoted by [-].

4   The structure as commutative L-monoid

In this section we illustrate the formal properties of the structure, that can be summarized saying that it is commutative l-monoid [15, 17, 19].

Some preliminary results are the following:

Proposition 1:

(i) The operation of composition $\Diamond$ is both commutative and associative and its unit element is U^{NC}.

(ii) The triple ($S(U)$, $\Diamond$, U^{NC}) is a commutative monoid.

Proof [15].

Our string consists of subsets of U, thus in order to define an ordering criterion among them, one has first to introduce an ordering criterion among sets.

Let us suppose that U is a totally ordered set, i.e. there is an ordering relation $\leq_U$ defined on U such that for every $x \in U$ and for every $y \in U$ one has either $x \leq_U y$ or $y \leq_U x$.

Let $X \subseteq U$, we define:

$max_1(X)$: the maximum of X as regards the relation $\leq_U$;

$max_2(X)$: the second maximum of X;

$max_3(X)$: the third maximum of X;

....

$max_{|X|}(X)$ the minimum of X.

It is now possible to introduce the following

Definition 3

Let X and Y be two subsets of U, then we say that $X \leq_S Y$ if holds true either $X = Y$ or $|X| < |Y|$ or $|X| = |Y|$ and for some i, $1 \leq i \leq |X|$, one has: $max_j(X) = max_j(Y)$ for every $j < i$ and $max_i(X) \leq_U max_i(Y)$

Proposition 2

The ordering relation $\leq_S$ on P(U) is total.

Proof [15].

In order to define the relation $\leq$ for the attribute strings, we note that the ordering is based on the information contained in each string and it can be evaluated according to the coefficients k in the strings. The greater is k, for a specific string, the higher is the number of strings whose composition has generated the string. Thus the greater is the content of information, the higher is the possibility for it to precede other strings in the ordering.

Definition 4

Let

$$A = a_n^{\alpha n}\ a_{n-1}^{\alpha n-1} \dots a_2^{\alpha 2}\ a_1^{\alpha 1}\ (k_A,\ d_{A,n},\dots,d_{A,2},d_{A,1})\ and$$

$$B = b_m^{\beta m} b_{m-1}^{\beta m-1} \dots b_2^{\beta 2} b_1^{\beta 1}\ (k_B,\ d_{B,m},\dots,d_{B,2},d_{B,1})$$

be two attribute strings of the universe U,

then $A \leq B$ if either $A=B$ or $k_A>k_B$ or ($k_A=k_B$ and there is an index i ($1 \leq i \leq min(n,m)$) such that for every $j<i$ $a_j=b_j$, $\alpha_j=\beta_i e$ $d_{A,j}=d_{B,j}$, whereas $a_i<_S b_i$ or $a_i=b_i$ and $\alpha_i< \beta_i$ or $a_i=b_i$ and $\alpha_i=\beta_i$ and $d_{A,i}>d_{B,i}$. Moreover for every $A \in S(U)$ we have $U^{NI} \leq A \leq U^{NC}$.

It is apparent that the strings U^{NC} and U^{NI} represent the minimum and the maximum, respectively. In general, this relation can be viewed as an ordering carried out with respect to the information contained in each string, and a string having a higher content of information than another one precedes the latter in the ordered sequence.

The set S(U) is endowed with both minimum and maximum elements, moreover the ordering of strings takes place by comparing the components of the strings themselves, namely:

- subsets of U,

- fuzzy numbers in the interval [0,1], in turn consisting of real numbers in [0,1],

- integers (coefficients).

Since one has that:

- subsets of U are finite,

- the interval [0,1] is complete,

- the infinite subsets of strings whose coefficients grow to infinite have as g.l.b. the string U^{NI},

the following proposition holds:

Proposition 3.

The structure $(S(U),\ \Diamond,\ U^{NC}, \leq)$ is a commutative l-monoid.

Proof [15].

5 Data analysis

5.1 The weight of the attributes

In order to get a practical tool, it is worth noting that not all attributes have the same relevance. However, in our structure each attribute contributes to the final result in the same measure. So it is desirable to have a mechanism that allows to give different weights to different attributes. For example, suppose that some individuals apply for a post of salesman. The individuals could be classified according to the following features: appearance, ability to persuade, punctuality. It is reasonable that the second feature is more important than the third one, so it is necessary to introduce a weighting mechanism within the structure. The mechanism that attaches a weight to a string should allow the elements to be singled out, beginning with their classification. Moreover, the mechanism should not alter the generation of clusters.

A possible solution is modify, with the help of an human expert, the values of the labels in the strings related to the most relevant attributes [18, 20], however we are interested in an automatic mechanism so that the human intervention is not required. Thus we could increment of a fixed quantity m the linguistic labels of the strings corresponding to the most relevant attributes. However, in such way all objects belonging to this string will have an improved evaluation in the final string. It would be better that most relevant elements improve their position in the classification, whereas least relevant elements move to lower positions in the classification. To this aim we give a positive increment to "high" labels and a negative increment to "low" labels.

However, modifying just the labels is not enough. In fact, beyond modifying the labels, new classes should be generated within the first parts of the attribute string. By introducing these new classes, we can shift the elements of the appropriate number of places. Moreover, suitable intermediate labels should be generated for the new classes.

We describe now the weighting mechanism [20], that modifies the position of the objects according to the value taken with respect to the weighted attribute.

Let $A = a_n^{\alpha n} \; a_{n-1}^{\alpha n-1} \ldots a_2^{\alpha 2} \; a_1^{\alpha 1}$ be the attribute to be weighted, and let m be its weight (a real positive). Then for each class a_i with label α_i :

a. We create on its left N_{α_i} new empty classes, with $N_{\alpha_i} = \mathrm{int}(m_{\alpha_i} * m)$, where m_{α_i} is the mean value of the fuzzy numbers represented by the label $\alpha_{i,;}$

b. We move on the left the first N_{α_i} labels α_i together with their corresponding labels α_i;

c. For each i, we assign to the new N_{α_i} classes the intermediate labels, between α_{i-1} and α_i which satisfy the following formula:

$$s_{ij} = \alpha_{i-1} + \frac{j*(\alpha_i - \alpha_{i-1})}{N_{\alpha_i}+1} \qquad j = 1,..,N_{\alpha_i} \text{ for each i}$$

Because the basic objects are fuzzy numbers, we use extended fuzzy operations in the formula.

Example 4

Consider the strings of Example 2:

brightness=$\{w,lb,y\}$ ^{high} $\{r\}$ ^{average} $\{v,b\}$ ^{low}
warmth=$\{r,y\}$ ^{high} $\{b,v\}$ ^{average} $\{lb,w\}$ ^{low}

Suppose that we aim to paint a painting with warm colors and carrying a very high level of brightness. Thus we can compose "brightness" and "warmth" with ratio 3-1 and give weight=3 to "brightness".

By applying the weighing mechanism, we get:

$N_{high}= int(1*3) = 3$ $N_{average} = int(0.5 * 3) = 1$ $N_{low} = int(0.3* 3) = 0$

and this means that the string becomes:

brightness $= [y,lb,w]^{high}$ $[-]^{s3,3}$ $[-]^{s3,2}$ $[-]^{s3,1}$ $[r]^{average}$ $[-]^{s2,1}$ $[b,v]^{low}$, where we have to single out the intermediate labels s_{33}, s_{32}, s_{31} e s_{21}. By applying the formula we get:

ss_{33}=[0.62,0.87,0.95]	ss_{32}=[0.55,0.75,0.9]
ss_{31}=[0.47,0.62,0.85]	ss_{21}=[0.2,0.4,0.65]

Thus we can compute brightness $\Diamond$ warmth, and we have, for the first parts:

$(c_9, c_8, c_7, c_6, c_5, c_4, c_3, c_2, c_1) = [y]\ [-]\ [lb,w]\ [-]\ [r]\ [-]\ [-]\ [b,v]\ [-]$

and for the second parts:

$\gamma_9 = [0.7, 1, 1] \approx high$ $\qquad\qquad$ $\gamma_8 = [0.6, 0.84, 0.93]$
$\gamma_7 = [0.49, 0.73, 0.85] \approx ib[average, high]$ $\gamma_6 = [0.45, 0.67, 0.88]$
$\gamma_5 = [0.42, 0.61, 0.8] \approx nt[average]$ $\qquad$ $\gamma_4 = [0.36, 0.55, 0.76]$
$\gamma_3 = [0.28, 0.5, 0.7]$ $\qquad\qquad\qquad$ $\gamma_2 = [0.15, 0.37, 0.61] \approx ib[low,average]$
$\gamma_1 = [0, 0.3, 0.5]$

where we have omitted the approximation for labels corresponding to empty sets. By comparing the weighted and not weighted strings:

brightness $\Diamond$ warmth =

$= [\text{y}]^{high} \; [\text{r}]^{ib[average,high]} \; [\text{lb},\text{w}]^{nt[average]} \; [\text{v},\text{b}]^{ib[low,average]}$

brightness' $\Diamond$ warmth =

$= [\text{y}]^{high} \; [\text{lb},\text{w}]^{ib[average,high]} \; [\text{r}]^{nt[average]} \; [\text{v},\text{b}]^{ib[low,average]}$

We note that, the increased weight of "brightness" with respect to "warmth" has improved the position of most bright colors (light blue and white), whereas the opposite has happened for colors warmer but less bright (red).

5.2   The relevance

When two strings A and B are composed into A$\Diamond$B in order to get the string C, essentially we "shake up" the information in A and B and put it into C. The "relevance" [21] measures the ability of A in influencing B (and viceversa) in order to get C. It is worth introducing two measures of relevance, as the operation of composition affects both first and second parts.

5.2.1 The relevance for the first parts

Let $A = a_n^{\alpha n} \ldots a_2^{\alpha 2} a_1^{\alpha 1}$ be an attribute string, and let C be the result of the composition of A with other attributes (e.g. $C = A \Diamond B \Diamond D$). Then the relevance $\mu^C A$ of A with respect to C is defined as follows

$$\mu_A^C = \sum_{j=1,\ldots n} \sum_{k=1,\ldots m} (p_{jk} - p_j)^2$$

where $p_{jk} = \#[A^{-1}(\alpha_j) \cap C_k]$ and $p_j = \#[A^{-1}(\alpha_j)]/m$ and $\{C_1, C_2, \ldots, C_m\}$ are the first parts of sthe string C

From the conceptual point of view the relevance of the attribute A represents the quantity of information conveyed by A into the attribute C or, equivalently, the ability of A in singling out the elements for the generation of the attribute C. The formula at first glance appears similar to the one for the variance, in fact we compute the extent to which the distribution of the classes of A in C differs from that in which the distribution of A is uniform.

Example 5

Let {CT,T,AT,F} be the values taken by a linguistic variable and suppose that we have:

$A= \{c, g\}^{CT} \{d\}^T \{b, e, f\}^{AT} \{a\}^F$ (1,1,1,1,1) and
$B = \{g\}^{CT} \{a,d,e,f\}^{AT} \{b,c\}^F$ (1,1,1,1), where
$C=A\lozenge B =\{g\}^{CT} \emptyset^{AT} \{c,d\}^{ib[AT,T]}\{e,f\}^{AT}\{a,b\}^{ib[F,AT]} \emptyset^F$ (2,1,2,3,3,2,1).

We compute $\mu_A^C= \sum_{j=1,...,4} \sum_{k=1,...,6}(p_{jk} - p_j)^2$, one has:

$p_1=\#[\{a\}]/6=1/6$ $\quad$ $p_2=\#[\{b,e,f\}]/6=3/6=1/2$ $\quad$ $p_3=1/6$ $\quad$ $p_4=1/3$	

$p_{1,1}=\#[\{a\}\cap\emptyset]=0$	$p_{2,1}=0$	$p_{3,1}=0$	$p_{4,1}=0$
$p_{1,2}=\#[\{a\}\cap\{a,b\}]=1$	$p_{2,2}=1$	$p_{3,2}=0$	$p_{4,2}=0$
$p_{1,3}=\#[\{a\}\cap\{e,f\}]=0$	$p_{2,3}=2$	$p_{3,3}=0$	$p_{4,3}=0$
$p_{1,4}=0$	$p_{2,4}=0$	$p_{3,4}=1$	$p_{4,4}=1$
$p_{1,5}=0$	$p_{2,5}=0$	$p_{3,5}=0$	$p_{4,5}=0$
$p_{1,6}=0$	$p_{2,6}=0$	$p_{3,6}=0$	$p_{4,6}=1$

thus

$\mu_A^C =(0-1/6)^2+(1-1/6)^2+(0-1/6)^2+(0-1/6)^2+(0-1/6)^2+(0-1/6)^2+(0-1/2)^2+$
$(1-1/2)^2+(2-1/2)^2+(0-1/2)^2+(0-1/2)^2+(0-1/2)^2+(0-1/6)^2+(0-1/6)^2+(0-1/6)^2+$
$(1-1/6)^2+(0-1/6)^2+(0-1/6)^2+(0-1/3)^2+(0-1/3)^2+(0-1/3)^2+(1-1/3)^2+(0-1/3)^2+$
$(1-1/3)^2 = 1/36+25/36+1/36+1/36+1/36+1/36+1/4+1/4+9/4+1/4+1/4+$
$1/4+1/36+1/36+1/36+25/36+1/36+1/36+1/9+1/9+1/9+4/9+$
$1/9+4/9=5/6+7/2+ 5/6+4/3=6.5.$

In a similar way one gets $\mu_B^C= 5.5$.

The conclusion is that A affects the composition of C more than B.

5.2.2 The relevance for the second parts

We note that the relevance does not take into account the labels, since it operates just on partitions. However, it is possible to evaluate how labels affect the operation $\lozenge$. We remember that the operation for second parts is based on the mean among fuzzy numbers, and in case a string consists of many clusters their labels will appear often in the computation of the means and the more the labels will be important the more they are nearby.

Let us consider how the computation of the labels
$\gamma_i={}^1/_{(kA+kB)*d_{c,i}} \sum_j d_{B,j} * d_{A,i-j+1} * (k_A * \alpha_{i-j+1} + k_B * \beta_j)$ is carried out.

The higher is the relevance of the label α_i in the string A, the higher are the coefficients k_A and d_{Ai}. Thus the relevance of a string is to be directly proportional to $k_A + \sum d_{Ai}$, and inversely proportional to the dispersion of the labels on the interval $[0,1]$, that can be represented as follows:

$$\frac{\sum_{i=1...n-1} \sum_{j=i+1..n} |\alpha_i - \alpha_j|}{n*(n-1)/2}$$

where if $\alpha_i=[\alpha_{i1}, \alpha_{i2}, \alpha_{i3}]$ and $\alpha_j=[\alpha_{j1}, \alpha_{j2}, \alpha_{j3}]$ then $|\alpha_i - \alpha_j|=|\alpha_{i1} - \alpha_{j1}|+|\alpha_{i2} - \alpha_{j2}|+|\alpha_{i3} - \alpha_{j3}|$.

The sum is divided by $n*(n-1)/2$ to find the mean of all these differences. Given a string $A = a_n^{\alpha n} \, a_{n-1}^{\alpha n-1} ... \, a_2^{\alpha 2} \, a_1^{\alpha 1}$ we can define the relevance for the second parts as follows:

$$\rho_A = \begin{cases} 0 & if A = U^{NC} \\ 1 & if A = U^{NI} \\ 2/\pi \; arctg \; \frac{n*(n-1)\,(k_A + \sum_{i=1..n} d_{Ai})}{2*\sum_{i=1...n-1}\sum_{j=i+1..n} |\alpha_i - \alpha_j|} & otherwise \end{cases}$$

The strings U^{NC} and U^{NI} attain, respectively, the minimum and maximum value of the relevance. The relevance of U^{NC} equals zero because this string does not affect the string to which it is applied ($U^{NC} \lozenge A=A$ for every A), whereas the relevance for U^{NI} equals 1 because "absorbs" any string ($U^{NI} \lozenge A=U^{NI}$ for every A).

It is worth emphasizing that the relevance ρ_A is computed beginning from only the information present in A. In other words, the relevance μ_A^C is the relevance of A with respect to a resulting string C, whereas ρ_A is the relevance of A in absolute terms. Thus μ is the *relative relevance* and ρ is *the absolute relevance*.

5.2.3 The total relevance
The *total relevance* of A with respect to C is $R_A^C= \mu_A^C * \rho_A$.

Example 6

In the previous example we have computed $\mu_A^C=6.5$ and $\mu_B^C=5.5$.

We note that F=[0,0,0.2], AT=[0.2,0.4,0.6], T=[0.5,0.7,0.9], CT= [0.8,1,1], thus we get for

$\rho_A=$ (|0-0.2| + |0 − 0.4| + |0.2 − 0.6|) + (|0 − 0.5| + |0 − 0.7| + |0.2 − 0.9|) + (|0-0.8|+|0-1|+|0.2-1|)+(|0.2-0.5|+|0.4-0.7|+|0.6-0.9|)+(|0.2-0.8|+|0.4-1|+|0.6-1|)+(|0.5-0.8|+|0.7-1|+|0.9-1|) = 1+1.9+2.6+0.9+1.6+0.7=8.7.

Thus $(4*3)*5 / 2*8.7 = 3.4$ and $\rho_A=(2/\pi)*\text{arctg}(3.4)=0.817$. We get $R^C_A=\mu^{C}_A* \rho_A=6.5 * 0.817=5.310$.

On gets that $\rho_B=(2/\pi) \, \text{arctg}(2.3)=0.73$.

Thus we finally get $R^C_B= \mu^{C}_B* \rho_B= 5.5*0.73 = 4.01$.

5.3   The similarity index

Given a string A and a weight w, the higher is the value w for A, the more the conservation feature holds true for the products A $\varDelta$Aw and B $\varDelta$ A $\varDelta$Aw, where B is another string. Moreover, the conservation effect attains its maximum value, for each value of w, the more the string B "resembles" the string A.

Let $\{A_1, A_2, A_3, \ldots, A_n\} \cup \{B\}$ be a set of strings and let $\{w_1, w_2, w_3, \ldots, w_n\}$ be a set of positive real numbers, how do we select the string Ai for which

$$B \, \varDelta \, Ai = B \, \varDelta Ai \, \varDelta Awi \, ?$$

Unfortunately, this event is very unlikely, thus we have to introduce a *similarity index* [7] for strings to be used when strings are multiplied.

In order to accomplish this task it is necessary to track the changes undergone by the labels. Let A and B denote two strings, then we build a list L containing differences between the string A and B. Selecting an element of L in position j, and denoting with αb and αa the labels in A and in B respectively in the position j of the list L, the value of the element is defined as follows:

$$0 \quad \text{if } \alpha_a=\beta_b$$
$$|P(\alpha_a)\text{-}P(\beta_b)| \text{ Otherwise,}$$

where the function P returns the position of the label in the ordered sequence of the original labels, and in those labels produced by the linguistic approximation. With n labels we have $|P(\alpha_a)\text{-}P(\beta_b)|_{max=4*n-4}$.

Example 7

$$A = [a, b]^{ci} \, [c, d]^{i} \, [e, f]^{ai} \, [g]^{ni}$$

$$B = [a, d]^{ci} \, [c, g]^{i} \, [e]^{IB[ai,i]} \, [f]^{NT[qi]} \, [b]^{ni}$$

Then one gets:

	a	b	c	d		e		f	g
A	ci	ci	i	i		ai		ai	ni
B	ci	ni	i	ci		$IB[ai,i]$		$NT[ai]$	i
L(A,B)	0	12	0	4		2		1	8

Let $\Sigma(A,B)$ be the sum of numeric values present in the list L(A,B). Let i and j denote the number of non empty clusters in A and B respectively and let $n_c=\min(i, j)$. We define the quantity: $d(A,B) = \Sigma(A,B) / n_c$ as the *dissimilarity index* for the strings A and B. Of course, d=0 if and only if the two strings are equal.

It is worth noting that it is not possible to define such an index as an algebraic sum of the elements of the string L. In fact, some values can cancel each other out as happens in this case: $A= [a,b,c]^{ci} [-]^i [-]^{ai} [d,e]^{ni}$, $B=[d,e]^{ci} [-]^i [-]^{ai} [a,b,c]^{ni}$. Moreover, the symmetric property would not hold for such an index. The maximum index value is attained by strings such as: $A=[a,b,c]^{ci} [-]^i [-]^{ai} [-]^{ni}$, $B=[-]^{ci} [-]^i [-]^{ai} [a,b,c]^{ni}$. If the denominator were not present, the index for the previous pairs of strings would take the same value.

It is easy to see that, if n is the number of labels present in the system and #D denotes the cardinality of the universe of discourse, one has $d_{MAX} = \#D * 4 * (n-1)$

Example 8

$A = [a,b,e,f]^{ci} [g,h]^i [i,j,k,c,d]^{ai} [l]^{ni}$, $B=[i,b,e,h]^{ci} [g,h,c,k]^i [-]^{ai} [a,l,d,f]^{ni}$

	a	b	c	d	e	f	g	h	i	j	k	l
A	ci	ci	Ai	ai	ci	ci	i	i	ai	ai	ai	ni
B	ni	ci	l	ni	ci	ni	i	i	ci	ci	i	ni
L(A,B)	12	0	4	4	0	12	0	0	8	8	4	0

$d(A, B) = 52/3 = 17.33333.....$

Thus the *similarity index* for the strings A and B can be defined as follows:

$$\delta(A,B) = 1 - d(A,B) / (\#D * 4 * (n-1))$$

It is worth noting that $\delta(A,A) = 1$ and $\delta(A,B) = \delta(B,A)$. In the last example, one has $\delta = 1- (52/3)/144 = 0.880...$

Now we are able to state our *selection criterion:* we single out, among the n strings, the A_k such that

$$\delta \ (B \ \Delta \ A_k \ , B \ \Delta \ A_k \ \Delta \ A_{w_k}) = \max_i \ \{B \ \Delta \ A_i, B \ \Delta \ A_i \ \Delta A_{w_i}\}$$

The following properties hold true:

1. If $B = A_j$ then $\max_i \{B \ \Delta A_i, B \ \Delta \ A_i \ \Delta A_{w_i}\} = A_j$.

2. If B and A_j carry a high index, this similarity is enhanced by the weighting mechanism.

3. If the difference between the indices related to B and A_i is similar in size to that related to B and A_j then the weight allows for discrimination between the two strings A_i and A_j.

4. If B is next to A_j and far from A_k then, by choosing a suitable value for the weight for A_k, A_k could be singled out.

6   Applications

6.1   A fuzzy user model in adaptive hypermedia systems

In [7] a fuzzy-based user model has been presented. This model presents a high level of adaptivity as regards both contents and presentation [5]. The basic idea underlying the stereotype approach is that a real user can be represented and approximated by a set of *stereotype users,* which differ with regard to the variety and intensity of cognitive interest in the knowledge domain of the system. Such representation will match a suitable decomposition of the knowledge base in order to evaluate which stereotype best approximates the behavior of the real user.

Basically, at a specific instant t_n, we have a string $U_r(t_n)$ that represents the real user by utilizing the nodes of the hypermedia space and the values of a fuzzy linguistic variable, representing the user's preferences and goals. Similar strings are used to represent stereotypes. Selecting a navigation element induces selection of a stereotype U_i and updating of user representation: $U_r(t_{n+1}) = U_r(tn) \ \Delta \ U_i$.

Each user's choice, each move is suitably "weighted" by the structure and then used to classify the real user starting from a set of stereotype users. The numerical values express the weight that each stereotype user assigns to that choice. One has small weights if the choice slightly affects the stereotype, whereas one has high weights if the user's choice singles out a specific stereotype. Beginning from the link activated by the user, the weights related to the stereotypes, the string representing the real user and the user's current position, the procedure selects the stereotype and leads the user to the next position. The selection of stereotype user that best fit the real user is obtained by means of the above introduced similarity function. Figure 1 depicts the model.

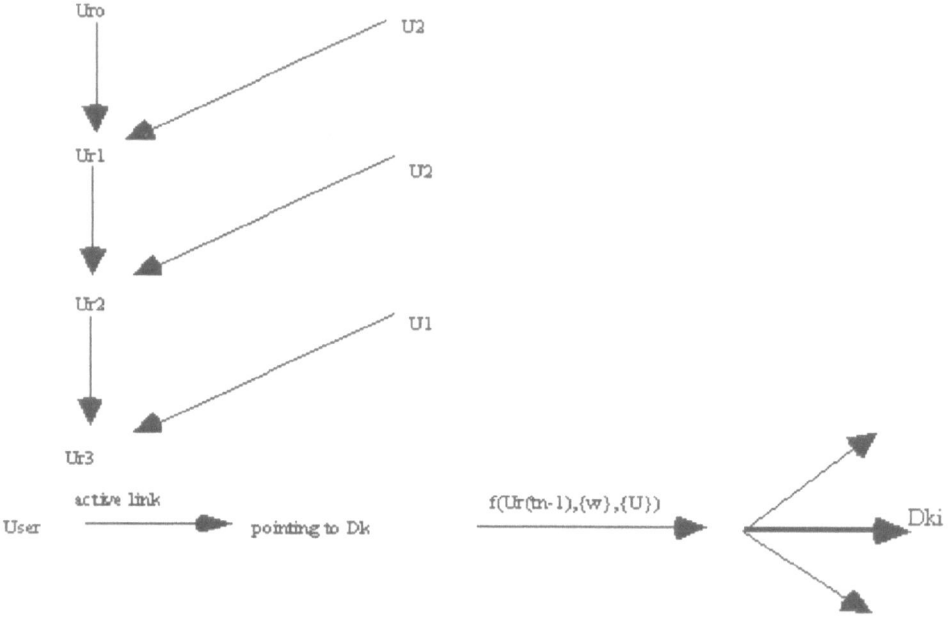

Fig. 1. Stereotype Selection

6.2   A model for computer aided medical diagnosis

In [12] we have presented a fuzzy-based technique for diagnosing the diabetic neuropathy, whose pathogenesis so far is not well known. In fact, the problem of diagnosing neuropathies is a very difficult task to be tackled by physicians.

The analysis is carried out by means of patient's anagraphical and clinical data, e.g. age, sex, duration of the disease, insulinic needs, severity of diabetes, possible presence of complications, referred to two samples of patients, both hospitalized at the Istituto Scientifico San Raffaele in Milan, Italy.

We aim at developing a tool which can give, for each patient under examination, the risk factor related to this disease. The results obtained by our methodology fully agree with those achieved clinically, also as regards possible origins of the disease.

In fact, by analyzing suitable relevance factors one gets reasonable information about the etiology of the disease: our results agree with most credited clinical hypotheses. It is shown how pathogenic factors can be discriminated against those not directly relevant for the insurgence of the neuropathy. Pathogenic factors are equal to those singled out in [6, 23].

6.3 Youngsters classification according to their constitution

In [7] we have chosen a clustering problem tackled by means of a *3-way data fuzzy clustering* in [25]. The problem involves classification of 38 youngsters (13, 14 and 15 years old) according to their constitution. By means of the 3-way data fuzzy clustering the boys have been partitioned into 4 classes. Cluster A includes the boys who preserve their strong constitution along the years, whereas in cluster B there are boys having rather weak constitution. Clusters C and D include boys with medium constitution, but those in C have a greater growth factor than those in D.

By T13, W13, C13 and S13 we denote, respectively, the strings related to tallness, weight, chest, and tallness when seated concerning boys who are 13 old. The same notation is adopted for boys who are 14 and 15. Beginning from data present in the strings, we aim at obtaining classifications related to the constitution of boys for each age. We carry out the following compositions:

| C13=T13◊W13◊C13 ◊S13 | C14=T14◊W14◊C14◊S14 | C15=T15◊W15◊C15◊S15 |

where C13, C14, C15 are the attribute strings related to the constitution at 13, 14 and 15 years old.

We have compared our results with those reported in [25]. We note that for almost all individuals the results are identical. In conclusion we can adfirm that our clustering methodology is satisfactory since grades adequately all individuals and attains the same levels of fuzzy clustering with additional advantage that results are immediately understandable thanks to linguistic labels.

Moreover we have investigated how the constitutions of the boys at the different ages have affected the global assessment. We attain the result that the final constitution is mostly affected by their constitution at 13 and 14. We could also use our method in order to investigate how their average constitution can be beneficially affected.

6.4   Educational taxonomies and learning evaluation

A method for students' evaluation, based on the space of linguistic variables representing the student's learning, is presented in [8, 9, 10]. This method allows representing adequately the student's mastery of a concept.

Using the notions of linguistic variable and fuzzy variable, one can in a natural way (1) model the inherent vagueness related to the learning level of a concept, (2) construct dynamically new terms of the linguistic variable Evaluation using suitable modifiers and operations and (3) achieve a rich description of the student's cognitive and psychological states. These states are described by Bloom taxonomies [4].

The Bloom taxonomies of educational goals classify students' behavior according to their learning process and include three main areas: cognitive, affective, psychomotory. The cognitive area includes all goals related to the mastery of concepts and skills whereas the affective area takes into account goals related to changes occurring in student's interests, behaviors and emotions.

The algebraic computing [29] on the space of formative goals and concepts is performed thanks to the commutative l-monoid, whose operations can be interpreted in terms of student's proficiency and mastery levels.

We note that the evaluation process involves many hedges like *satisfactory, very good, excellent*, etc. which are inherently vague and thus carry a substantial amount of fuzziness.

For example, we suppose that a teacher has singled out the following goals: Clarity of expression (K_1), Consistency of the answerscripts (K_2), Ability in singling out the optimal strategy (K_3), Completeness of the answerscripts (K_4). By our method we get string like the following: $[K_1]^{IB(f,g)}$ $[K_3, K_4]^{B(f)}$ $[K_2]^{NT(s)}$, from which we conclude that: *The student has achieved a very fair clarity of expression, the consistency of the answerscripts and his/her ability in singling out the optimal strategy are almost fair, the completeness of the answerscritps is more than satisfactory.* We can also get information about the evaluation of single questions. *The student has achieved in question 35 a more than good performance, in question 7 a very satisfactory performance, in question 8 a satisfactory performance.*

Finally, the level of performance as regards the educational goals can be summarized by strings from which it is possible to deduce, for example, that *The student has achieved an almost excellent level of synthesis, his/her ability in singling out basic principles is more than good, mastery of terminology, conventions and applications are almost fair, the ability in translating information into other forms is more than satisfactory.* The level of conceptualization, in turn, can be as well obtained by our method.

The method for evaluation can be extended to deal with items of the affective area in Bloom taxonomy so that the student's psychological profile can be evaluated.

The results of our model are similar to those attained by Biswas [3]. However, we note that our model can be applied also to structured questions.

6.5 Decision Making and Fuzzy Screening

Decision Making (DM) techniques [24] have been introduced in early seventies and can be defined as the process aiming at selecting one or more particular choices among a set of them, provided that some constraints are satisfied. In particular, Bellman and Zadeh [2] define fuzzy DM as a decision process where goals and constraints, and possibly the controlled system, are inherently fuzzy.

In [13] we have shown that the problems of DM and Fuzzy Screening can be profitably dealt with the commutative monoid illustrated in the previous sections. A suitable model has been compared with methods belonging to the class max-min and to the class using aggregation operators with weighted mean. In both cases the results are very near to those presented in [1, 26, 27]. We can adfirm that the algebraic structure allows to get more precise results provided that fuzzy numbers are suitably set beginning from data at disposal.

6.6 Financial investments

In [20] we have interpretated some financial indicators in terms of fuzzy values and tested our mechanism by using data from a sample of firms whose securities are exchanged on the Boston Stock Exchange.

Our case study has been developped using data available from the balance sheet information and the financial ratios about firms issuing securities. Having obtained this infomation, we provvide a judgement on its reliability in terms of linguistics labels VH (very high), H (high), M (medium), L (low), treated operationally with triangular numbers.

6.7 FuzzyEvoAgent and evolutionary laws

In [16] we extend the actor-based model of concurrent and distributed programming toward a framework in which the agents can learn emerging behaviours through a fuzzy evolutionary structure. The framework is based on the notion of FuzzyEvoAgent, i.e. an entity that exploiting the basic issues of actors (asynchronous message passing, concurrent computation) is submitted to evolutionary laws that reinforce the most suitable behaviours respecting the environment. We propose a formal definition as well as an implementation model of FuzzyEvoAgents that has been verified via simple simulation of Artificial Life.

7   A case study

Graphologists by examining handwriting is able to detect, to some extent, writer's psychological features. Handwriting reveals intelligence quotient, specific attitudes, somatic features, some diseases. The underlying idea is that handwriting is the projection of the writer. In which way our structure can help graphologist's task ? We note that graphology bases handwriting interpretation on the form of each stroke denoted as graphic sign. Examples of signs are the width of letters, writing speed, length of strokes, distance between letters and words, and so on. Graphologist evaluate the degree of each sign. For example, handwriting presenting small letters denotes intelligence and deep feelings. Thus our structure is suitable for the task since the graphologist says that letters are "quite small", speed is "not very high". Of course the problem of interpretation is not easy at all since the presence of several signs single out general aspects of the personality, such as creativity, emotionalism, and so on.

In the following case study we aim to classify some individuals with respect to their "intelligence" as evaluated by combining the signs related to intelligence. Graphology singles out signs *quantitative* and *qualitative* with respect to intelligence. Quantitative signs are:

1) Fluid/static. Fluid handwriting is characterized by signs that move more horizontally than vertically. Fluidity denotes self confidence, quick mental mechanisms, high level of creativity. On the contrary, staticity (signs extend vertically) denotes that the individuals is not courageous, does not trust in himself, and is slow in reasoning.

2) Quick/calm: Quick writing is characterized by irregular strokes that change the usual aspect of letters, its opposite is calm handwriting. Quick writing denotes impatience and ability to take decisions in short times whereas the calm one denotes patience, and ability to dominate passions but also not having quick reflexes.

3) Moderate/exuberant. Moderate writing use the minimum of signs to write letters, whereas exuberant individual exceed in signs. Moderate handwriting denotes the ability to focus the attention on the essential part of a problem, whereas the exuberant one denotes people unable to concentrate on the essential aspects of a problem.

4) Small/large: Small letters denote seriousness, discipline whereas large letters indicate abruptness and little criticism.

5) Robust/weak: Robust handwriting is characterised by vigorous strokes, whereas the weak one presents strokes not clearly impressed. Robustness denotes strong mental abstraction, and ability to take decisions, whereas weakness indicates the opposite.

6) Wide/narrow: Wide handwriting presents closed letters (such as "o", "a", "d") that extend more horizontally than vertically, narrow handwriting the opposite. Wide writing indicates the ability to analyze ideas in all its aspects, and thus scientific mentality, whereas the narrow one denotes inclination towards irrationality.

7) Open/closed. Open writing if ascending and descending strokes, typical those of letters "m" and "n" tend to open wide. Instead, if strokes are, in some measure, superimposed the handwriting is closed. Open writing denotes conceptual precision and ability to discern whereas the other one indicates people with scarce attitude to discern.

8) Attached/separated: Attached handwriting is characterized by continuous strokes, and the writer does not separate letters and sometimes even words. Separated writing shows letters and words separated each other. Attached handwriting denotes ability in logical synthesis, continuity and perseverance, the separated one indicates the reverse.

Other signs are related to the quality of intelligence, and they can be summarized as follows:

a) Ascending/descending. Ascending writing tends to go up the horizontal row and denotes ambition and optimism. On the contrary descending writing indicates lack of ambition and, to some extent, pessimism.

b) Blanks between words/No blanks: Lot of blank space between words denote prudence and high level of criticism. The opposite indicates imprudence and inclination to depression.

c) Clear/obscure: Clear writing denotes clarity in behaviors whereas the opposite indicates an inclination towards approxiiation.

d) Vivacious/flat: Vivacious writing presents jumps of letters whereas in the flat one all letters are lined up. Vivacious writing denotes high levels of creativity, the flat one the opposite.

e) Tortuous: In this case no parallelism among the axes of the letters is present. This feature denotes versatility and investigation abilities.

f) Animated: This type of writing is characterized by excessive strokes and denotes mental vigor along with hyperactivity of fantasy.

Of course, we have summarised some basic elements of graphology, a discipline very complex and whose results should always placed in a wider framework in order to get reliable assessments. Now suppose that the graphologist has to analyze the handwriting of 20 individuals, then our universe of discourse is U={1, 2, 3, 4, 5, 6, ..., 20}.

Then the expert examines the 20 handwritings and states, for each writing, to what extent each of the signs 1-8 and a-f is satisfied:

Very high ⇒ **VH**=[0.8,1,1]	high ⇒ **H**=[0.7,0.8,0.9]
Almost high ⇒ **AH**=[0.5,0.6,0.7]	Avarage ⇒ **A**=[0.3,0.4,0.5]
Low ⇒ **L**=[0.1,0.2,0.3]	Very low ⇒ **VL**=[0,0,0.1]

Then each attribute 1-8 and a-f induces on U a classification with the above-mentioned labels. For short, attributes are denoted by the following names:

fluid **FL**	ascending **AS**	quick **QU**
blanks between words **BW**	moderate **MO**	clear **CL**
small **SM**	vivacious **VI**	robust **RO**
Toutuous **TO**	wide **WI**	animated **AN**
open **OP**	attached **AT**	

Thus, after graphologist's report, we have that the attribute strings are:

• Strings related to the quantity of intelligence:

FL= $[1,4,9,17]^{VH}$ $[5,7,14,19]^{H}$ $[2,10,15,18,20]^{AH}$ $[8]^{A}$ $[3,12,13,16]^{L}$ $[6,11]^{VL}$

QU= $[5,13,20]^{VH}$ $[4,14,15]^{H}$ $[1,2,12,18,19]^{AH}$ $[6,7,8,11]^{A}$ $[3,9,17]^{L}$ $[10,16]^{VL}$

MO= $[4]^{VH}$ $[1,3,5,14]^{H}$ $[9,10,15,17]^{AH}$ $[6,8,11,12,13,18,19]^{A}$ $[2,7,16,20]^{L}$ $[-]^{VL}$

SM= $[1,6,7]^{VH}$ $[9,10,12,13,14]^{H}$ $[8,16]^{AH}$ $[3,4,5,11,20]^{A}$ $[15,17]^{L}$ $[2,18,19]^{VL}$

RO= $[-]^{VH}$ $[-]^{H}$ $[2,6,10,15,17,20]^{AH}$ $[1,9,11]^{A}$ $[3,4,12,13,14,18,19]^{L}$ $[5,7,8,16]^{VL}$

WI= $[12,18]^{VH}$ $[1,2,3,13,19]^{H}$ $[4,8,14,20]^{AH}$ $[5,6,9,15]^{A}$ $[7,10,16]^{L}$ $[11,17]^{VL}$

OP= $[8,10,16]^{VH}$ $[1,6,15,17,18,19]^{H}$ $[12,13,14]^{AH}$ $[2,3,20]^{A}$ $[4,5,7]^{L}$ $[9,11]^{VL}$

AT= $[1,2,13,14,18,19]^{VH}$ $[6,10,15,20]^{H}$ $[3,5,8,11,16,17]^{AH}$ $[4]^{A}$ $[7,12]^{L}$ $[9]^{VL}$

• Strings related to the quality of intelligence:

AS= $[6,16,20]^{VH}$ $[7,8,9,10,15,19]^{H}$ $[1,3,13]^{AH}$ $[2,4,5,11,14,17]^{A}$ $[12,18]^{L}$ $[-]^{VL}$

BW= $[1,11]^{VH}$ $[2,4,12]^{H}$ $[3,6,13,17,18,19]^{AH}$ $[5,8,14,16]^{A}$ $[7,10,15]^{L}$ $[9,20]^{VL}$

CL= $[\text{-}]^{VH}$ $[\text{-}]^{H}$ $[1,6,8,13,15,16,17,18,19,20]^{AH}$ $[5,7,12]^{A}$ $[4,9,10,11]^{L}$ $[2,3,14]^{VL}$

VI= $[1,2]^{VH}$ $[3,6]^{H}$ $[4,7]^{AH}$ $[5,9]^{A}$ $[8,10,11,12,13,16,18,20]^{L}$ $[14,15,17,19]^{VL}$

TO= $[\text{-}]^{VH}$ $[2,4,6]^{H}$ $[1,7]^{AH}$ $[3,5,8,15,16,17]^{A}$ $[9,10,13,18,19]^{L}$ $[11,12,14,20]^{VL}$

AN= $[7]^{VH}$ $[1,6,13,15,20]^{H}$ $[4,12,14,16]^{AH}$ $[2,3,11,18,19]^{A}$ $[5,8,9,17]^{L}$ $[10]^{VL}$

Now we can carry out the first level of classification with respect to the "quantity of intelligence" by combining the first eight attributes by means of the operator $\Diamond$, and by combining the remaining string with respect to the "quality of intelligence". In order to express the results of the composition new fuzzy labels are to be defined related to the degree of intelligence:

Very high' $\Rightarrow$ **VH'**=[0.8,1,1]	High' $\Rightarrow$ **H'**=[0.5,0.6,0.8]
Almost high' $\Rightarrow$ **AH'**=[0.2,0.3,0.5]	Not high' $\Rightarrow$ **NH'**=[0,0,0.2]

In such way we get:

QUANTINT=FL$\Diamond$QU$\Diamond$MO$\Diamond$SM$\Diamond$RO$\Diamond$WI$\Diamond$OP$\Diamond$AT=

= $[1]^{ib(H',VH')}$ $[14]^{nt(H')}$ $[4,10,13,15,18,19,20]^{H'}[2,5,6]^{bt(H')}$ $[3,8,9,12,17]^{ib(AH',H')}$ $[7,16]^{nt(AH')}$ $[11]^{AH'}$

QUALINT=AS$\Diamond$BW$\Diamond$CL$\Diamond$VI$\Diamond$TO$\Diamond$AN=

$[1,6]^{ib(H',VH')}[7]^{H'}$ $[2,4,16]^{bt(H')}$ $[3,8,13,15,19,20]^{ib(AH',H')}$ $[5,11,12,17,18]^{nt(AH')}$ $[9]^{AH'}$ $[10,14]^{bt(AH')}$.

QUANTINT and QUALINT represent, respectively, the classifications of the individuals with respect to their quantitative and qualitative degree of intelligence. In order to obtain an overall measure, the two strings can be composed, imposing that the quantitative intelligence have a weight slightly higher than the qualitative in the final classification. To this aim we give weight m=0.5 to the string QUANTINT.

We get:

INTEL = $0.5*$QUANTINT$\Diamond$QUALINT = $[1]^{ib(H',VH')}$ $[6]^{nt(H')}$ $[4,13]^{H'}$ $[2,15,19,20]^{bt(H')}$ $[3,5,7,8,10,12,14,16,17,18]^{ib(AH',H')}$ $[9]^{nt(AH')}$ $[11]^{AH'}$

Thus the final graphologist's report can be summarized as follows:

With respect to the degree of intelligence, the individual 1 ranks first, the individuals 6, 4, 13, 2, 15, 19, 20 have a good level 3, 5, 7, 8, 10, 12, 14, the individuals 16, 17, 18 show an average level; whereas the individuals 9, 11 rank last.

8   Concluding remarks

In this paper we have illustrated the basic properties of an algebraic structure and we have applied the structure to case studies well known in the literature. Our approach can be summarized as follows:

1. A *classification* is a particular string, and more classifications can be *multiplied* by applying the operation ◊.

2. A *linear ordering* is present in the structure, so that strings can be compared as regards their contents of information.

3. The structure includes special linguistic labels: *Not compatible* and *No information*. Introducing these labels allows that the structure be endowed with a unit element with respect to the operation ◊ and a minimum element with respect to the ordering. In such way the structure becomes an *integral commutative l-monoid*.

4. We have introduced a *similarity index* between two strings or classifications and a mechanism that allows to give *different weights* to different attributes.

5. Finally, the concept of *relevance of a component* with respect to a resulting classification and the concept of *absolute relevance* of a string have been introduced. The two measures of relevance give rise to the *total relevance*.

Our approach introduces a suitable framework in the field of soft computing and it appears as a tool flexible and powerful for data processing in fuzzy environments.

References

1. Baldwin J. F., (1994) *"FRIL-Fuzzy and Evidential Reasoning"* in AI, Research Studies Press.
2. Bellman R. E. and Zadeh L. A., (1970) *"Decision-making in a fuzzy environment"*, Management Sci., 17(4) 141-164 .
3. Biswas R., (1995) *"An application of fuzzy sets in students' evaluation"*, Fuzzy Sets and Systems 74, 187-195.
4. Bloom B.S. et al., (1984) *"Taxonomy of educational objectives: Handbook 1 - 2"*, Addison Wesley.
5. Brusilovsky P., (1996) *"Methods and techniques of adaptive hypermedia"*, User Modeling and User-Adapted Interaction, 6, (2-3), 87-129.
6. Comi G., Galardi G., Amadio S., Bianchi E., Secchi A., Martinenghi S., Caldara R., Pozza G., Canal N., (1991) *"Neurophysiological study of the effect of combined kidney and pancreas transplantation ondiabetic neuropathy: a 2-year follow-up evaluation"*, Diabetologia 1991;34 Suppl 1:S103-7.
7. Di Lascio L., Fischetti E., Gisolfi A., (1999)*" A Fuzzy-Based Approach To Stereotype Selection In Hypermedia"*, User Modeling and User-Adapted Interaction, 9:285-320.

8. Di Lascio L., Fischetti E., Gisolfi A., (2000) " *Learning Evaluation= Computing with Words*", Techical Report, DMI-0043, 2000.

9. Di Lascio L., Fischetti E., Gisolfi A., (2001) "*Fuzzy modeling and evaluation of an ITM user*", Computacion and Systemas, in press.

10. Di Lascio L., Fischetti E., Gisolfi A., (2000) "*Molto attento x abbastanza buono = piú che discreto*", Atti DIDAMATICA-AICA, Cesena, Italy, 2000.

11. Di Lascio L., Gisolfi A., (1998) "*Averaging linguistic truth values in fuzzy approximate reasoning*", International Journal of Intelligent Systems, 13, 301-318.

12. Di Lascio L., Gisolfi A., Albunia A., Galardi G., Meschi F., (2000) "*A fuzzy-based methodology for the analyis of diabetic neuropathy*", submitted,

13. Di Lascio L., Gisolfi A., (2000) "*Decision Making and Fuzzy Screeneng: a new computational approach*", Techical Report, DMI-0045, 2000.

14. Di Lascio L., Gisolfi A., Loia V., (1998) "*Uncertainty processing in user-modeling activity*", Information Sciences, 106, 25-47.

15. Di Lascio L., Gisolfi A., Rosa G., (2000) "*A commutative l-monoid for classification with fuzzy attributes*", to appear in Int. J. Of Approx. Reasoning.

16. Di Nola A., Gisolfi A., Loia V., Sessa S., "*Emerging Behaviors in Fuzzy Evolutionary Agents* ", Proceedings of 7th European Congress on Intelligent Techniques and Soft, Aachen, Germany, Sept. 13-16, 1999.

17. Gisolfi A., (1992) "*An Algebraic Fuzzy Structure For The Approximate Reasoning*", Fuzzy Sets And Systems 45, 37-43.

18. Gisolfi A., (1995), "*Classifying Through An Algebraic Fuzzy Structure: The Relevance Of The Attributes*", International Journal Of Intelligent Systems, Vol. 10, pp. 715- 734.

19. Gisolfi A., Cicalese F., (1996) "*Classifying through a fuzzy algebraic structure*", Fuzzy Sets and systems 78, 317-331.

20. Gisolfi A., Loia V., (1995) "*A Complete, Flexible Fuzzy-Based Approach To The Classification Problem*", International Journal Of Approximate Reasoning, 13, 151-183.

21. Gisolfi A., Nunez G., (1993) "*An Algebraic Approximation to the Classification with Fuzzy Attributes*", International Journal of Approximate Reasoning, 9, 75-95.

22. Klir G. J., Yuan B., (1995) "*Fuzzy sets and fuzzy logic: theory and applications*", Prentice - Hall, London.

23. Martinenghi S., Comi G., Galardi G., Di Carlo V., Pozza G., Secchi A., (1997) "*Amelioration of nerve conduction velocity following simultaneous kidney/pancreas transplantation is due to the glycaemic control provided by the pancreas*", Diabetologia 1997;40(9):1110-2.

24. Ribeiro R. A., (1996) "*Fuzzy multiple attribute decision making: A review and new preference elicitation techniques*", Fuzzy Set and Systems, 78, 155-181.

25. Sato. M, Sato Y., Jain L.C., (1997) "*Fuzzy Clustering Models and Applications*", Studies in Fuzzyness and Soft Computing n. 9, Physica-Verlag, New York.

26. Yager R. R., (1977) "*Multiple objective decision-making using fuzzy sets*", Int. J. Man-Machine Studies 9 , 375-382.

27. Yager R. R., (1987) "*Fuzzy decision making including unequal objectives*", Fuzzy Set and Systems, 87-95.

28. Zadeh L. A., (1975) "*The concept of a Linguistic Variable and Its Application to Approximate Reasoning*", I, II, III, Information Sciences, 1975 (8) 199-249, (8) 301-357, (9) 43-89.

29. Zadeh L. A., (1996) *"Fuzzy Logic = Computing with Words""*, IEEE Transactions on Fuzzy Systems, (4):2, 103-111.
30. Zadeh, L. A., (1965) *"Fuzzy sets"*, Information and Control, (8): 3, 338-353.

Free BL_Δ Algebras

Franco Montagna

Dipartimento di Matematica, Universitá di Siena
Via del Capitano 15, 53100 Siena - montagna@unisi.it

Abstract. We prove that the BL algebra consisting of the ordinal sum of $n + 1$ copies of the standard MV-algebra on $[0, 1]$ generates the variety generated by all n generated BL algebras.

1   Introduction

In [H98], fuzzy logics are presented as the logics of continuous t-norms and their residua. There, a fuzzy logic is presented, named BL (*Basic Logic*), and it is conjectured that BL is the logic of continuous t-norms and their residua. This concjecture has been proved to be true in [CEGT99]. In [H98], the notion of *ordinal sum* is presented, and it is shown that any linearly ordered BL algebra (i.e., any linearly ordered algebraic model of BL) satisfying suitable conditions is the ordinal sum of an ordered family of product algebras and of MV algebras (cf [H98] for definitions). In [AFM99] the authors use some techniques from [BF96] and a slightly different notion of ordinal sum, and obtain an improvement of [CEGT99]: BL is the logic of the class of all continuous t-norms that are ordinal sums of a finitely many copies of Lukasievicz t-norms (with their residua, of course). It follows that any infinite! ordinal sum of copies of the
Giving a good description of free algebras is a rather important problem both in Universal Algebra and in Fuzzy Logic. Free MV algebras have been described by McNaughton in [MN51]. The paper [CT98] contains a description of free product algebras, where [MP99] contains a description of free algebras of varieties of MV algebras with additional operators. Thus a good description of free BL algebras would be an important contribution to the developement of the algebraic aspects of fuzzy logic. We tried unsuccessfully to reach this goal. Our idea was the following. It is well-known that given a variety $\mathbf{V}$ and an algebra if $\mathcal{A} \in \mathbf{V}$ which generates precisely the variety generated by the class of n generated algebras of $\mathbf{V}$, then the free algebra of $\mathbf{V}$ on n generators is the subalgebra of $\mathcal{A}^{\mathcal{A}^n}$ generated by the projections.
Now we can prove that the algebra $(n + 1)[0, 1]$ consisting of the ordinal sum of $n + 1$ copies of the standard MV algebra on $[0, 1]$ generates the variety generated by all n generated BL algebras. Thus in order to describe the free BL algebra on n generators it is sufficient to characterize the subset of $((n + 1)[0, 1])^{((n+1)[0,1])^n}$ generated by the projections.

Unfortunately, this is not an easy task. Our functions behave like Mc-Naughton's functions (cf. [MN51]) when all arguments lie in $[0,1)$ (the first copy of the standard MV algebra on $[0,1]$, 1 excluded), but they may behave in a very complicate way in the other intervals. For example, if $x, y \in [0,1)$, then $x \to y = \neg y \to \neg x = \neg\neg x \to y$, but if $1 < x < 2$ and $1 \leq y < x$, then $y < x \to y < 2$, $\neg y \to \neg x$ is the top element $n+1$ of $(n+1)[0,1]$, whereas $\neg\neg x \to y = y$. Thus, different binary functions may have the same restriction to the first copy of $[0,1)^2$. On the other hand, the behavior of a function in $[0,1)$ induces some restrictions to its behavior in the other intervals: e.g. if $f(x) = x$ in $[0,1)$, then in $[1,2)$ one must have: either $f(x) = x$ or $f(x) = n+1$.

Due to the above mentioned difficulties, we did not succeed giving a full characterization of functions which are in the free BL algebra on n generators. Nevertheless, we have been able to perform an intermediate step, that is, to characterize free BL_Δ algebras. These are the free algebras of the variety of BL algebras with an additional operator Δ (cf. [Ba96]) whose intended meaning is: $\Delta(x) = 1$ if $x = 1$, and $\Delta(x) = 0$ otherwise (we will give a precise definition in Section 1). The operator Δ allows for definitions by cases, and makes it possible to obtain more functions, which can be defined independently in different intervals. This makes characterization of free BL_Δ algebras much easier, but still far from trivial, as we shall see in Theorem 3.

2   Preliminaries about basic hoops

We review some definitions and results from [BO], [BP94], [H98], [BF96] and [AFM99].

Definition 1. [BO], [BP94] A *hoop* is a structure $\langle A, \cdot, \to, 1 \rangle$, such that, letting $x \leq y \equiv x \to y = 1$, the following conditions hold:

- $\langle A, \cdot, 1 \rangle$ is a commutative monoid.
- $x \leq x$.
- $x \to (y \to z) = (x \cdot y) \to z$.
- $x \cdot (x \to y) = y \cdot (y \to x)$.

A *bounded hoop* is a hoop A equipped with a constant 0 satisfying $0 \leq x$ for all $x \in A$.
A *basic hoop* is a hoop satisfying the equation
$$((x \to y) \to z) \cdot ((y \to x) \to z) \leq z$$
A *BL algebra* is bounded basic hoop.

Hájek's definition of BL algebra in [H98] is slightly different, but it can be shown that it is equivalent to ours. In [H98] it is shown that a basic hoop is a distributive lattice with respect to $\leq$, with lattice operations $\wedge$ and $\vee$

defined by $x \wedge y = x \cdot (x \rightarrow y)$, and $x \vee y = ((x \rightarrow y) \rightarrow y) \wedge ((y \rightarrow x) \rightarrow x)$, and that every BL algebra is isomorphic to a subdirect product of a family of lienarly ordered BL algebras. A shorter proof of these results can be found in [AFM99].

Definition 2. Let $\langle I, \leq \rangle$ be a totally ordered set with minimum i_0. Let for all $i \in I - \{i_0\}$, $\mathcal{A}_i$ be a linearly ordered basic hoop, and let $\mathcal{A}_{i_0}$ be a linearly ordered BL algebra such that for $i \neq j \in I$ $\mathcal{A}_i \cap \mathcal{A}_j = \{1^{\mathcal{A}_i}\} = \{1^{\mathcal{A}_j}\}$. Then, $\bigoplus_{i \in I} \mathcal{A}_i$ is the structure whose domain is $\bigcup_{i \in I} \mathcal{A}_i$, $0 = 0^{\mathcal{A}_{i_0}}$, $1 = 1^{\mathcal{A}_{i_0}}$, and

- If for some i $x, y \in \mathcal{A}_i$, then $x \rightarrow y = x \rightarrow_i y$, and $x \cdot y = x \cdot_i y$.
- If $x \in \mathcal{A}_i - \{1\}$ and $y \in \mathcal{A}_j - \{1\}$, with $i < j$ then $x \rightarrow y = 1$, $y \rightarrow x = x$, and $x \cdot y = y \cdot x = x$.

In [AM00], the following is shown:

Theorem 1. *Every linearly ordered BL hoop $\mathcal{A}$ is the ordinal sum of a family of Wajsberg hoops. Every linearly ordered BL algebra is the ordinal sum of a family of Wajsberg hoops, whose first component is a Wajsberg algebra.* ∎

3 BL_Δ algebras

Definition 3. (cf [H98]). A BL_Δ *algebra* is a BL algebra with an additional operator Δ satisfying the following equations:

(i) $\Delta(x) \vee \neg\Delta(x) = 1$.
(ii) $\Delta(1) = 1$
(iii) $\Delta(x \rightarrow y) \leq \Delta(x) \rightarrow \Delta(y)$
(iv $\Delta(x) \leq x$
(v) $\Delta(\Delta(x)) = \Delta(x)$

In [H98] it is shown that every subdirectly irreducible BL_Δ algebra is linearly ordered, and that in every linearly ordered BL_Δ algebra one has:

(•) $\Delta(x) = \begin{cases} 1 \text{ if } x = 1 \\ 0 \text{ otherwise.} \end{cases}$

It follows that every linearly ordered BL_Δ algebra is a structure whose BL reduct is the ordinal sum of a family of Wajsberg hoops whose first component is a BL algebra, and such that Δ is defined by condition (•).
In [AFM99] it is shown that the the variety of BL algebras is generated by all finite ordinal sums of finite Wajsberg chains. The result goes through without substantial changes for BL_Δ algebras. Moreover, we can improve it slightly. Let $BL(n)$ ($BL_\Delta(n)$ respectively) denote the class of all n-generated BL (BL_Δ respectively) algebras, and let $Var(BL(n))$ ($Var(BL_\Delta(n))$ respectively) denote the variety generated by $BL(n)$ ($BL_\Delta(n)$ respectively).

Theorem 2. *Var(BL(n)) is generated by all ordinal sums of $\leq n+1$ finite Wajsberg chains, and $Var(BL_\Delta(n))$ is generated by all ordinal sums of $\leq n+1$ finite Wajsberg chains, with Δ defined by the formula ($\bullet$).*

Proof. Let $e(x_1, \cdots, x_k)$ be an identity which is not valid in some n generated BL algebra. Using Birkhoff's subdirect decomposition theorem and the fact that every subdirectly irreducible BL algebra is linearly ordered, we can find a linearly ordered n-generated BL algebra $\mathcal{A}$ and elements $a_1, \cdots, a_k \in \mathcal{A}$ such that $\mathcal{A} \not\models e(a_1, \cdots, a_k)$. Since $\mathcal{A}$ is n-generated, there are terms $t_i(x_1, \cdots, x_n)$ and elements $b_1, \cdots, b_n \in \mathcal{A}$ such that for $i = 1, \cdots, k$, $a_i = t_i(b_1, \cdots, b_n)$. Thus $e(t_1(b_1, \cdots, b_n), \cdots, t_k(b_1, \cdots, b_n))$ is not true in $\mathcal{A}$. Now we recall ([AFM99]) that every finite linarly ordered partial can be embedded into a finite linearly ordered BL algebra, which is the ordinal sum of finitely many linearly ordered Wajsberg algebras. Let T be the set of all subterms of terms occurring in $e(t_1(x_1, \cdots, x_n), \cdots, t_n(x_1, \cdots, x_n))$, and let $\mathcal{P} = \{t(b_1, \cdots, b_n) : t(x_1, \cdots, x_n) \in T\}$. Then there is a finite linearly ordered BL algebra $\mathcal{F}$ in which $\mathcal{P}$ embeds. It follows that $e(t_1(b_1, \cdots, b_n), \cdots, t_k(b_1, \cdots, b_n))$ is not true in $\mathcal{F}$. Moreover $\mathcal{F}$ is the ordinal sum of finitely many linearly ordered Wajsberg algebras. Clearly, only the components which contain either 0 or some b_i are relevant to the evaluation of $e(t_1(b_1, \cdots, b_n), \cdots, t_k(b_1, \cdots, b_n))$. Let $\mathcal{W}_1, \ldots, \mathcal{W}_h$ ($h \leq n+1$) be these components. It is clear that $e(t_1(x_1, \cdots, x_n), \cdots, t_k(x_1, \cdots, x_n))$ fails to hold identically in $\mathcal{B} \equiv \bigoplus_{i=1}^{h} \mathcal{W}_i$. Thus $e(x_1, \cdots, x_k)$ fails to hold in $\mathcal{B}$, which is the ordinal sum of at most $n+1$ finite linearly ordered Wajsberg algebras.

The proof for BL_Δ algebras is similar: recall that the variety of BL_Δ algebras is generated by its linearly ordered members. So, any invalid identity $e(x_1, \cdots, x_k)$ fails to hold in some linearly ordered BL_Δ algebra $\mathcal{A}$. In $\mathcal{A}$, Δ is defined according to the conditions ($\bullet$). Now repeat the construction of Theorem 2, replacing $\Delta(a)$ by 0 if $a \neq 1$, and by 1 if $a = 1$. We obtain a finite BL algebra $\mathcal{B}$ which is the ordinal sum of $\leq n+1$ finite Wajsberg chains such that $e(x_1, \cdots, x_k)$ fails to hold identically in the BL_Δ algebra resulting from $\mathcal{B}$ by defining Δ according to condition ($\bullet$). ∎

Corollary 1. *$Var(BL_\Delta(n))$ is generated by the algebra (still denoted, by abuse of language, $(n+1)[0,1]$) whose BL reduct is $\underbrace{[0,1] \oplus \cdots \oplus [0,1]}_{n+1 \text{ times}}$, (where $[0,1]$ denotes the standard Wajsberg algebra on the real unit interval), and in which Δ is defined according to condition ($\bullet$).*

Proof. Since every finite Wajsberg chain can be embedded into $[0,1]$, every ordinal sum of $\leq n+1$ finite Wajsberg chains can be embedded into the BL reduct of $(n+1)[0,1]$. ∎

4   Free BL_Δ algebras

Let $B_\Delta(n)$ denote the free BL_Δ algebra on n generators. Clearly $B_\Delta(n)$ is also the free $Var(BL_\Delta(n))$ algebra on n generators. Now as noted in the introduction, if a variety $\mathbf{V}$ is generated by a single algebra $\mathcal{A}$, then the free algebra of $\mathbf{V}$ on n generators is the subalgebra of $(\mathcal{A})^{\mathcal{A}^n}$ generated by the projections. Taking $\mathbf{V} = Var(BL_\Delta(n))$, we obtain:

Corollary 2. $B_\Delta(n)$ *is the subalgebra of* $(n+1)[0,1]^{(n+1)[0,1]^n}$ *generated by the projections.*

Our plan is to give a good description of this algebra in terms of Mc Naughton functions. First recall that the Lukesiewicz t norm $\odot$ in $[0,1]$ and its residuum $\rightarrow_L$ are defined by:

$$x \odot y = max\{x + y - 1, 0\} \qquad x \rightarrow_L y = min\{y + 1 - x, 1\}.$$

Notation 1. For all $a \in \mathbf{R}$, $i(a)$ denotes the maximum integer b such that $b \leq a$. For all $i \leq n$, let

$$x \odot_i y = i \vee (x + y - i - 1), \text{ and } x \rightarrow_i y = (i + 1) \wedge (y + i + 1 - x).$$

The algebra $(n+1)[0,1]$ can be described as the algebra whose domain is $[0, n+1]$, where the bottom is 0, the top is $n+1$, and the operations $\cdot$, $\rightarrow$ and Δ are defined as follows:

$$x \cdot y = \begin{cases} x \odot_{i(x)} y & \text{if } i(x) = i(y) \leq n \\ min\{x, y\} & \text{otherwise.} \end{cases}$$

$$x \rightarrow y = \begin{cases} n + 1 & \text{if } x \leq y \\ y & \text{if } i(y) < i(x) \\ x \rightarrow_{i(x)} y & \text{otherwise.} \end{cases}$$

$$\Delta(x) = \begin{cases} n + 1 \text{ if } x = n + 1 \\ 0 & \text{otherwise.} \end{cases}$$

Notation 2. For typographic reasons, in the sequel we write $\vec{x}$, $\vec{a}$, etc. for $x_1, \cdots, x_n$, $a_1, \cdots, a_n$, etc.

Lemma 1. *Let* $\vec{a}$, $\vec{b}$ *be n-tuples of elements of* $(n+1)[0,1]$ *such that for* $h, j = 1, \cdots, n$ *the following conditions hold:*

(a) $a_h - b_h$ *is an integer.*
(b) $i(a_h) < i(a_j)$ *iff* $i(b_h) < i(b_j)$.
(c) $i(a_h) > 0$ *iff* $i(b_h) > 0$.
(d) $a_h = n + 1$ *iff* $b_h = n + 1$.

Then for every pair $t(\vec{x})$, $s(\vec{x})$ *of terms one has:*

(i) $t(\vec{a}) - t(\vec{b})$ *and* $s(\vec{a}) - s(\vec{b})$ *are integers.*

(ii) $i(t(\vec{a})) < i(s(\vec{a}))$ *iff* $i(t(\vec{b})) < i(s(\vec{b}))$, *and* $i(s(\vec{a})) < i(t(\vec{a}))$
iff $i(s(\vec{b})) < i(t(\vec{b}))$.

(iii) $t(\vec{a}) = n + 1$ *iff* $t(\vec{b}) = n + 1$, *and* $s(\vec{a}) = n + 1$ *iff* $s(\vec{b}) = n + 1$.

Proof. Induction on the sum k of the complexities (number of occurrences of $\cdot$, $\rightarrow$ and Δ) of s and t.

If $k = 0$, the claim is an immediate consequence of (a), (b), (c) and (d).

Suppose that the claim holds for $k \leq m$, and let us prove it for $k = m + 1$. Let $s(\vec{x})$ and $t(\vec{x})$ be terms such that the sum of their complexities is $m + 1$. Without loss of generality we can assume that t has positive complexity. We distinguish the following cases:

- $t(\vec{x}) = u(\vec{x}) \cdot v(\vec{x})$.

  By the induction hypothesis, the claim holds for the pairs $s(\vec{x})$, $u(\vec{x})$ and $s(\vec{x})$, $v(\vec{x})$. Claim (i) for $s(\vec{x})$ follows from the induction hypothesis, and claim (i) for $t(\vec{x})$ follows from the induction hypothesis and from the definition of $\cdot$.

  As regards to claim (ii), one has:

  $i(s(\vec{a})) < i(t(\vec{a}))$ iff $i(s(\vec{a})) < i(u(\vec{a}))$ and $i(s(\vec{a})) < i(v(\vec{a}))$ iff (induction hypothesis) $i(s(\vec{b})) < i(u(\vec{b}))$ and $i(s(\vec{b})) < i(v(\vec{b}))$ iff $i(s(\vec{b})) < i(t(\vec{b}))$.

  Moreover, $i(t(\vec{a})) < i(s(\vec{a}))$ iff either $i(u(\vec{a})) < i(s(\vec{a}))$ or $i(v(\vec{a})) < i(s(\vec{a}))$ iff (induction hypothesis) either $i(u(\vec{b})) < i(s(\vec{b}))$ or $i(v(\vec{b})) < i(s(\vec{b}))$ iff $i(t(\vec{b})) < i(s(\vec{b}))$.

  Finally, claim (iii) follows immediately from the induction hypothesis.

- $t(\vec{x}) = u(\vec{x}) \rightarrow v(\vec{x})$. Claims (i) and (iii) for $s(\vec{x})$ follow from the induction hypothesis. Moreover the induction hypothesis for $u(\vec{x})$ and $v(\vec{x})$ implies that $u(\vec{a}) \leq v(\vec{a})$ iff $u(\vec{b}) \leq v(\vec{b})$, therefore claims (i) and (iii) for $t(\vec{x})$, as well as claim (ii) for $t(\vec{x})$ and $s(\vec{x})$ are immediate if $u(\vec{a}) \leq v(\vec{a})$.

  If $i(v(\vec{a})) < i(u(\vec{a}))$, then by induction hypothesis $i(v(\vec{b})) < i(u(\vec{b}))$, $t(\vec{a}) = v(\vec{a})$, and $t(\vec{b}) = v(\vec{b})$. Thus, claims (i), (ii) and (iii) follow from the induction hypothesis.

  Finally, if $v(\vec{a}) < u(\vec{a})$, but $i(v(\vec{a})) = i(u(\vec{a}))$, then by induction hypothesis $i(v(\vec{b})) = i(u(\vec{b}))$, and since $u(\vec{a}) - u(\vec{b})$ and $v(\vec{a}) - v(\vec{b})$ are integers, claim (i) follows easily from the definition of $\rightarrow$. Moreover, $i(t(\vec{a})) = i(u(\vec{a}))$, $i(t(\vec{b})) = i(u(\vec{b}))$, and both $t(\vec{a}) - t(\vec{b})$, $s(\vec{a}) - s(\vec{b})$

are integers. Thus, claim (ii) follows easily from the induction hypothesis. Finally, $t(\overrightarrow{a}) \neq n+1$, and $t(\overrightarrow{b}) \neq n+1$. Thus claim (iii) follows.

- $t(\overrightarrow{x}) = \Delta(u(\overrightarrow{x}))$. Claims i), (ii), and (iii) follow immediately from the induction hypotheis.

This concludes the proof. ∎

Definition 4. We say that $\overrightarrow{a}$ and $\overrightarrow{b}$ are *equivalent* iff conditions (a), (b), (c) and (d) of Lemma 1 hold. We say that $\overrightarrow{a}$ is *initial* if the set $\{i(a_j) : j = 1, \cdots, n\} \cup \{0\} - \{n+1\}$ is an initial segment of the set of natural numbers. The set of initial elements of $(n+1)[0,1]$ is denoted $I(n)$.

Clearly, for every n-tuple $\overrightarrow{a} \in (n+1)[0,1]$ there is an initial n-tuple $\overrightarrow{b}$ such that $\overrightarrow{b}$ is equivalent to $\overrightarrow{a}$. Moreover, again by Lemma 1, for every term $t(\overrightarrow{x})$ such that $t(\overrightarrow{a}) \neq n+1$, one has $t(\overrightarrow{b}) \neq n+1$. Thus if some term $t(\overrightarrow{x})$ is not identically equal to $n+1$ in $(n+1)[0,1]$, then there is an initial n-tuple $\overrightarrow{b}$ such that $t(\overrightarrow{b}) \neq n+1$. It follows that if we replace every element of $B_\Delta(n)$ (thought of as a map from $(n+1)[0,1]^n$ into $(n+1)[0,1]$) by its restriction to $I(n)$, we obtain an isomorphic algebra. Thus we will regard $B_\Delta(n)$ as the set of functions from $I(n)$ into $(n+1)[0,1]$ generated by the projections (with operations defined pointwise).

Let for $j = 0, \cdots, n$, $I_j = [j, j+1)$, and let $I_{n+1} = \{n+1\}$. It is readily seen that $I(n)$ is the disjoint union of all sets of the form $I_{k_1} \times \cdots \times I_{k_n}$, where $\{0, k_1, \cdots, k_n\} - \{n+1\}$ is an initial segment of $\{0, 1, \cdots, n\}$ (repetitions are allowed, i.e., possibly $k_i = k_j$ for $i \neq j$). Such sets will be called *cells*.

Definition 5. We define, for $x, y \in (n+1)[0,1]$: $x \prec y$ iff $i(x) < i(y)$, and $x \equiv y$ iff $i(x) = i(y)$

Lemma 2. *The folowing conditions hold:*

(i) $\equiv$ *is an equivalence, and $\prec$ is transitive and irreflexive.*

(ii) *For all $a, b \in (n+1)[0,1]$, either $a \prec b$, or $b \prec a$, or $a \equiv b$, and these conditions are mutually exclusive.*

(iii) *For all $a, b, c \in (n+1)[0,1]$, if either $a \equiv b$ and $b \prec c$, or $a \prec b$ and $b \equiv c$, then $a \prec c$.*

(iv) *For all $a \in (n+1)[0,1]$, $a \not\prec 0$, and $n+1 \not\prec a$.*

(v) *Given two relations R and S on $X_n = \{x_1, \cdots, x_n, 0, n+1\}$ such that conditions (i),...,(iv) are satisfied with R in place of $\prec$, S in place of $\equiv$, and X_n in place of $(n+1)[0,1]$, there are a (uniquely determined) cell C such that for all $(\overrightarrow{a}) \in C$, and for $i, j = 1, \cdots, n$ one has: $x_i R x_j$ iff $a_i \prec a_j$; (hence, $x_i S x_j$ iff $a_i \equiv a_j$); $x_i R n+1$ iff $a_i \prec n+1$; $0 R x_i$ iff $0 \prec x_i$ (hence, $x_i S 0$ iff $a_i \equiv 0$, and $x_i S n+1$ iff $a_i = n+1$).*

Proof. Almost obvious.   ∎

Definition 6. A *McNaughton function* is the truth function of a *BL* formula in the Wajsberg algebra $[0, 1]$.

It is well-known that McNaughton functions in $x_1, \cdots, x_n$ are precisely those functions $g(\vec{x})$ that can be represented as

$$g(\vec{x}) = 0 \vee (1 \wedge \bigvee_{i=1}^{k} \bigwedge_{j=1}^{h_i} g_{i,j}(\vec{x})), \text{ where } g_{i,j}(\vec{x}) \text{ are linear functions with integer}$$
coefficients.

Definition 7. Let $y_1, \cdots, y_k$ be variables among $\vec{x}$, let $f(y_1, \cdots, y_k)$ be a McNaughton function, and let $i \leq n + 1$. We define a function $f_{i,n}$ from I_i^k into $[i, i+1) \cup \{0, n+1\}$ as follows:

- If $i = 0$, then
$$f_{i,n}(y_1, \cdots, y_k) = \begin{cases} f(y_1, \cdots, y_k) \text{ if } f(y_1, \cdots, y_k) \neq 1 \\ n+1 \qquad \text{otherwise.} \end{cases}$$
- If $0 < i \leq n$, if $f(1, \cdots, 1) = 1$, and $f(y_1 - i, \cdots, y_k - i) \neq 1$, then
$f_{i,n}(y_1, \cdots, y_k) = f((y_1 - i), \cdots, (y_k - i)) + i$.
- If $0 < i \leq n$, if $f(1, \cdots, 1) = 1$, and $f(y_1 - i, \cdots, y_k - i) = 1$, then
$f_{i,n}(y_1, \cdots, y_k) = n + 1$.
- If $i = n + 1$ and $f(1, \cdots, 1) = 1$, then $f_{i,n}(y_1, \cdots, y_k) = n + 1$.
- If $0 < i \leq n + 1$ and $f(1, \cdots, 1) = 0$, then $f_{i,n}(y_1, \cdots, y_k) = 0$.

Note that if $i(x) = i(y) \leq n$, then $x \odot_{i(x),n} y = x \odot_{i(x)} y$, and that if, in addition, $y < x$, then $x \to_{i(x),n} y = x \to_{i(x)} y$ (cf. Notation 1 for the definitions of $\odot_i$ and $\to_i$).

Lemma 3. *Let $f(y_1, \cdots, y_k)$ be a McNaughton function in k variables. Then, there is a term $t_f(y_1, \cdots, y_k)$ of the language of BL algebras such that for $i = 0, \cdots, n+1$ and for $(y_1, \cdots, y_k) \in I_i^k$, one has: $t_f(y_1, \cdots, y_k) = f_{i,n}(y_1, \cdots, y_k)$.*

Proof. We prove by induction on g that if $g(y_1, \cdots, y_k)$ is any term of MV algebras, then there are a terms $t_g(y_1, \cdots, y_k)$, $s_g(y_1, \cdots, y_k)$ such that for $i = 0, \cdots, n+1$ and for $(y_1, \cdots, y_k) \in I_i^k$,

$t_g(y_1, \cdots, y_k) = g_{i,n}(y_1, \cdots, y_k)$, and $s_g(y_1, \cdots, y_k) = (\neg g)_{i,n}(y_1, \cdots, y_k)$.
Since every McNaughton function is the truth function of a term of MV algebras, this will suffice.
If g is atomic, let $t_g = g$, and $s_g = \neg g$ (remind that the interpretation of 1 in $(n + 1)[0, 1]$ is $n + 1$, and that if $x \geq 1$, then $\neg x = 0$).
Now suppose $g = h \odot k$. As regards to t_g it is sufficient to define: $t_g = t_h \cdot t_k$.

As regards to s_g, note that $(\neg g) = (\neg h \rightarrow_L (\neg h \odot \neg k)) \rightarrow_L \neg k = h \rightarrow_L \neg k = k \rightarrow_L \neg h$.

Distinguish the following cases:

- If $h(1, \cdots, 1) = k(1, \cdots, 1) = 1$, then it suffices to define $s_g = \neg t_g$. (Note that if $i \neq 0$, then both s_g and $\neg g_{i,n}$ are 0 $([i, i+1))^k)$.
- If $h(1, \cdots, 1) = k(1, \cdots, 1) = 0$, then define $s_g = (s_h \rightarrow (s_h \cdot s_k)) \rightarrow s_k$.
- If $h(1, \cdots, 1) = 1$ and $k(1, \cdots, 1) = 0$, define: $s_g = t_h \rightarrow s_k$.
- If $h(1, \cdots, 1) = 0$ and $k(1, \cdots, 1) = 1$, define: $s_g = t_k \rightarrow s_h$.

An easy computation shows that s_g satisfies our requirements.

Next suppose that $g(y_1, \cdots, y_k) = h(y_1, \cdots, y_k) \rightarrow_L k(y_1, \cdots, y_k)$. First of all, we define t_g. Again, we distinguish several cases:

- If $h(1, \cdots, 1) = k(1, \cdots, 1) = 1$, then define: $t_g = t_h \rightarrow t_k$.
- If $h(1, \cdots, 1) = k(1, \cdots, 1) = 0$, then define: $t_g = s_k \rightarrow s_h$.
- If $h(1, \cdots, 1) = 0$ and $k(1, \cdots, 1) = 1$, then define: $t_g = s_h \rightarrow t_k$.
- If $h(1, \cdots, 1) = 1$ and $k(1, \cdots, 1) = 0$, then define $t_g = t_h \rightarrow t_k$ (note that in this case, using the induction hypothesis, we see that if $i > 0$, then in $([i, i+1))^k$ both $g_{i,n}$ and t_g are equal to 0).

As regards to s_g, it is sufficient to define: $s_g = t_h \cdot s_k$.

Once again, an easy computation shows that this definition meets our requirements. ∎

Definition 8. Let $f(\vec{x})$ be any function from a cell C into $[0, n+1]$, and let $i \leq n$. We say that f *does not depend on* x_i if for all $(a_1, \cdots, a_n) \in C^n$ and for all $b \in C$, one has:

$f(\vec{a}) = f(a_1, \cdots, a_{i-1}, b, a_{i+1}, \cdots, a_n)$.

Otherwise, we say that f *depends on* x_i. If $y_1, \cdots, y_k$ are variables among $x_1, \cdots, x_n$, we say that $f(\vec{x})$ *depends only on* $y_1, \cdots, y_k$ if f does not depend on the other variables.

Definition 9. Let C be a cell. A *linear semialgebraic subset of* C is a subset Y of C for which there is a set E of linear inequations of the form $f(\vec{x}) \lhd g(\vec{x})$, where $\lhd$ is either $\leq$ or $<$, such that the following conditions hold:

- For any inequation $e \equiv f(\vec{x}) \lhd g(\vec{x})$ in E there is a natural number $i(e) \in \{0, 1, \cdots, n+1\}$ such that for all $j \leq n$, if either f or g depends on x_j, then for all $(\vec{a}) \in C$, $i(a_j) = i(e)$.
- For every inequation $e \equiv f(\vec{x}) \lhd g(\vec{x})$ in E, there are variables $y_1, \cdots, y_k$ among $\vec{x}$ and McNaughton's functions $h(y_1, \cdots, y_k)$ and $r(y_1, \cdots, y_k)$ such that $f(\vec{x})$ and $g(\vec{x})$ depend only on $y_1, \cdots, y_k$, and for all

$(x_1, \cdots, x_n) \in C$ one has:
$f(\vec{x}) = h_{i(e),n}(y_1, \cdots, y_k)$, and $g(\vec{x}) = k_{i(e),n}(y_1, \cdots, y_k)$. Moreover, if $i(e) > 0$, then
$h(1, \cdots, 1) = r(1, \cdots, 1) = 1$

- Y is the set of all elements of C which are solutions of all inequations in E.

Lemma 4. *For every linear semialgebraic subset Y of a cell C, there is a term $p_Y(\vec{x})$ of BL_Δ algebras such that for all $\vec{a} \in ((n+1)[0,1])^n$ one has: $(\vec{a}) \in Y$ iff $p_Y(\vec{a}) = 1$*

Proof. First of all, we prove the claim in the case where Y is the whole of C. Recall that C is uniquely determined by the relations R and S on $X_n \equiv \{\vec{x}, 0, n+1\}$ defined according to Lemma 2, (v). Let $L(x,y)$ denote the term $\neg\Delta(x) \wedge \Delta((y \to x) \to x)$. It is readily seen that $L(x,y) = n+1$ if $x_j R x_j$, and $L(x,y) = 0$ otherwise. Now let $E(x,y)$ denote the term $\neg(L(x,y) \wedge \neg L(y,x))$. Then $E(x,y) = n+1$ if xSy, and $E(x,y) = 0$ otherwise. Now define

$$p_Y(\vec{x}) = p_C(\vec{x}) = (\bigwedge_{x,y \in X_n : xRy} L(x,y)) \wedge (\bigwedge_{x,y \in X_n : xSy} E(x,y)).$$

Now let $e \in E$. By Lemma 3, if e has the form $f_{i(e),n} \leq g_{i(e),n}$, then the set of elements of C which satisfy e is the set of all $(\vec{a})$ such that $p_C(\vec{a}) \wedge \Delta(t_f(\vec{a}) \to t_g(\vec{a})) = n+1$ (where t_f and t_g are related to f and g according to Lemma 3). Moreover, if e has the form $f_{i(e),n} < g_{i(e),n}$, then the set of elements of C which satisfy e is the set of all $(\vec{a})$ such that $p_C(\vec{a}) \wedge \neg\Delta(t_g(\vec{a}) \to t_f(\vec{a})) = n+1$. Let p_e denote $p_C \wedge \Delta(t_f \to t_g)$ if e is $f_{i(e),n} \leq g_{i(e),n}$, and $p_C \wedge \neg\Delta(t_g \to t_f)$ if e is $f_{i(e),n} < g_{i(e),n}$. Then, it suffices to define:

$$p_Y(\vec{x}) = p_C(\vec{x}) \wedge \bigwedge_{e \in E} p_e(\vec{x}). \qquad \blacksquare$$

Definition 10. A BL_Δ *partition* of $I(n)$ is a partition $\mathcal{P}$ of $I(n)$ into semialgebraic subsets of cells.
An *elementary BL_Δ function* is a function f whose domain is an algebraic subset Y of some cell C, and such that there are an $i(f) \in \{1, \cdots, n+1\}$ and a Mc Naughton function $g(\vec{x})$ satisfying the following conditions:

- For $j = 1, \cdots, n$, if g depends on x_j, then for all $(\vec{a}) \in C^n$ $i(a_j) = i(f)$.
- $f = g_{i(f),n}$

If an elementary BL_Δ function f defined on Y is constant on Y, then f does not depend on any variable, and $i(f)$ might be defined arbitrarily. However, in this case we *define* $i(f)$ to be the constant value (0 or $n+1$) of f. We are ready to prove the central result of this paper.

Theorem 3. *The free BL_Δ algebra on n generators $\mathcal{B}_\Delta(n)$ can be represented as follows:*

- *The domain of $\mathcal{B}_\Delta(n)$ is the set $D(n)$ of all functions from $I(n)$ into $[0, n+1]$ for which there are a BL_Δ partition $\mathcal{P} = \{P_1, \cdots, P_m\}$ of $I(n)$ and elementary BL_Δ functions $f_1, \cdots, f_m$ with the following properties:*
  (a) For $i = 1, \cdot, m$, the domain of f_i is P_i.
  (b) For $i = 1, \cdot, m$, the restriction of f to P_i is f_i.
- *The operations $\star, \Rightarrow$, and Δ on $\mathcal{B}_\Delta(n)$ are defined as follows:*
  (i) $(f \star g)(\vec{x}) = f(\vec{x}) \cdot g(\vec{x})$.
  (ii) $(f \Rightarrow g)(\vec{x}) = f(\vec{x}) \to g(\vec{x})$.
  (iii) $(\Delta(f))(\vec{x}) = \begin{cases} n+1 & \text{if } f(\vec{x}) = n+1 \\ 0 & \text{otherwise.} \end{cases}$

Proof. That the operations are defined according to (i), (ii) and (iii) depends on the fact that $\mathcal{B}_\Delta(n)$ is a subalgebra of $(n+1)[0, 1]^{I(n)}$. Thus, we only have to prove that the domain of $\mathcal{B}_\Delta(n)$ is $D(n)$.

If $f \in D(n)$, then there are a BL_Δ partition $\mathcal{P} = \{P_1, \cdots, P_m\}$ of $I(n)$ and elementary BL_Δ functions $f_1, \cdots, f_m$ satisfying properties (a) and (b) of the present theorem. It follows that

$$f(\vec{x}) = \bigvee_{j=1}^{m} (p_{P_j}(\vec{x}) \wedge t_{f_j}(\vec{x})).$$

where for $j = 1, \cdots, n$, p_{P_j} is defined according to Lemma 4, and t_{f_j} is defined according to Lemma 3. Thus every function in $D(n)$ is an element of the subalgebra of $((n+1)[0, 1])^{I(n)}$ generated by the projections.

To the other direction, we prove that $D(n)$ contains the projection functions and the constants, and is closed under the operations $\star, \Rightarrow$ and Δ. That projection functions and constants are in $D(n)$ is trivial. Let $f(\vec{x}), g(\vec{x}) \in D(n)$, and let $\mathcal{P} = \{P_1, \cdots, P_m\}$, $\mathcal{Q} = \{Q_1, \cdots, Q_h\}$ be BL_Δ partitions, and $f_1, \cdots f_m, g_1, \cdots, g_h$ be BL_Δ elementary functions satisfying condition (a) of the present theorem, and such that for $j = 1, \cdots, m$, and for $k = 1, \cdots, h$, f coincides with f_j in P_j, and g coincides with g_k in Q_k. Let $R_{j,k} = P_j \cap Q_k$. Then $\{R_{j,k} : j = 1, \cdots, m; k = 1, \cdots, h\}$ is a BL_Δ partition of $I(n)$. Now let

$$U(f_j) = \{(\vec{x}) \in P_j : f_j(\vec{x}) = n+1\}, \, U(g_k) = \{(\vec{x}) \in Q_k : g_k(\vec{x}) = n+1\}$$

$$E(j, k) = \{(\vec{x}) \in R_{j,k} : f_j(\vec{x}) \leq g_k(\vec{x})\}, \, E_1(j, k) = R_{j,k} - U(f_j),$$

$$E_2(j, k) = R_{j,k} - U(g_k), \, E_3(j, k) = R_{j,k} \cap U(f_j) \cap U(g_j),$$

$$E_4(j, k) = R_{j,k} - U(f_j) - U(g_k), \text{ and } E_5(j, k) = R_{j,k} - E(j, k) - U(f_j).$$

It is easily seen that $U(f_j), U(g_k), E_1(j, k), E_2(j, k)$ and $E_3(j, k), E_4(j, k)$ are linear semialgebraic sets. Moreover, if $i(f_j) = i(g_k)$, then $E(j, k)$ and $E_5(j, k)$ are also linear semialgebraic.

Now it is easily seen that if f and g are elementary BL_Δ functions defined on a semialgebraic set Y, if $f \neq n+1$ and $g \neq n+1$ in Y, and $i(f) = i(g)$, then $f \odot_{i(f)} g$ is also elementary. If in addition $g < f$ in Y, then $f \rightarrow_{i(f)} g$ is in turn elementary. (Cf. Notation 1 for the definitions of $\odot_{i(f)}$ and $\rightarrow_{i(f)}$)

At this point, if $(\vec{x}) \in R_{j,k}$ and $i(f_j) = i(g_k)$, then:

$$f(\vec{x}) \cdot g(\vec{x}) = \begin{cases} f_j(\vec{x}) \odot_{i(f_j)} g(\vec{x}) & \text{if } (\vec{x}) \in E_4(j,k) \\ f_j(\vec{x}) & \text{if } (\vec{x}) \in U(g_k) \\ g_k(\vec{x}) & \text{if } (\vec{x}) \in U(f_j) \end{cases}$$

$$f(\vec{x}) \rightarrow g(\vec{x}) = \begin{cases} n+1 & \text{if } (\vec{x}) \in E(j,k) \\ g_k(\vec{x}) & \text{if } (\vec{x}) \in U(f_j) \\ f_j(\vec{x}) \rightarrow_{i(f_j)} g_k(\vec{x}) & \text{if } (\vec{x}) \in E_5(j,k) \end{cases}$$

If $i(f_j) < i(g_k)$, then

$$f(\vec{x}) \cdot g(\vec{x}) = \begin{cases} f_j(\vec{x}) & \text{if } (\vec{x}) \in E_1(j,k) \\ g_k(\vec{x}) & \text{if } (\vec{x}) \in U(f_j) \end{cases}$$

$$f(\vec{x}) \rightarrow g(\vec{x}) = \begin{cases} n+1 & \text{if } (\vec{x}) \in E_1(j,k) \\ g_k(\vec{x}) & \text{if } (\vec{x}) \in U(f_j) \end{cases}$$

If $i(g_k) < i(f_j)$, then

$$f(\vec{x}) \cdot g(\vec{x}) = \begin{cases} g_k(\vec{x}) & \text{if } (\vec{x}) \in E_2(j,k) \\ f_j(\vec{x}) & \text{if } (\vec{x}) \in U(g_k) \end{cases}$$

$$f(\vec{x}) \rightarrow g(\vec{x}) = g_k(\vec{x})$$

Finally, in every case

$$\Delta(f(\vec{x})) = \begin{cases} n+1 & \text{if } (\vec{x}) \in U(f_j) \\ 0 & \text{if } (\vec{x}) \in E_1(j,k) \end{cases}$$

Thus for every element f of $\mathcal{B}_\Delta(n)$ there are a partition $\mathcal{P} = \{P_1, \cdots, P_m\}$ of $(n+1)[0,1]^I(n)$ and elementary BL_Δ functions $f_1, \cdots, f_m$ satisfying conditions (a) and (b) of the present theorem, i.e. every element of $\mathcal{B}_\Delta(n)$ is in $D(n)$. ∎

Examples.

- The cells corresponding to $\mathcal{B}_\Delta(2)$ are: I_0^2, $I_0 \times I_1$, $I_0 \times I_3$, I_1^2, $I_1 \times I_0$, $I_1 \times I_2$, $I_2 \times I_1$, $I_1 \times I_3$, $I_3 \times I_1$, and $I_3^2 = \{3,3)\}$.
- The function corresponding to $x \rightarrow y$ is equal to 3 in $I_0 \times I_j$, $(j = 1,3)$, in $I_1 \times I_h$, $(h = 2,3)$, and in the linear algebraic subsets of I_0^2 and of I_1^2 defined by the inequation $x \leq y$. $x \rightarrow y = y$ in $I_j \times I_0$ $(j = 1,2)$ and in $I_1 \times I_h$ $(h = 2,3)$. Finally in the linear semialgebraic subset of I_0^2 defined by $y < x$, $x \rightarrow y = y + 1 - x$, and in the linear semialgebric subset of I_1^2 defined by $y < x$, $x \rightarrow y = y + 2 - x = x \rightarrow_1 y$.
- There is no term whose interpretation in $\mathcal{B}_\Delta(2)$ depends on both x, y in $I_0 \times I_1$, or in $I_0 \times I_2$.
  There is no term whose interpretation $f(x,y)$ in $I_0 \times I_1$ is, say, 0 if $3x \leq y$, and $n+1$ otherwise (the inequation $3x \leq y$ does not define a linear

semialgebraic subset of $I_0 \times I_1$).
There is no term whose interpretation in $B_\Delta(2)$ is equal to $2 - x$ in I_1^2
($\neg x$ is 0 if $x \geq 1$).

References

[AFM99] P. AGLIANÓ, I.M.A. FERREIRIM, F. MONTAGNA *Basic hoops: an algebraic study of continuous t-norms*, Preprint 1999.

[AM00] P. AGLIANÓ, F. MONTAGNA *Varieties of basic algebras*, Preprint 2000.

[BO] J. R. BÜCHI AND OWENS, *Complemented monoids and groups*, unpublished manuscript.

[Ba96] M. BAAZ: Infinite-valued Gödel logics with 0-1 projections and relativizations. In GÖDEL'96 - Logical foundations of mathematics, computer science and physics. Lecture Notes in Logic 6 (1996), P. Hájek (Ed.), Springer Verlag, pp.23–33.

[BF96] W.J. BLOK, I.M.A. FERREIRIM *On the structure of hoops*, to appear in Algebra Universalis.

[BF93] W.J. BLOK, I.M.A. FERREIRIM, *Hoops and their implicational reducts*, Algebraic Logic, Banach Center pub. 28, Warsaw 1993.

[BP94] W.J. BLOK, D. PIGOZZI, *On the structure of varieties with equationally definable principal congruences III*, Algebra Universalis 32, 545-608.

[Ch58] C. C. CHANG *Algebraic analysis of many-valued logics*, Trans Amer. Math. Soc. 88, 467-490.

[Ch59] C. C. CHANG *A new proof of the completeness of Lukasiewicz axioms*, Trans Amer. Math. Soc. 93, 74-80.

[CEGT99] R. CIGNOLI, F. ESTEVA, L. GODO, A. TORRENS, *Basic fuzzy logic is the logic of continuous t-norms and their residua*, to appear in Soft Computing.

[COM95] R. CIGNOLI, I.M.L. D'OTTAVIANO, D. MUNDICI, *Algebras das Logicas de Lukasiewicz* , Centro de Logica, Epistemologia e Historia de Ciencia UNICAMP, 1995, Campinas, (Brazil).

[CT98] R. CIGNOLI, A. TORRENS, *An algebraic analysis of Product Logic*, to appear in Multiple-valued Logic.

[H98] P. HÁJEK, *Metamathematics of Fuzzy Logic*, Kluwer, 1988.

[Ha98] P. HÁJEK, *Basic fuzzy logic and BL algebras*, Soft Computing 2 (1998), 124-128.

[MMT] R. MACKENZIE, G. MCNULTY, W. TAYLOR, *Algebras, Lattices, Varieties*, Vol I, Wadsworth and Brooks/Cole, Monterey CA, 1987.

[MN51] R. MCNAUGHTON, *A theorem about infinite-valued sentential logic*, Journ. Symb. Log. 16 (1951), 71-83.

[MP99] F. MONTAGNA, G. PANTI, *Adding structure to MV algebras*, Preprint 1999.

[Mu86] D. MUNDICI, *Interpretation of AFC* algebras in Lukasiewicz sentential calculus*, Journal of Functional Analysis, 65 (1986) 15-63.

Natural Duality as a Tool to Study Algebras Arising from Logics

Philippe Niederkorn

Algebra and Logic, University of Liège

Abstract. MV-algebras are the algebraic counterpart of Łukasiewicz's infinite-valued logic, just as Boolean algebras correspond to the classical propositional calculus. The finitely generated subvarieties of the variety $\mathcal{M}$ of all MV-algebras are generated by a finite number of finite chains.

We present Davey and Werner's theory of natural duality, illustrated by its application to a few classes of algebras arising from classical and non-classical logics. We insist on the subvarieties of $\mathcal{M}$ generated by one finite chain, for which some simple applications of the dualities are proposed.

Introduction

The theory of natural dualities has been introduced by B. A. Davey and H. Werner in a huge foundation paper ([5]), and further developed by several authors, including D. M. Clark, B. A. Davey, H. A. Priestley and H. Werner. It turns out to be a very powerful tool for the study of many classes of algebras, including those algebras arising from classical and non-classical logics. The main purpose of the present paper is to provide some evidence to sustain this claim.

To this end, we present the well-known duality discovered by Stone in 1936 between Boolean algebras and certain topological spaces known as Boolean spaces. We then proceed to show how this particular example can be generalised to other classes of algebras, and we illustrate the theory, along with some sample applications of it, by applying it to the varieties of Post algebras of order n, and to the varieties of MV-algebras generated by one finite chain.

MV-algebras were introduced by C. C. Chang ([2]) as the algebraic counterpart of Łukasiewicz's many-valued propositional calculus. Let us recall that an *MV-algebra* is an algebra $\langle A; \oplus, \odot, \neg, 0, 1 \rangle$ of type $(2, 2, 1, 0, 0)$ such that $\langle A; \oplus, 0 \rangle$ is an Abelian monoid, and satisfying the following identities : $\neg\neg x = x$, $x \oplus 1 = 1$, $\neg 0 = 1$, $x \odot y = \neg(\neg x \oplus \neg y)$, $(x \odot \neg y) \oplus y = (y \odot \neg x) \oplus x$. For instance, the unit interval $[0, 1]$ is an MV-algebra when endowed with the operations $x \oplus y := \min(x+y, 1)$, $x \odot y := \max(x+y-1, 0)$, $\neg x := 1-x$. For each positive integer n we denote by $\underline{L}_n$ the subalgebra $\{0, \frac{1}{n}, \ldots, \frac{n-1}{n}, 1\}$ of $[0, 1]$.

The subvarieties $\mathbb{HSP}(\underline{L}_n)$ of the variety $\mathcal{M}$ of all MV-algebras, henceforth denoted by $\mathcal{M}_n$, will be our main illustration of the theory of natural duality. We describe their injective and finite projective members, their free algebras, and their algebraically and existantially closed members. To end with, we give two results about the automorphism group and the endomorphism monoid of the finite MV-algebras. Most results are not new, but the simple proofs we provide all rely on the natural dualities we establish, and will therefore hopefully contribute to demonstrate the usefulness of such dualities.

1 Boolean algebras

A *Boolean algebra* is an algebra $\langle B; \vee, \wedge, ', 0, 1 \rangle$ of type $(2, 2, 1, 0, 0)$ such that $\langle B; \vee, \wedge, 0, 1 \rangle$ is a bounded distributive lattice and a' is the (necessarily unique) complement of a in B. They are intimately connected with the classical propositional calculus, since the Lindenbaum algebras of this calculus are Boolean algebras under obvious operations.

We first recall the duality, discovered by Stone in 1936, between the class of Boolean algebras and a class of topological spaces known as Boolean spaces. A subset F of a Boolean algebra B is called a *filter* provided F is increasing (if $a \in F$ and $b \geq a$, then $b \in F$) and closed under $\wedge$. A filter F of a Boolean algebra B is called *proper* provided it is properly included in B. A maximal proper filter (with respect to set inclusion) is called an *ultrafilter*. Now comes Stone's idea. We denote by $\mathrm{Spec}(B)$ the *spectral space* of a Boolean algebra B, that is the set of its ultrafilters endowed with the topology whose subbasis is given by the sets.

$$r(b) := \{U \mid U \text{ is an ultrafilter of } B \text{ and } b \in U\}$$

for each element b of B. Since $r(b') = \mathrm{Spec}(B) \setminus r(b)$, each $r(b)$ is a *clopen* set, that is a set which is simultaneously closed and open. Actually they are the only clopen sets in $\mathrm{Spec}(B)$, and B can be recovered from its spectral space, as is stated in the following result.

Theorem 1 (Stone Representation Theorem [15]). *Let B be a Boolean algebra, and $\mathrm{Spec}(B)$ its spectral space. Then the algebra $cl(\mathrm{Spec}(B))$ of clopen subsets of $\mathrm{Spec}(B)$ is a Boolean algebra when endowed with the two binary operations of set union $\cup$ and set intersection $\cap$, the unary operation of set complementation with respect to $\mathrm{Spec}(B)$, and the two constants $\emptyset$ and $\mathrm{Spec}(B)$. Moreover, $B \cong cl(\mathrm{Spec}(B))$.*

The previous result thus gives a *representation* of each Boolean algebra as an algebra of clopen subsets of its spectral space. But much more can be said. First of all, the class of spectral spaces of Boolean algebras can be caracterized topologically. They are exactly the *Booelan spaces*, that is the compact Hausdorff spaces having a basis of clopen subsets.

More important maybe is the fact that the homomorphisms between two Boolean algebras can also be interpreted. Let B_1 and B_2 be two Boolean algebras, and $h : B_1 \to B_2$ be a homomorphism. Define $h^* : \mathrm{Spec}(B_2) \to \mathrm{Spec}(B_1)$ by $h^*(U) := h^{-1}(U)$. Then h^* is a continuous map. This lifts up the representation obtained above to a *duality* (known as *Stone's duality*), and $\mathrm{Spec}(B)$ is called the *dual* of B. Notice that the duality reverses arrows, since a homomorphism from B_1 into B_2 gives rise to a continuous map from $\mathrm{Spec}(B_2)$ into $\mathrm{Spec}(B_1)$.

To end with, the correspondance between Boolean algebras and Boolean spaces is complete. Every Boolean space is the spectral space of a Boolean algebra : if X is a Boolean space, then X is homeomorphic to $\mathrm{Spec}(\mathrm{cl}(X))$. Every continuous map between two Boolean spaces arises from a homomorphism between the corresponding Boolean algebras. Moreover, ono-to-one (resp. onto) homomorphisms correspond to onto (resp. one-to-one) continuous maps.

2   Natural dualities

Stone's duality for Boolean algebras, along with the important historical example of Priestley's duality for bounded distributive lattices [12,?] (see also [4]), has led Davey and Werner [5] to discover a very general theory of duality. The so-called natural duality theory can be applied to many classes of algebras, particularly to classes of algebras arising from the algebraisation of non-classical logics. The essential reference on the subject is Clark and Davey's monography [3].

To expalin the theory, let us come back again to Stone's duality, and restate it in a different way. We denote by $\mathbf{2}$ the two-element Boolean algebra, and by $\underset{\sim}{\mathbf{2}}$ the two-element discrete topological space. Starting from a Boolean algebra B, let $H(B)$ be the set of homomorphisms from B into $\mathbf{2}$, endowed with the topology inherited from the product topology on $\mathbf{2}^B$. The function mapping a homomorphism $h : B \to \mathbf{2}$ to its cokernel $h^{-1}(1)$ is obviously a bijection from $H(B)$ onto $\mathrm{Spec}(B)$. Less obvious, but still true, is the fact that this function is a homeomorphism, so that $H(B)$ can replace $\mathrm{Spec}(B)$ as the dual of B.

Symetrically, let $C(X)$ be the set of continuous functions from a Boolean space X into $\underset{\sim}{\mathbf{2}}$, endowed with the componentwise operations inherited from the Boolean algebra operations on $\mathbf{2}$. Again, the function mapping a continuous function $\alpha : X \to \underset{\sim}{\mathbf{2}}$ to its cokernel $\alpha^{-1}(1)$ is not only a bijection from $C(X)$ onto $\mathrm{cl}(X)$, but also a Boolean algebra isomorphism.

Finally, the theorem 1 can be rephrased : if B is a Boolean algebra, then $B \cong C(H(B))$. An explicit isomorphism is given by the *evaluation map*

$$e_B : B \to C(H(B)) : b \mapsto e_B(b) \quad \text{where} \quad e_B(b) : H(B) \to \underset{\sim}{\mathbf{2}} : h \mapsto h(b).$$

Moreover, if X is a Boolean space, then $X \cong H(C(X))$ and an explicit isomorphism is given by

$$\varepsilon_X : X \to H(C(X)) : x \mapsto \varepsilon_X(x) \text{ where } \varepsilon_X(x) : C(X) \to \underline{2} : \alpha \mapsto \alpha(x).$$

What makes things work is the fact that $\underline{2}$ and $\widetilde{2}$ are two structures defined on the same universe (a two-element set). This object is said to have a *schizophrenic nature* : we consider it sometimes as a two-element Boolean algebra, sometimes as a two-element discrete topological space.

This formulation has a major advantage over the one presented in the previous section. Instead of using the concepts of ultrafilter (which is specific to Boolean algebras) and clopen set (typical of Boolean spaces), the duality is described in term of homomorphisms (resp. continuous maps) into a particular Boolean algebra (resp. Boolean space). Being no longer specific to Boolean algebras, the theory can be extended to various other algebras. We now proceed to present it.

First we start from an algebra $\underline{M}$ (to play the same role as $\underline{2}$ in Stone's duality). If we are to represent some algebras in a way that mimics what we have described above, then those algebras will be isomorphic to a subalgebra of a power of $\underline{M}$ (in Stone's duality, a Boolean algebra B is isomorphic to $C(H(B))$, which is a subalgebra of $\underline{2}^{H(B)}$). Therefore we consider the class $\mathcal{A} := \mathbb{ISP}(\underline{M})$. We also need another structure $\underline{M}$ (corresponding to $\widetilde{2}$ in Stone's duality) defined on the universe M of $\underline{M}$. It won't be enough just to endow M with the discrete topology (as in the Boolean case) : the resulting theory would only apply to a very narrow range of algebras $\underline{M}$, known as *primal algebras* (see later). The solution is to define $\underline{M}$ as a relational topological structure. This means that we endow M, not only with the discrete topology, but also with finitary relations. To sum up we have $\underline{M} = \langle M; R, \tau \rangle$, where R is a (possibly infinite) set of finitary relations on M (a n-ary relation on M is just a subset of M^n).

Note that a general presentation of natural duality should include operations and partial operations in the type of $\underline{M}$. This would not cause any major modification in the presentation, but since we don't need them for our purposes we prefer to avoid the notational complications that their inclusion would incur.

Now that we have our schizophrenic pair $\underline{M}$ and $\widetilde{M}$, we can try to mimic the definitions of H and C in the Boolean case. Therefore we define D on $\mathcal{A}$ so that $D(A)$ has as its universe the set of $\mathcal{A}$-homomorphisms from A into $\underline{M}$, that we denote by $\mathcal{A}(A, \underline{M})$. As in Stone's duality we endow this set with the topology inherited from the product topology on $\underline{M}^A$. But since $\widetilde{M}$ also includes relations, we endow $\mathcal{A}(A, \underline{M})$ with the componentwise relations inherited from the relations on $\widetilde{M}$. More explicitly, if r is a n-ary member of R then the corresponding relation $r_{D(A)}$ on $\mathcal{A}(A, \underline{M})$ contains those n-tuples $(h_1, \ldots, h_n)$ of homomorphisms from A into $\underline{M}$ such that for all a in A we have that $(h_1(a), \ldots, h_n(a))$ is a member of r. With this definition, the *dual* $D(A)$ of an algebra A of $\mathcal{A}$ is a topological structure of the same type as $\underline{M}$, in symbols

$$D(A) = \langle \mathcal{A}(A, \underline{M}); \{r_{D(A)} \mid r \in R\}, \tau \rangle,$$

and it is not very hard to check that $D(A)$ is a (topologically) closed substructure of $\underline{M}^A$. Therefore the class $\mathcal{X} := \mathbb{IS}_c\mathbb{P}(\underline{M})$ of topological structures of the same type as $\underline{M}$ which are isomorphic to a closed substructure of a power of $\underline{M}$ appears as a natural candidate to 'house' our dual structures.

If $\mathbf{X}$ and $\mathbf{Y}$ are two topological structures of $\mathcal{X}$, then a map $\varphi : X \to Y$ is a *morphism* from $\mathbf{X}$ into $\mathbf{Y}$ provided it is continuous and preserves their relations. This last condition means that for each n-ary relation r of R we have that $(\varphi(x_1), \ldots, \varphi(x_n))$ belongs to $r_\mathbf{Y}$ whenever $(x_1, \ldots, x_n)$ belongs to $r_\mathbf{X}$. An *isomorphism* in $\mathcal{X}$ is a bijective morphism φ such that φ^{-1} is also a morphism. An *embedding* in $\mathcal{X}$ is an isomorphism onto a closed substructure. If $\mathbf{X}$ and $\mathbf{Y}$ are two members of $\mathcal{X}$, then the set of morphisms from $\mathbf{X}$ into $\mathbf{Y}$ is denoted by $\mathcal{X}(\mathbf{X}, \mathbf{Y})$.

As one could expect, we now define E on $\mathcal{X}$ so that $E(\mathbf{X})$ has $\mathcal{X}(\mathbf{X}, \underline{M})$ as its universe. As in the Boolean cas we endow this set with componentwise operations inherited from the operations on $\underline{M}$. Of course we hope that if $\mathbf{X}$ is a member of $\mathcal{X}$, then $E(\mathbf{X})$ will be an algebra in $\mathcal{A}$. This will only be true if we impose some conditions on the set R of relations that defines $\underline{M}$. A n-ary relation on M (i.e. a subset of M^n) is called *algebraic* (over $\underline{M}$) provided it is an $\mathcal{A}$-sublagebra of $\underline{M}^n$, and $\underline{M}$ is called *algebraic over* $\underline{M}$ provided each relation r of R is algebraic. This last condition is equivalent to the fact that $E(\mathbf{X})$ belongs to $\mathcal{A}$ for each member $\mathbf{X}$ of $\mathcal{X}$. We now have the following.

Theorem 2 ([5]). *Assume that the structure on $\underline{M}$ is algebraic over $\underline{M}$:*

(i) *for all $A \in \mathcal{A}$ define $D(A) := \mathcal{A}(A, \underline{M})$ and for all $A, B \in \mathcal{A}$ and each $u \in \mathcal{A}(A, B)$ define $D(u) : D(B) \to D(A) : x \mapsto x \circ u$,*

(ii) *for all $\mathbf{X} \in \mathcal{X}$ define $E(\mathbf{X}) := \mathcal{X}(\mathbf{X}, \underline{M})$ and for all $\mathbf{X}, \mathbf{Y} \in \mathcal{X}$ and each $\varphi \in \mathcal{X}(\mathbf{X}, \mathbf{Y})$ define $E(\varphi) : E(\mathbf{Y}) \to E(\mathbf{X}) : \alpha \mapsto \alpha \circ \varphi$,*

(iii) *for all $A \in \mathcal{A}$ and $\mathbf{X} \in \mathcal{X}$ define the evaluations $e_A : A \to ED(A)$ by $e_A(a)(x) := x(a)$ and $\varepsilon_\mathbf{X} : \mathbf{X} \to DE(\mathbf{X})$ by $\varepsilon_\mathbf{X}(x)(\alpha) := \alpha(x)$.*

Then $D : \mathcal{A} \to \mathcal{X}$ and $E : \mathcal{X} \to \mathcal{A}$ are contravariant functors which are adjoint to each other with e and ε as the natural transformations which act as the units of the adjunction. Moreover for all $A \in \mathcal{A}$ and $\mathbf{X} \in \mathcal{X}$ the evaluations e_A and $\varepsilon_\mathbf{X}$ are embeddings.

In order to have a duality, we need that the evaluations be not only embeddings, but isomorphisms. Unfortunately, this will in general not be the case, and it will heavily depend on the choice we make of the structure $\underline{M}$. More precisely, we will say that $\underline{M}$ (or R) yields a *natural duality* on $\mathcal{A}$ provided the evaluation map e_A is an isomorphism for every algebra A in $\mathcal{A}$. Whenever this occurs we say that A (or $\underline{\mathrm{M}}$) is *dualisable*. The duality will be called *full* provided the evaluation map $\varepsilon_{\mathbf{X}}$ is also an isomorphism for every structure $\mathbf{X}$ in $\mathcal{X}$. A *strong* duality is a full duality obtained in a special way. The formal definition doesn't matter to us : all we need to know is that a strong duality is a full duality such that $\underline{\mathrm{M}}$ is injective in $\mathcal{X}$ (see later). One of the main purposes of the theory has been to give sufficient conditions on the structure of $\underline{\mathrm{M}}$, depending on the algebra $\underline{\mathrm{M}}$, for the existence of a natural duality (resp. a strong duality) yielded by $\underline{\mathrm{M}}$ on $\mathcal{A}$. Not every finite algebra is dualisable, while some of them admit a very simple strong duality.

3   Dualities for Post algebras and MV-algebras

The first family of dualisable algebras we are going to consider are the above-mentioned *primal* algebras. Let us recall that an algebra is called primal provided it is a finite algebra such that every finitary operation on its universe is a term function. A well-known example of primal algebra is $\mathbf{2}$, the two-element Boolean algebra. More generally we consider the finite Post chains : for each positive integer n we denote by $\underline{\mathrm{P}}_n$ the $(n+1)$-element Post algebra

$$\underline{\mathrm{P}}_n = \langle \{0, \frac{1}{n}, \ldots, 1\}; \vee, \wedge, {}^*, D_1, \ldots, D_n, c_0, \ldots, c_n \rangle$$

where $\vee$ and $\wedge$ correspond to the chain order $0 < \frac{1}{n} < \ldots < 1$, the constants c_i satisfy $c_i = \frac{i}{n}$, and the additional unary operations are given by

$$a^* = \begin{cases} 1 \text{ if } a = 0 \\ 0 \text{ if } a \neq 0 \end{cases} \quad \text{and} \quad D_i(a) = \begin{cases} 1 \text{ if } a \geq \frac{i}{n} \\ 0 \text{ if } a < \frac{i}{n} \end{cases}.$$

Remark that every primal algebra is obviously term-equivalent to one of the $\underline{\mathrm{P}}_n$ (depending on its cardinality). We refer to [1] and [14], and to the bibliography therein, for a detailled discussion on Post algebras. We just would like to point out that, extending Stone's duality, T. K. Hu proves in [10] that the discrete $(n+1)$-element topological space $\mathbf{P}_n$ yields a strong duality on the variety $\mathcal{P}_n := \mathbb{I}\mathbb{S}\mathbb{P}(\underline{\mathrm{P}}_n)$ of Post algebras of order $n+1$. It turns out that for each positive integer n the class $\mathbb{I}\mathbb{S}_c\mathbb{P}(\mathbf{P}_n)$ is the class of Boolean spaces, that we henceforth denote by $\mathcal{Z}$. The corresponding functors will be denoted by $S_n : \mathcal{P}_n \to \mathcal{Z}$ and $T_n : \mathcal{Z} \to \mathcal{P}_n$. Notice that $\underline{\mathrm{P}}_1$ is $\mathbf{2}$, the 2-element Boolean algebra, and that S_1 and T_1 actually define Stone's duality (i.e. $S_1 = H$ and $T_1 = C$). These examples are special indeed : the simplest possible topological structure (namely $R = \emptyset$) yields a strong duality on the given algebra. They actually are the only ones, since an algebra for which it is true has to be primal.

Next we turn to a slightly more general class of algebras for which a strong duality is easily obtained : the *semi-primal* algebras. Let us recall that an algebra $\underline{M}$ is called a semi-primal algebra provided it is finite, the ternary discriminator operation $t(x,y,z) := \begin{cases} x \text{ if } x \neq y \\ z \text{ if } x = y \end{cases}$ is a term function on $\underline{M}$, and there are no isomorphisms between subalgebras of $\underline{M}$ with more than one element other than the identity maps. As noted in [3, 3.3.14], those algebras admit a simple strong duality, and the corresponding dual classes are easy to axiomatize.

To illustrate this result, we now consider the above-mentioned varieties $\mathcal{M}_n=\mathbb{HSP}(\underline{L}_n)$ of MV-algebras. As it is proven in [11], the algebra $\underline{L}_n$ is a semi-primal algebra for each positive integer n, so that $\mathcal{M}_n=\mathbb{ISP}(\underline{L}_n)$. We define the topological structure

$$\underline{L}_n = \langle L_n; \mathbb{S}(\underline{L}_n), \tau_d \rangle,$$

where τ_d is the discrete topology, and R contains each subalgebra of $\underline{L}_n$ as a unary relation. The structure $\underline{L}_n$ is obviously algebraic over $\underline{L}_n$. We denote by $\mathcal{X}_n$ the topological quasi-variety $\mathbb{IS}_c\mathbb{P}(\underline{L}_n)$, and by $D_n : \mathcal{M}_n \to \mathcal{X}_n$ and $E_n : \mathcal{X}_n \to \mathcal{M}_n$ the corresponding hom-defined functors. We then have the following result.

Theorem 3 ([11]). *For each positive integer n, the structure $\underline{L}_n$ yields a strong duality on $\mathcal{M}_n$. A topological structure $\langle X; \{r_m \mid m \in div(n)\}, \tau \rangle$ of the same type as $\underline{L}_n$ (i.e. whose relations are all unary) belongs to $\mathcal{X}_n$ iff*

(i) *$\langle X; \tau \rangle$ is a Boolean space,*
(ii) *for each $m \in div(n)$, the relation r_m is a closed subspace of $\langle X; \tau \rangle$,*
(iii) *for all $m, m' \in div(n)$, we have $r_m \cap r_{m'} = r_{gcd(m,m')}$.*

where $div(n)$ denotes the set of positive divisors of n and $gcd(m, m')$ denotes the greatest common divisor of m and m'.

Remark that when describing a member of $\mathcal{X}_n$ explicitly by

$$\mathbf{X} = \langle X; \{r_m \mid m \in div(n)\}, \tau \rangle,$$

we mean that r_m is the (unary) relation of $\mathbf{X}$ corresponding to the relation $\underline{L}_m$ of $\underline{L}_n$.

Since the dualities of Theorem 3 are strong, they turn products in $\mathcal{M}_n$ (resp. $\mathcal{X}_n$) to coproducts in $\mathcal{X}_n$ (resp. $\mathcal{M}_n$) and vice versa. Moreover, finite coproducts in $\mathcal{X}_n$ are given by disjoint union (as it is the case in $\mathcal{Z}$). This allows a direct computation of the duals of finite members of $\mathcal{M}_n$. If m divides n, then the only homomorphism from $\underline{L}_m$ into $\underline{L}_n$ is the canonical embedding. Therefore, $D_n(\underline{L}_m)$ is a one-point discrete space such that the relation r_k on it corresponding to $\underline{L}_k$ contains this only point if and only if m divides k (and is empty otherwise).

More generally, each finite member of $\mathcal{M}_n$ is isomorphic to a direct product of finitely many (non-trivial) subalgebras of $\underline{L}_n$. Consequently, the dual of such an algebra A is a finite discrete space containing one point for each factor in this product. Again, the point corresponding to a factor $\underline{L}_m$ belongs to the relation r_k of $D_n(A)$ corresponding to $\underline{L}_k$ if and only if m is a divisor of k.

Conversely, each finite member $\mathbf{X}$ of $\mathcal{X}_n$ has a dual $E_n(\mathbf{X})$ which is isomorphic to a direct product of as many subalgebras of $\underline{L}_n$ as there are points in X. The factor corresponding to a point x of X is $\underline{L}_m$, where m is the greatest common divisor of the integers k such that x belongs to the relation r_k of $\mathbf{X}$ corresponding to $\underline{L}_k$.

When considering the finite members of $\mathcal{X}_n$, i.e. the duals of finite members of $\mathcal{M}_n$, it can prove useful to define the following derived unary relations. If

$$\mathbf{X} = \langle X; \{r_m \,|\, m \in \operatorname{div}(n)\}, \tau \rangle$$

is a finite member of $\mathcal{X}_n$, then for each positive divisor m of n define the unary relation s_m by

$$s_m := r_m \setminus \bigcup_{k|m,\, k \neq m} r_k.$$

Indeed, in these conditions we have that

$$E_n(\mathbf{X}) \cong \prod_{m|n} \underline{L}_m^{|s_m|},$$

where $|s_m|$ denotes the number of elements in the unary relation s_m. Of course, we can recover the relations r_m from the derived relations s_m : for each positive divisor m of n we have that $r_m = \bigcup_{k|m} s_k$. Notice also that the derived relations s_m form a partition of X.

One property of the members of $\mathcal{M}_n$ that can easily be expressed in term of their dual structures is their belonging to subvarieties of $\mathcal{M}_n$. Indeed, if m divides n, then an algebra A of $\mathcal{M}_n$ belongs to the subvariety $\mathcal{M}_m$ of $\mathcal{M}_n$ if and only if the relation r_m of its dual $D_n(A)$ satisfies $r_m = \mathcal{M}_n(A, \underline{L}_n)$. In particular, A is a Boolean algebra if and only if $r_1 = \mathcal{M}_n(A, \underline{L}_n)$. More generally, if $m_1, \ldots, m_k$ divide n, then an algebra A of $\mathcal{M}_n$ belongs to the subvariety $\mathbb{HSP}(\underline{L}_{m_1}, \ldots, \underline{L}_{m_k})$ of $\mathcal{M}_n$ if and only if the relations $r_{m_1}, \ldots, r_{m_k}$ of its dual $D_n(A)$ satisfy $\bigcup_{i=1}^k r_{m_i} = \mathcal{M}_n(A, \underline{L}_n)$.

A result of a similar nature is the following. If A is a member of $\mathcal{M}_n$, then A is the reduct of a Post algebra of order $n+1$ if and only if every relation of its dual $D_n(A)$ is empty.

Finally, we would like to point out one last important feature of the dualities of Theorem 3 : they turn embeddings in $\mathcal{M}_n$ (resp. $\mathcal{X}_n$) to surjection in $\mathcal{X}_n$ (resp. $\mathcal{M}_n$), as it is the case for Stone's duality.

4   Special members of $\mathcal{M}_n$

In this section we use the dualities obtained above to describe some remarkable classes of members of $\mathcal{M}_n$. We denote by $\mathcal{F}_{\mathcal{A}}(\kappa)$ the *free algebra* generated in $\mathcal{A}$ by κ generators (κ is a cardinal). If $\mathcal{A} = \mathbb{ISP}(\underline{\mathbf{M}})$ admits a natural duality, then $\mathcal{F}_{\mathcal{A}}(\kappa) \cong E(\underline{\mathbf{M}}^{\kappa})$.

An object A of a class $\mathcal{A}$ is *injective* in $\mathcal{A}$ if, for every embedding $e : B \to C$ and morphism $u : B \to A$ in $\mathcal{A}$, there is a morphism $v : C \to A$ such that $v \circ e = u$. Dually, an object A is *projective* in $\mathcal{A}$ if, for every surjective morphism $s : B \to C$ and morphism $u : A \to C$, there is a morphism $v : A \to B$ such that $s \circ v = u$. Under Hu's dualities, as well as under the dualities of Theorem 3, the injective members of $\mathcal{A}$ (resp. $\mathcal{X}$) correspond to the projective members of $\mathcal{X}$ (resp. $\mathcal{A}$), which will help describe them.

The last classes we want to consider are *algebraically closed* and *existentially closed* algebras. Generalizing the classical definition in field theory, an algebra A of a class $\mathcal{A}$ is algebraically closed provided every conjunction of atomic formulae with parameters in A that is satisfiable in some extension of A in $\mathcal{A}$ is also satisfiable in A. An algebra A is existentially closed provided the same is true for every quantifier-free first-order formula with parameters in A. The class of algebraically (resp. existentially) closed members of $\mathcal{A}$ is denoted by $\mathcal{A}^{ac}$ (resp. $\mathcal{A}^{ec}$). Section 5.3 of [3] shows that, under rather mild hypotheses on $\mathcal{A}$ — which are satisfied by the varieties $\mathcal{P}_n$ of Post algebras and the varieties $\mathcal{M}_n$ of MV-algebras —, the algebraically and existentially closed members of $\mathcal{A}$ can be characterized in term of diagram completion properties, and some further properties of $\mathcal{A}^{ec}$ can be proved.

We now proceed to describe all these classes of algebras in the varieties $\mathcal{P}_n$ of Post algebras (remember that $\mathcal{P}_1$ is the variety of Boolean algebras). Most of these results either belong to the folklore or are easily proven, so that we omit the proofs.

First we give a description of the finitely generated free algebras in $\mathcal{P}_n$: if k is a nonnegative integer, then

$$\mathcal{F}_{\mathcal{P}_n}(k) = \underline{\mathbf{P}}_n^{(n+1)^k}.$$

The free algebra in $\mathcal{P}_n$ over $\aleph_0$ generators is more easily described via its dual :

$$S_n(\mathcal{F}_{\mathcal{P}_n}(\aleph_0)) \cong \underline{\mathbf{P}}_n^{\aleph_0}.$$

One way to visualize $\underline{\mathbf{P}}_n^{\aleph_0}$ is to consider it as the subspace of the Euclidean unit interval [0,1] containing those real numbers whose decimal expansion written in base $2n+1$ has only even digits. More precisely, define Γ_n in the following way. Starting from $\Gamma_{n,0} = [0,1]$, define $\Gamma_{n,k+1}$ from $\Gamma_{n,k}$ by cutting each interval composing $\Gamma_{n,k}$ into $2n+1$ intervals of the same lenght, and removing every second of these smaller intervals. In other words, number the $2n+1$ intervals from 0 to $2n$, and keep those having an even number. Then $\Gamma_n := \bigcap_{k \geq 1} \Gamma_{n,k}$. Endow Γ_n with the topology induced from the Euclidean topology on [0,1], and you obtain a topological space which turns out to be homeomorphic to $\underset{\sim}{\mathbf{P}}_n^{\aleph_0}$. An explicit homeomorphism is given by

$$\varphi : \underset{\sim}{\mathbf{P}}_n^{\aleph_0} \to \Gamma_n : x \mapsto \sum_{i=0}^{+\infty} \frac{2n.x_i}{(2n+1)^{i+1}}.$$

Note that Γ_1 is often denoted by Γ, and known as the *Cantor space*. To end with, remark that although they might look quite different from one another, all the Γ_n actually are homeomorphic, because they all are perfect metric Boolean spaces.

The injective members of $\mathcal{P}_n$ correspond to the projective Boolean spaces, which have been characterized by Gleason ([8]). He proved that a Boolean space $\mathbf{X}$ is projective if and only if it is extremally disconnected, which means that the closure of every open set in $\mathbf{X}$ is (cl)open. Since these spaces correspond under Hu's dualities to the complete Post algebras (as ordered sets), it follows that an algebra is injective in $\mathcal{P}_n$ if and only if it is complete. As for injective Boolean spaces (and projective members of $\mathcal{P}_n$), no characterization is known to date. We shall nevertheless point out that every finite Boolean space is injective, so that every finite member of $\mathcal{P}_n$ is projective.

Finally, we note that every algebra in $\mathcal{P}_n$ is algebraically closed. Moreover, an algebra A in $\mathcal{P}_n$ is existentially closed if and only if it is atomless. This is equivalent to saying that $S_n(A)$ has no isolated point, i.e. no point x such that $\{x\}$ is (cl)open. The class $\mathcal{P}_n^{ec}$ is $\aleph_0$-categorical, i.e. it has, up to isomorphism, exactly one countably infinite member. Its unique countably infinite member, which turns out to be $\mathcal{F}_{\mathcal{P}_n}(\aleph_0)$, is $\mathcal{P}_n$-universal, i.e. every non-trivial member of $\mathcal{P}_n$ of cardinality $\leq \aleph_0$ can be inbedded in it.

Now we turn to the varieties $\mathcal{M}_n$ of MV-algebras, and proceed to characterize these special classes of algebras in the varieties $\mathcal{M}_n$. Again, we start by giving a description of the finitely generated free algebras in $\mathcal{M}_n$. In the sequel we denote by $P(n)$ the set of prime divisors of a positive integer n.

Proposition 1 ([9,6,11]). *For all positive integers k and n we have*

$$\mathcal{F}_{\mathcal{M}_n}(k) \cong \prod_{m \mid n} \underline{L}_m^{f(k,m)},$$

where

$$f(k,m) = \sum_{X \subseteq P(m)} (-1)^{|X|}.(\frac{m}{\prod X} + 1)^k.$$

Proof. We know that

$$\mathcal{F}_{\mathcal{M}_n}(k) \cong E_n(\underline{L}_n^k) \cong \prod_{m|n} \underline{L}_m^{|s_m|},$$

where the s_m are the derived unary relations corresponding to the relations r_m of $\underline{L}_n^k$. Since the s_m form a partition of $\underline{L}_n^k$, we have that

$$\sum_{m|n} |s_m| = (n+1)^k.$$

Therefore, using the Möbius inversion formula, we obtain

$$|s_m| = \sum_{j|m} \mu(j).(\frac{m}{j} + 1)^k,$$

where μ is the *Möbius function*, whose value is $(-1)^{|P(n)|}$ if n is square-free, and 0 otherwise. But each nonzero term in this sum arises from a square-free divisor j of m, i.e. a product of prime factors of m, whence

$$|s_m| = \sum_{X \subseteq P(m)} (-1)^{|X|}.(\frac{m}{\prod X} + 1)^k.$$

The proposition follows at once.$\square$

In $\mathcal{M}_n$, as in $\mathcal{P}_n$, the dual $D_n(\mathcal{F}_{\mathcal{M}_n}(\aleph_0))$ of the free algebra over $\aleph_0$ generators can be pictorially described. For each positive divisor m of n, define I_m to be the set of positive integers i such that $\frac{i}{n}$ belongs to $\underline{L}_m$. We also define $2I_m$ to be the set whose members are twice the members of I_m. Now consider the subset, denoted by $r_{n,m}$, of the unit interval $[0, 1]$ containing those real numbers whose decimal expansion written in base $2n + 1$ has only digits in $2I_m$. An inductive definition of $r_{n,m}$ can also be given. It differs from the one given for Γ_n in that only those intervals whose number belongs to $2I_n$ are kept at each step. The following result should then come as no surprise.

Proposition 2 ([11]). *The member Γ'_n of $\mathcal{X}_n$, obtained by endowing Γ_n with the unary relations $r_{n,m}$, is isomorphic to $D_n(\mathcal{F}_{\mathcal{M}_n}(\aleph_0))$.*

Proof. It is straightforward to check that

$$\varphi : \underline{L}_n^{\aleph_0} \to \Gamma_n : x \mapsto \sum_{i=0}^{+infty} \frac{2n.x_i}{(2n+1)^{i+1}}$$

is an $\mathcal{X}_n$-isomorphism from $\underline{L}_n^{\aleph_0}$ onto Γ'_n.$\square$

Next we caracterize the injective and the finite projective members of $\mathcal{M}_n$. Of course, this is achieved through a caracterization of the projective and the finite injective members of $\mathcal{X}_n$.

Proposition 3 ([11,?]). *If A is a member of $\mathcal{M}_n$, then*

(i) *A is injective in $\mathcal{M}_n$ if and only if A is the reduct of a complete Post algebra of order $n+1$;*

(ii) *A is a finite projective member of $\mathcal{M}_n$ if and only if $A \cong \underline{L}_1 \times B$, where B is any finite member of $\mathcal{M}_n$.*

Proof. Let **X** be a projective member of $\mathcal{X}_n$. Define **Y** to be the one-point topological space endowed with empty relations (except for the relation corresponding to $\underline{L}_n$), and **Z** to be the same space endowed with relations all containing its only element. Let α be the only morphism from **Y** onto **Z**, and β be the only morphism from **X** onto **Z**. Since **X** is projective in $\mathcal{X}_n$, there must exist a morphism γ from **X** to **Y** such that $\beta = \alpha \circ \gamma$. But this is possible only if every relation of **X** is empty, excepted the one corresponding to $\underline{L}_n$.

On the reverse, if this last condition is fulfilled, then every morphism from **X** is just a continuous map (no further condition is imposed by the empty relations), whence such an **X** is projective in $\mathcal{X}_n$ if and only if $\langle X; \tau \rangle$ is a projective Boolean space. As a consequence, a member A of $\mathcal{M}_n$ is injective in $\mathcal{M}_n$ if and only if it is the reduct of a complete Post algebra of order $n+1$.

Let us now prove the second part of the proposition. Let **X** be a finite injective member of $\mathcal{X}_n$. Define **Y** to be the one-point topological space endowed with empty relations (except for the relation corresponding to $\underline{L}_n$), and **Z** to be the two-point discrete space endowed with relations all containing exactly one of its two elements (excepted the relation corresponding to $\underline{L}_n$). Let α be the only embedding from **Y** into **Z**, and β be any morphism from **Y** to **X**. Since **X** is injective in $\mathcal{X}_n$, there must exist a morphism γ from **Z** to **X** such that $\beta = \gamma \circ \alpha$. Since the relation corresponding to $\underline{L}_1$ is empty on **Y** and nonempty on **Z**, the existence of γ is possible only if the relation corresponding to $\underline{L}_1$ on **X** is nonempty.

On the reverse, suppose there is at least one point x_0 in the relation corresponding to $\underline{L}_1$ on **X**. Let α be an embedding from **Y** into **Z** and β be a morphism from **Y** to **X**. Since α and β are continuous maps, there is a continuous map δ from $\langle Z; \tau \rangle$ to $\langle X; \tau \rangle$ such that $\beta = \delta \circ \alpha$, because every finite Boolean space is projective in $\mathcal{Z}$. Of course, $\delta|_{\alpha(\mathbf{Y})}$ is a morphism, because α is an embedding. For each x in **X**, define m_x to be the unique natural number m such that x belongs to s_m, and

$$F_x := \delta^{-1}(x) \cap \bigcup \{ r_m \mid m_x \text{ does not divide } m \},$$

where the r_m are the relations of **Z**. Finally, define $F := \bigcup_{x \in \mathbf{X}} F_x$. We have that F is closed because the relations r_m are closed and δ is continuous, and that $F \cap \alpha(\mathbf{Y})$ is empty because $\delta|_{\alpha(\mathbf{Y})}$ is a morphism. Let ω be a clopen subset of **Z** containing F and such that $\omega \cap \alpha(\mathbf{Y})$ is empty. Now define γ

from $\mathbf{Z}$ to $\mathbf{X}$ by

$$\gamma(x) = \begin{cases} x_0 & \text{if } x \in \omega \\ \delta(x) & \text{otherwise} \end{cases}.$$

Obviously, γ is a morphism such that $\beta = \gamma \circ \alpha$. This ends the proof. $\square$

Finally we describe the algebraically and existantially closed members of $\mathcal{M}_n$. As announced earlier, we use the results of [3, 5.3], which express these properties in term of diagram completion properties satisfied by the duals of the algebras.

Proposition 4. *If A is a member of $\mathcal{M}_n$, then*

(i) *A is algebraically closed if and only if A is the reduct of a Post algebra of order $n + 1$;*
(ii) *A is a existentially closed if and only if A is the reduct of an atomless Post algebra of order $n + 1$.*

Proof. Let $\mathbf{X}$ be the dual $D_n(A)$ of an algebraically closed member A of $\mathcal{M}_n$. By [3, 5.3], for every finite members $\mathbf{Y}$ and $\mathbf{Z}$ of $\mathcal{X}_n$ and every surjections φ from $\mathbf{X}$ to $\mathbf{Z}$ and ψ from $\mathbf{Y}$ to $\mathbf{Z}$, there exists a morphism λ from $\mathbf{X}$ to $\mathbf{Y}$ such that $\varphi = \psi \circ \lambda$. As in Proposition 3, define $\mathbf{Y}$ to be the one-point topological space endowed with empty relations (except for the relation corresponding to $\underline{L}_n$), and $\mathbf{Z}$ to be the same space endowed with relations all containing its only element. Let φ be the only morphism from $\mathbf{X}$ onto $\mathbf{Z}$, and ψ be the only morphism from $\mathbf{Y}$ onto $\mathbf{Z}$. The existence of a morphism λ such that $\varphi = \psi \circ \lambda$ is possible only if every relation of $\mathbf{X}$ is empty, excepted the one corresponding to $\underline{L}_n$.

On the reverse, if this last condition is fulfilled, then every morphism from $\mathbf{X}$ is just a continuous map (no further condition is imposed by the empty relations), whence such an $\mathbf{X}$ satisfies the diagram completion property in $\mathcal{X}_n$ if and only if $\langle X; \tau \rangle$ does so in $\mathcal{Z}$. As a consequence, a member A of $\mathcal{M}_n$ is algebraically closed if and only if it is the reduct of a Post algebra of order $n + 1$.

Similarly, because of [3, 5.3], a member A of $\mathcal{M}_n$ is existentially closed if and only if it is the reduct of an existentially closed Post algebra of order $n + 1$, i.e. an atomless Post algebra of order $n + 1$. Indeed, the diagram completion property corresponding to existantial closure is the same as for algebraic closure, except that λ is required to be a surjection. $\square$

Some more can be said about the class $\mathcal{M}_n^{ec}$ of existantially closed members of $\mathcal{M}_n$. As in the Post case, $\mathcal{M}_n^{ec}$ is $\aleph_0$-categorical, i.e. it has, up to isomorphism, exactly one countably infinite member. Its unique countably infinite member, which turns out to be the reduct of $\mathcal{F}_{\mathcal{P}_n}(\aleph_0)$, is $\mathcal{M}_n$-universal, i.e. every non-trivial member of $\mathcal{M}_n$ of cardinality $\leq \aleph_0$ can be inbedded in it.

5   Further applications

The last application of natural duality we wish to illustrate concerns automorphisms and endomorphisms of algebras. If A is an algebra, we denote by Aut(A) the set of automorphisms of A (i.e. of isomorphisms from A into itself), considered as a group under functional composition. Likewise, we denote by End(A) the set of endomorphisms of A (i.e. of homomorphisms from A to itself), considered as a monoid under functional composition.

The natural dualities reverse arrows, so that the automorphism group and endomorphism monoid of an algebra A are anti-isomorphic respectively to the automorphism group and endomorphism monoid of its dual $D(A)$. When the latter are straightforward to compute, they provide an easy access to the former. This will be the case for the finite MV-algebras. In the sequel we denote by Sym(n) the symmetric group on an n-element set (we consider that the symmetric group on an empty set is the trivial group).

Proposition 5 ([6,?]). *If $A \cong \prod_{m|n} \underline{L}_m^{i_m}$, then*

$$Aut(A) \cong \prod_{m|n} Sym(i_m).$$

Proof. Denote by r_m the relation of $D_n(A)$ corresponding to $\underline{L}_m$, and by s_m the corresponding derived unary relations. Consider the dual $D_n(h)$ of an automorphism h of A. It has to be an isomorphism from $D_n(A)$ onto itself, i.e. a permutation of $D_n(A)$ which sends each point in s_m to another point in s_m. Since $|s_m| = i_m$, the result follows at once.$\square$

Concerning the endomorphism monoid of the finite MV-algebras, we won't be able to be so precise. But we can still compute their size.

Proposition 6. *If $A \cong \prod_{m|n} \underline{L}_m^{i_m}$, then*

$$|End(A)| = \prod_{m|n} \left(\sum_{k|m} i_k \right)^{i_m}.$$

Proof. Again, denote by r_m the relation on $D_n(A)$ corresponding to $\underline{L}_m$, and by s_m the corresponding derived unary relations. Consider the dual $D_n(h)$ of an endomorphism h of A. It has to be a morphism from $D_n(A)$ to itself, i.e. a mapping which sends each point in s_m to a point in r_m. Since $|s_m| = i_m$ and $|r_m| = \sum_{k|m} |s_k|$, the result follows at once.$\square$

References

1. R. Balbes and P. Dwinger, "Distributive lattices", University of Missouri Press, 1974.
2. C. C. Chang, Algebraic analysis of many-valued logics, *Trans. Amer. Math. Soc.* **88** (1958), 467–490.
3. D. M. Clark and B. A. Davey, "Natural Dualities for the Working Algebraist", Cambridge University Press, 1998.
4. B. A. Davey and H. A. Priestley, "Introduction to Lattices and Order", Cambridge University Press, 1990.
5. B. A. Davey and H. Werner, Dualities and equivalences for varieties of algebras, *in* "Colloquia mathematica societatis János Bolyai", Vol. 33, North-Holland, 1983.
6. A. Di Nola, R. Grigolia and G. Panti, Finitely generated free MV-algebras and their automorphism groups, *Studia Logica* **61** (1998), 65–78.
7. A. Di Nola and R. Grigolia, Projective MV-algebras and their automorphism groups, *to appear*
8. A. M. Gleason, Projective topological spaces, *Ill. J. Math.* **2** (1958), 482–489.
9. R. Grigolia, "Free algebras of non-classical logics", Metsniereba, Tbilisi, 1987.
10. T. K. Hu, On the topological duality for primal algebra theory, *Alg. Univ.* **1** (1971), 152–154.
11. P. Niederkorn, Natural dualities for varieties of MV-algebras. I, to appear in *J. Math. Anal. Appl.*
12. H. A. Priestley, Representation of distributive lattices by means of ordered Stone spaces, *Bull. London Math. Soc.* **2** (1970), 186–190.
13. H. A. Priestley, Ordered topological spaces and the representation of distributive lattices, *Proc. London Math. Soc.* **24** (1972), 507–530.
14. H. Rasiowa, "An Algebraic Approach to Non-Classical Logics", Studies in logic and the foundations of mathematics (volume 78), North-Holland / American Elsevier, 1974.
15. M. H. Stone, The Theory of representations for Boolean Algebras, *Trans. Amer. Math. Soc.* **40** (1936), 37–111.

The Principles of Fuzzy Logic: Its Mathematical and Computational Aspects*

Vilém Novák and Irina Perfilieva

University of Ostrava
Institute for Research and Applications of Fuzzy Modeling
30. dubna 22, 701 03 Ostrava 1, Czech Republic
NovakV@dynami.osu.cz

1 Introduction

Our aim in this chapter is to give a brief overview of the main aspects of fuzzy logic. We introduce the concept of fuzzy logic and discuss its philosophical background. We argue that people encounter a phenomenon of indeterminacy which has two complementary facets, namely uncertainty and vagueness. Fuzzy logic is then considered as a mathematical model useful for modelling of the latter. Furthermore, we outline the theory of special structures, which are suitable for representation of the structure of truth values.

We distinguish fuzzy logic in narrow sense (FLn) and that in broader sense (FLb). Our further explanation continues by an overview of some main concepts of FLn. In Section 6, we discuss important questions of the correspondence between formulas of FLn and special functions defined in $[0, 1]$, as well as the problem of approximation of continuous functions using functions derived from special formulas of FLn. The material is finished by a brief introduction to FLb, which is an extension of FLn enabling to capture the meaning of special natural language expressions, namely the so called evaluating syntagms, which characterize position on an ordered scale and are used especailly in the characterization of the behaviour of dynamic systems.

2 Vagueness, uncertainty and fuzzy logic

2.1 Vagueness and uncertainty

The phenomena of *uncertainty* and *vagueness* characterize situations in which we regard the world surrounding us; they are concerned with the amount of knowledge we have at disposal (or can have at disposal). We argue that

* The research has been supported by the grant A1187901/99 of the GA AV ČR and the project VS96037 of the MŠMT of the Czech Republic.

uncertainty and vagueness are two, rather complementary facets of a more general phenomenon called the *indeterminacy*.

The phenomenon of uncertainty emerges due to the *lack of knowledge* about all the influences which may lead to *occurrence* of some *event*. Uncertainty is encountered when an experiment (process, test, etc.) is to proceed, the result of which is not known to us. Let us stress that there is no uncertainty after the experiment was realized and the result is known to us. Recall that a specific form of uncertainty is *randomness*. The mathematical model (i.e. quantified characterization) of the uncertainty phenomenon is provided especially by the *probability theory*. There are also other mathematical theories addressing the uncertainty phenomenon, for example the possibility theory, belief measures and others.

The vagueness phenomenon raises during the process of grouping together objects having some property φ (of objects). The result of this process will be called a *grouping* of objects. In a slightly formal way, the grouping can be written as

$$X = \{x \mid x \text{ has the property } \varphi\} \tag{1}$$

where x varies over objects.

It is important to stress that, in general, the grouping X *cannot be taken as a set* since the property φ may not make us possible to characterize the grouping X precisely and unambiguously; there can exist *borderline* elements x for which it is unclear whether they have the property φ (and thus belong to X), or not.

Example 1. A typical vague property is *to be a small natural number*. Can we imagine all the small natural numbers? Clearly, 0 is small, 1 is small as well, etc. But where does this sequence finish? The only sure fact is that there exists a number, say $1,000,000,000$, which is not small.

An attempt at sharp (exact) explanation would mean to be able to find a small natural number n, $0 < n < 1,000,000,000$ such that $n + 1$ is not small. However, such a conclusion can hardly be defended. If n is small then $n + 1$ must be small, as well. Hence, there is *no last small* number before $1,000,000,000$ and *no first number* $n < 1,000,000,000$ which is *not small*. We can distinguish small numbers from big ones but we are not able to say unambiguously about *each* number whether it is small or not. The property of "small" is vague and small numbers disappear inside the sequence of numbers ranging from 0 to 1,000,000,000. Consequently, they form a vague grouping of numbers.

Vagueness is an opposite to exactness and we argue that it cannot be avoided in the human way of regarding the world. Any attempt to explain an extensive detailed description necessarily leads to using vague concepts since precise description contains abundant number of details. To understand it, we must group them together — and this can hardly be done precisely. It is likely that the explanation would significantly rely on the use of natural language since vagueness is often connected with its use. However, the problem lays deeper, in the way how people regard the phenomena around them. Vagueness is necessary to convey relevant information.

Vagueness should be distinguished from generality and from ambiguity. In our terms, to be more general means to take into account more (various) groupings of objects, while ambiguity occurs in the language when more alternative meanings are assigned to the same word expression.

A typical feature of vagueness is its *continuity*. This means that if an object has a vague property and another one differs very little from it, then it must have the same property. In other words, a small difference between objects cannot lead to abrupt change in the decision of whether either of them has, or has not a vague property. The transition from having a (vague) property to not having it is smooth. A mathematical theory of the vagueness phenomenon is most successfully provided by fuzzy logic.

Important is the principle of *graded (fuzzy) approach*, which is employed throughout fuzzy set theory and fuzzy logic. This means the use of a scale when characterizing a relation between object and its property. The graded approach will be mathematized by means of a special scale being an ordered set. In order to be sufficiently general and to be able to capture the continuity feature of vagueness, we will suppose that this set is uncountable.

3   What is fuzzy logic

In general, fuzzy logic can be characterized as the *many-valued logic with special properties aiming at modeling of the vagueness phenomenon and some parts of the meaning of natural language via graded approach.*

We will distinguish fuzzy logic in narrow and broader sense. *Fuzzy logic in narrow sense*, FLn, is a special many-valued logic, which aims at providing formal background for the graded approach to vagueness. *Fuzzy logic in broader sense*, FLb, is an extension of FLn and it aims at developing mathematical model of natural human reasoning, in which principal role is played by the natural language.

In general, fuzzy logic is the result of the graded approach to the formal logical systems. Due to the graded approach, fuzzy logic provides solution of some, classically non-solvable problems. For example, there are well known ancient paradoxes of *sorites* (heap) and *falakros* (bald men).

> One grain of wheat does not make a heap. Neither make it two grains, three, etc. Hence, there are no heaps.

These paradoxes rise if we understand the properties "to be a heap" and "to be bald" sharply, i.e. if we neglect their vagueness.

Let $\mathbb{FN}(x)$ denote the proposition "the number x is small" (x be natural). The problem lays in the truth verification of implication $\mathbb{FN}(x) \Rightarrow \mathbb{FN}(x+1)$, i.e. classical induction cannot be applied to vague property. If the latter is taken classically (i.e. absolute true) then starting from $\mathbb{FN}(0)$ we necessarily arrive to the above paradoxes. However, for different subsequents of x, the verification of the proposition $\mathbb{FN}(x) \Rightarrow \mathbb{FN}(x+1)$ is different. In general, verification that x is small does not imply that we will be able to verify that also $x+1$ is small with the same effort. For example, if we verify that 1000 is small by counting one thousand lines then verification that 1001 is also small means that we must count one line more, i.e. our effort to verify that 1001 is small is a little (imperceptibly) greater than that for 1000. Consequently, the above implication is not fully convincing. A solution offered by fuzzy logic is to assume that the implication $\mathbb{FN}(x) \Rightarrow \mathbb{FN}(x+1)$ is true only in some degree close to 1, say $1 - \varepsilon$ where $\varepsilon > 0$. Then sorites (as well as falakros) paradox disappears.

The formal apparatus of fuzzy logic provides us the following outcome.

1. Because the graded approach serves us for modeling of the vagueness phenomenon, we are able to deal with *any possible truth value* on both levels, namely syntactical (make graded syntactical derivations) as well as semantical (evaluate truth degrees of formulas in the interpretation). Note that unlike other multivalued logics, all truth values in FLn are *equal in their importance*, i.e. there are no designated truth values.
2. FLn and especially FLb as the extension of the latter seem to be working theories, which may be used as formal apparatus explicating the approximate reasoning schemes.
3. FLn is a generalization of classical logic. It opens different, more general way for explanation of, at least some of the classical problems. We may also expect new problems of classical mathematics influenced directly by FLn.

The system of FLn is basically truth functional. There are reasons to assume the truth values to form a complete, infinitely distributive residuated lattice, or an MV-algebra. FLn aims at *extension of the capability of classical logic* to the directions where the latter cannot provide satisfactory solution. Hence, FLn neither mimics classical logic by adding slight generalizations, nor denies it.

4 Algebraic structures for logical calculi

4.1 Residuated lattices

In order to have more than two values for evaluation of formulas, many attempts have been done to generalize the classical algebra for logic. On this way, two special structures took a distinguished role, namely residuated lattices and MV-algebras.

Suppose that the set L of truth values forms a lattice and contains the least 0 and the greatest 1 elements. Besides that we add two specific operations of multiplication $\otimes$ and addition $\oplus$. A residuated lattice can be placed between Boolean and MV-algebras. In the case that $L = \{0, 1\}$, these coincide with $\wedge$ and $\rightarrow$, respectively.

Definition 1. A residuated lattice is an algebra

$$\mathcal{L} = \langle L, \vee, \wedge, \otimes, \rightarrow, 0, 1 \rangle. \qquad (2)$$

with four binary operations and two constants such that

(i) $\langle L, \vee, \wedge, 0, 1 \rangle$ is a lattice with the ordering $\leq$ defined using the operations $\vee, \wedge$ as usual, and $0, 1$ are its least and the greatest elements, respectively;

(ii) $\langle L, \otimes, 1 \rangle$ is a commutative monoid, that is, $\otimes$ is a commutative and associative operation with the identity $a \otimes 1 = a$;

(iii) the operation $\rightarrow$ is a residuation operation with respect to $\otimes$, i.e.

$$a \otimes b \leq c \quad \text{iff} \quad a \leq b \rightarrow c. \qquad (3)$$

We see that in comparison with the ordinary lattice, this structure is enriched by a special couple $\langle \otimes, \rightarrow \rangle$ of operations called the *adjoint couple*. The defining relation (3) is called the *adjunction property*. We will call $\otimes$ the *multiplication* and $\rightarrow$ the *residuation*.

Example 2 (Boolean algebra for classical logic).

$$\mathcal{L}_B = \langle \{0, 1\}, \vee, \wedge, \rightarrow, 0, 1 \rangle$$

where $\rightarrow$ is the classical implication, is the simplest residuated lattice (the multiplication $\otimes = \wedge$). In general, every Boolean algebra is a residuated lattice when putting $a \rightarrow b = a' \vee b$.

Example 3 (Gödel algebra).

$$\mathcal{L}_G = \langle [0, 1], \vee, \wedge, \rightarrow_G, 0, 1 \rangle$$

where the multiplication $\otimes = \wedge$ and

$$a \rightarrow_G b = \begin{cases} 1 & \text{if } a \leq b, \\ b & \text{if } b < a. \end{cases} \qquad (4)$$

Gödel algebra is a special case of the *Heyting algebra* used in intuitionistic logic.

Example 4 *(Product algebra).*

$$\mathcal{L}_P = \langle [0, 1], \vee, \wedge, \odot, \rightarrow_P, 0, 1 \rangle$$

where the multiplication $\odot = \cdot$ is the ordinary product of reals and

$$a \rightarrow_P b = \begin{cases} 1 & \text{if} \quad a \le b, \\ \frac{b}{a} & \text{if} \quad b < a. \end{cases} \tag{5}$$

Example 5 *(Łukasiewicz algebra).*

$$\mathcal{L}_L = \langle [0, 1], \vee, \wedge, \otimes, \rightarrow_L, 0, 1 \rangle \tag{6}$$

where

$$a \otimes b = 0 \vee (a + b - 1), \qquad \text{(Łukasiewicz conjunction)} \tag{7}$$
$$a \rightarrow_L b = 1 \wedge (1 - a + b). \qquad \text{(Łukasiewicz implication)} \tag{8}$$

Example 6 *(Residuated lattice of $[0, 1]$-valued functions).* Let X be a non-empty set. For each two functions $f, g \in [0, 1]^X$ we put

$$(f \vee g)(x) = f(x) \vee g(x), \qquad (f \otimes g)(x) = f(x) \otimes g(x), \tag{9}$$
$$(f \wedge g)(x) = f(x) \wedge g(x), \qquad (f \rightarrow g)(x) = f(x) \rightarrow g(x) \tag{10}$$

for all $x \in X$ where $\vee, \wedge, \otimes, \rightarrow$ are the corresponding operations of the Łukasiewicz algebra (Example 5). Furthermore, let $\mathbf{1}$ and $\mathbf{0}$ be constant functions taking the values 1 and 0, respectively. Then

$$\mathcal{L}^{func}(X) = \langle [0, 1]^X, \vee, \wedge, \otimes, \rightarrow, \mathbf{0}, \mathbf{1} \rangle$$

is a residuated lattice.

Lemma 1. *Let $\mathcal{L}$ be a residuated lattice given by (2). Then for every $a, b, c \in L$ the following holds true.*

(a) $a \otimes b \le a,$ $a \otimes b \le b,$ $a \otimes b \le a \wedge b,$
(b) $b \le a \rightarrow b,$
(c) $a \otimes (a \rightarrow b) \le b,$ $b \le a \rightarrow (a \otimes b),$
(d) if $a \le b$ then

$$c \otimes a \le c \otimes b, \qquad c \rightarrow a \le c \rightarrow b,$$
$$a \rightarrow c \ge b \rightarrow c,$$

(e) $a \otimes (a \rightarrow \mathbf{0}) = \mathbf{0},$
(f) $a \le b$ iff $a \rightarrow b = 1,$ $a = 1 \rightarrow a,$
(g) $(a \vee b) \otimes c = (a \otimes c) \vee (b \otimes c),$

Lemma 2. *Let $\mathcal{L}$ be a residuated lattice given by (2). If*

$$1 = a \vee (a \rightarrow \mathbf{0})$$

holds for every $a \in L$ then $\mathcal{L}$ is a Boolean lattice and $\otimes = \wedge$.

Additional operations and their properties. To be used as an algebra for many-valued logic, residuated lattices should be completed by additional operations corresponding (among others) to the logical connectives. Analogously as in classical logic, they can be obtained as derived operations from the basic ones.

$$\neg a = a \to 0 \qquad \text{(negation)}, \qquad (11)$$

$$a \leftrightarrow b = (a \to b) \land (b \to a) \quad \text{(biresiduation)}, \qquad (12)$$

$$a \oplus b = \neg(\neg a \otimes \neg b) \qquad \text{(addition)}, \qquad (13)$$

$$a^n = \underbrace{a \otimes \cdots \otimes a}_{n-\text{times}} \qquad \text{(}n\text{-fold multiplication)}, \qquad (14)$$

$$na = \underbrace{a \oplus \cdots \oplus a}_{n-\text{times}} \qquad \text{(}n\text{-fold addition)}. \qquad (15)$$

first i.e.

Lemma 3. *Let $\mathcal{L}$ be a residuated lattice given by (2). Then the following holds true for every $a, b, c, d \in L$.*

(a) $a \leftrightarrow 1 = a, \qquad a = b \ \text{iff} \ a \leftrightarrow b = 1,$
(b) $(a \leftrightarrow b) \land (c \leftrightarrow d) \le (a \land c) \leftrightarrow (b \land d),$

Definition 2. Let $\mathcal{L}$ be a residuated lattice.

(i) $\mathcal{L}$ is *linearly ordered* if $a \le b$ or $b \le a$ holds for every $a, b \in L$.
(ii) $\mathcal{L}$ is *complete* if the underlying lattice is complete.

Lemma 4. *Let $\mathcal{L}$ be a complete residuated lattice. Then the following identities hold for every $a, b \in L$ and sets $\{a_i \mid i \in I\}$, $\{b_i \mid i \in I\}$ of elements from L over arbitrary set of indices I.*

(a) $a \to b = \bigvee \{x \mid a \otimes x \le b\},$
(b) $a \otimes b = \bigwedge \{x \mid a \le b \to x\}.$
(c) $(\bigvee_{i \in I} a_i) \otimes b = \bigvee_{i \in I} (a_i \otimes b),$
(d) $\bigwedge_{i \in I} (a \to b_i) = a \to (\bigwedge_{i \in I} b_i),$
(e) $\bigwedge_{i \in I} (a_i \to b) = (\bigvee_{i \in I} a_i) \to b,$
(f) $\bigvee_{i \in I} (a \to b_i) \le a \to (\bigvee_{i \in I} b_i),$
(g) $\bigvee_{i \in I} (a_i \to b) \le (\bigwedge_{i \in I} a_i) \to b.$

4.2 MV-algebras

The notion of MV-algebra was introduced by C. C. Chang. MV-algebras have been developed as generalizations of Boolean algebras. However, it turned out that MV-algebras stand for the algebraic structures of truth values for various non-classical logical calculi including fuzzy logics. From the algebraic point of view, the MV-algebra differs from the Boolean one by absence of the idempotency law for their algebraic operations (addition and multiplication) and also, by the lack of the law of excluded middle for the lattice operations.

Definition 3. An MV-algebra is an algebra

$$\mathcal{L} = \langle L, \oplus, \otimes, \neg, \mathbf{1}, \mathbf{0} \rangle \tag{16}$$

in which the following identities are valid.

$$
\begin{array}{lr}
a \oplus b = b \oplus a, & a \otimes b = b \otimes a, \tag{17} \\
a \oplus (b \oplus c) = (a \oplus b) \oplus c, & a \otimes (b \otimes c) = (a \otimes b) \otimes c, \tag{18} \\
a \oplus \mathbf{0} = a, & a \otimes \mathbf{1} = a, \tag{19} \\
a \oplus \mathbf{1} = \mathbf{1}, & a \otimes \mathbf{0} = \mathbf{0}, \tag{20} \\
a \oplus \neg a = \mathbf{1}, & a \otimes \neg a = \mathbf{0}, \tag{21} \\
\neg(a \oplus b) = \neg a \otimes \neg b, & \neg(a \otimes b) = \neg a \oplus \neg b, \tag{22} \\
a = \neg\neg a, & \neg \mathbf{0} = \mathbf{1}, \tag{23} \\
\neg(\neg a \oplus b) \oplus b = \neg(\neg b \oplus a) \oplus a. & \tag{24}
\end{array}
$$

The following are the examples of an MV-algebra.

Example 7 (Lukasiewicz algebra). The Lukasiewicz algebra $\mathcal{L}_L$ from Example 5 is an MV- algebra. It can be written as

$$\mathcal{L}_L = \langle [0, 1], \oplus, \otimes, \neg, 0, 1 \rangle \tag{25}$$

where $\otimes$ is the Lukasiewicz conjunction defined in (7), $\oplus$ is called the Lukasiewicz disjunction defined by

$$a \oplus b = 1 \wedge (a + b) \tag{26}$$

and $\neg$ is the negation operation defined by $\neg a = 1 - a$.

Example 8 (MV-algebra of $[0, 1]$-valued functions). The residuated lattice of $[0, 1]$-valued functions on a nonempty X from Example 6 is an MV-algebra.

Definition 4. Let $\mathcal{L}$ be an MV-algebra given by (16). $\mathcal{L}$ is locally finite if to every $a \neq \mathbf{1}$ there is $n \in \mathbb{N}$ such that $a^n = \mathbf{0}$, or equivalently, to every $a \neq \mathbf{0}$ there is $n \in \mathbb{N}$ such that $na = \mathbf{1}$. (In other terminology this property is called a nilpotentness.)

For example, Lukasiewicz MV-algebra $\mathcal{L}_L$ is locally finite.

Put

$$a \leq b \quad \text{iff} \quad \neg a \oplus b = \mathbf{1}. \tag{27}$$

It can be verified that (27) has the properties of partial ordering.

The lattice operations can be introduced by

$$a \vee b = \neg(\neg a \oplus b) \oplus b = (a \otimes \neg b) \oplus b, \tag{28}$$

$$a \wedge b = \neg(\neg a \vee \neg b) = (a \oplus \neg b) \otimes b. \tag{29}$$

Moreover, put $a \to b = \neg a \oplus b$ and see that this is a residuation w.r.t. $\otimes$.

Theorem 1.   *(a) Every MV-algebra is a residuated lattice.*
(b) A residuated lattice $\mathcal{L}$ is an MV-algebra iff

$$(a \to b) \to b = a \vee b \tag{30}$$

holds for every $a, b \in L$.

Corollary 1. *An MV-algebra is a Boolean algebra if the operation $\otimes$ (or, equivalently, $\oplus$) is idempotent.*

By Theorem 1 (b) both Gödel $\mathcal{L}_G$ as well product $\mathcal{L}_P$ algebras are not MV-algebras.

Lemma 5. *The following are identities in every MV-algebra.*

(a) $(a \to b) \vee (b \to a) = 1$,
(b) $a \to b = \neg b \to \neg a$.

As usual, an MV-algebra is complete if the underlying lattice is complete.

Lemma 6. *Let $\mathcal{L}$ be a complete MV-algebra. Then the following holds true for every $a, b \in L$ and sets $\{a_i \mid i \in I\}$, $\{b_i \mid i \in I\}$ of elements from L over arbitrary set of indices I.*

(a) $\bigvee_{i \in I} a_i = \neg \bigwedge_{i \in I} \neg a_i, \qquad \bigwedge_{i \in I} a_i = \neg \bigvee_{i \in I} \neg a_i,$
(b) $a \wedge (\bigvee_{i \in I} b_i) = \bigvee_{i \in I}(a \wedge b_i), \qquad a \vee (\bigwedge_{i \in I} b_i) = \bigwedge_{i \in I}(a \vee b_i),$
(c) $a \otimes (\bigwedge_{i \in I} b_i) = \bigwedge_{i \in I}(a \otimes b_i),$
(d) $\bigvee_{i \in I}(a_i \to b) = (\bigwedge_{i \in I} a_i) \to b,$
(e) $\bigvee_{i \in I}(a \to b_i) = a \to (\bigvee_{i \in I} b_i).$

4.3   Elements of the theory of t-norms

The previous section we have presented three distinguished examples, namely, Łukasiewicz, Gödel and product algebras, which are residuated lattices based on $[0, 1]$ differing from each other by the chosen multiplication (and residuation). However, the class of such examples is significantly wider. In this section we will study the general class of multiplications known as *triangular norms* (briefly t-norms). These are binary operations $t : [0, 1]^2 \longrightarrow [0, 1]$, which have been introduced by K. Menger. They became interesting for fuzzy logic because they preserve the fundamental properties of the conjunction 'and' (to hold at the same time), and thus, they serve as a natural generalization of the classical conjunction to many valued reasoning systems.

A concept associated with the t-norm is the triangular conorm (t-conorm) $s : [0, 1]^2 \longrightarrow [0, 1]$. This corresponds to the behaviour of truth values when joined by the connective 'or'. The other two concepts studied in this section are the residuation operation and the negation operation $n : [0, 1] \longrightarrow [0, 1]$ generalizing behaviour of the logical negation.

We will mostly confine to the properties of the continuous t-norms because these are especially interesting due to the continuity property of the vagueness phenomenon.

Basic properties of triangular norms and conorms.

Definition 5. A t-norm is a binary operation $t : [0, 1]^2 \longrightarrow [0, 1]$ such that the following axioms are satisfied for all $a, b, c \in [0, 1]$:

(i) commutativity
$$a \, t \, b = b \, t \, a,$$

(ii) associativity
$$a \, t \, (b \, t \, c) = (a \, t \, b) \, t \, c,$$

(iii) monotonicity
$$a \leq b \quad \text{implies} \quad a \, t \, c \leq b \, t \, c,$$

(iv) boundary condition
$$1 \, t \, a = a.$$

Note that due to the commutativity, we also have $a \, t \, 1 = a$. Moreover, every t-norm fulfils the additional boundary condition $0 \, t \, a = a \, t \, 0 = 0$ for all $a \in [0, 1]$. Note also that all t-norms coincide on $\{0, 1\}$.

Example 9. The most important t-norms are *minimum* '∧', *product* '·' and *Łukasiewicz conjunction* '⊗' (7). Other examples of t-norms are *drastic product*

$$a \, t_W \, b = \begin{cases} \min(a, b) & \text{if } \max(a, b) = 1, \\ 0 & \text{otherwise,} \end{cases}$$

or *nilpotent minimum* (also called Fodor t-norm)

$$a \, t_F \, b = \begin{cases} \min(a, b) & \text{if } a + b > 1, \\ 0 & \text{otherwise.} \end{cases}$$

The t-norms can be (partially) ordered as functions, i.e. for every t-norms t_1, t_2 we can put

$$t_1 \leq t_2 \quad \text{iff} \quad a\, t_1\, b \leq a\, t_2\, b, \qquad a, b \in [0, 1]. \tag{31}$$

If (31) holds then the t-norm t_1 is weaker than t_2 or, equivalently, t-norm t_2 is stronger than t_1. We write $t_1 < t_2$ if $t_1 \leq t_2$ and $t_1 \neq t_2$.

Lemma 7. *For every t-norm t it holds that*

$$t_W \leq t \leq \wedge.$$

This means that the drastic product t_W is the weakest and $\wedge$ the strongest t-norm. Furthermore, the following inequality holds for the t-norms introduced above:

$$t_W < \otimes < \cdot < \wedge,$$

Given a t-norm t, we put

$$a_t^n = \underbrace{a\, t \cdots t\, a}_{n-\text{times}}.$$

A dual concept to t-norm is that of t-conorm.

Definition 6. A t-conorm is a binary operation $s : [0, 1]^2 \longrightarrow [0, 1]$, which fulfils the axioms (i)–(iii) from Definition 5, and for all $a \in [0, 1]$ it fulfils the following boundary condition:

$$0\, s\, a = a. \tag{32}$$

A t-conorm is dual to the given t-norm t if

$$a\, s\, b = 1 - (1 - a)\, t\, (1 - b)$$

holds for all $a, b \in [0, 1]$.

Note that due to the commutativity, $a s 0 = a$ holds for all $a \in [0, 1]$. Moreover, every t-conorm fulfils the additional boundary condition $1\, s\, a = a\, s\, 1 = 1$.

Example 10. The most important t-conorms dual to the t-norms from Example 9 are the following. Dual to minimum is *maximum* '$\vee$', dual to product is *probabilistic sum*

$$a\, s_P\, b = a + b - a \cdot b,$$

dual to Lukasiewicz conjunction is *Lukasiewicz disjunction*

$$a \oplus b = \min(1, a + b),$$

dual to drastic product is *drastic sum*

$$a\, s_W\, b = \begin{cases} \max(a, b) & \text{if } \min(a, b) = 0, \\ 1 & \text{otherwise,} \end{cases}$$

and dual to nilpotent minimum is *nilpotent maximum*

$$a \, t_F \, b = \begin{cases} \max(a, b) & \text{if } a + b < 1, \\ 1 & \text{otherwise.} \end{cases}$$

Lemma 8. *For every t-conorm* s *it holds that*

$$\vee \leq s \leq s_W.$$

This means that maximum is the weakest and the drastic sum the strongest t-conorm. For the t-conorms introduced above, we have

$$\vee < s_P < \oplus < s_W.$$

Given a t-conorm s, we put

$$na_s = \underbrace{a \, s \cdots s \, a}_{n-\text{times}}.$$

Triangular norms with special properties. The set of t-norms can be classified into various, partly overlapping groups according to their specific properties. We will focus especially to three classes, namely continuous, Archimedean and non-Archimedean t-norms.

Example 11. The minimum, product and Lukasiewicz conjunction are continuous and, hence, left- as well as right-continuous. An example of a non-continuous left-continuous t-norm is the nilpotent minimum t_F.

Definition 7.   (i) A t-norm t is called strictly monotone if

$$a \, t \, b < c \, t \, b$$

holds for every $a, b, c \in (0, 1)$ and $a < c$. It is strict if it is strictly monotone and continuous.

 (ii) A t-norm t is Archimedean if for every $a, b \in (0, 1)$ there is $n \in \mathbb{N}$ such that

$$a_t^n < b.$$

(iii) A t-norm t is idempotent if $a \, t \, a = a$ holds for all $a \in [0, 1]$.
(iv) A t-norm t is nilpotent if for every $a \in (0, 1)$ there is n such that $a_t^n = 0$.

An element $a \in [0, 1]$ is called an *idempotent* if $a \, t \, a = a$. An element $a \in (0, 1)$ is *nilpotent* if there is n such that $a_t^n = 0$. We may see that a t-norm is Archimedean if it has no other idempotents except for 0 and 1.

Example 12. The product, Lukasiewicz conjunction, and drastic product are Archimedean while minimum is not. The Lukasiewicz conjunction is nilpotent and minimum is idempotent. The product is strict.

We obtain the following important classification of continuous t-norms: Every continuous Archimedean t-norm is either strict of nilpotent. Furthermore, every continuous t-norm is isomorphic either to product, or to Lukasiewicz t-norm, or it is equal to minimum '∧', or a kind of 'mixture' of all of them. Hence, when speaking about continuous t-norms, we may confine ourselves to these three t-norms. Tehrefore, we will call them the *basic t-norms*.

Residuation. We can introduce the residuation operation in the same way as is done in Lemma 4.

Definition 8. Let **t** be a t-norm. The residuation operation $\to_t: [0,1]^2 \longrightarrow [0,1]$ is defined by

$$a \to_t b = \bigvee \{z \mid a \, t \, z \leq b\}. \tag{33}$$

It can be proved that if **t** is continuous, then $a \to_t b = 1$ iff $a \leq b$.

Proposition 1. *The following are the residuation operations corresponding to the basic t-norms:*

(a) *Gödel implication $\to_G$ given in (4) is a residuation operation corresponding to minimum '∧'.*
(b) *Product implication $\to_P$ given in (5) is a residuation operation corresponding to multiplication '·'.*
(c) *Łukasiewicz implication $\to_L$ given in (8) is a residuation operation corresponding to Łukasiewicz conjunction '⊗'.*

Theorem 2. *Let **t** be a continuous t-norm and $\to_t$ the corresponding residuation. Then $a \, t \, (a \to_t b) = a \wedge b$.*

operation right

Negation. In the definition of t-conorm, we have used the operation $\neg a = 1 - a$, which has originally been used by L. A. Zadeh for the definition of the complement of a fuzzy set. His choice was good since it turned out that this operation is a convenient interpretation of the negation in fuzzy logic. However, it is an interesting question, what properties must be fulfilled by an operation to be considered as a many-valued negation. Since every generalization should preserve the original, it seems that the weakest requirement is merely to reverse 0 to 1 and vice-versa.

Definition 9. The negation is a non-increasing operation $\mathbf{n} : [0,1] \longrightarrow [0,1]$ such that $\mathbf{n}(0) = 1$ and $\mathbf{n}(1) = 0$. The negation is involutive if $\mathbf{n}(\mathbf{n}(a)) = a$ holds for every $a \in [0,1]$. The negation $\mathbf{n}$ is strict if it is continuous and $a < b$ implies $\mathbf{n}(a) > \mathbf{n}(b)$. It is strong if it is strict and involutive.

Example 13. A typical example of the involutive negation is $\mathbf{n}(a) = 1 - a$. This is obtained from Lukasiewicz implication by putting $\mathbf{n}_L(a) = a \to_L 0$. This negation is even strong.

The negation $\mathbf{n}_G(a) = a \to_G 0$ is non involutive since $\mathbf{n}_G(0) = 1$ but $\mathbf{n}_G(a) = 0$ for every $a \in (0,1]$.

Let $\mathbf{t}$ be a t-norm and $\mathbf{n}$ be a negation. Put

$$a \, \mathbf{s} \, b = \mathbf{n}(\mathbf{n}(a) \, \mathbf{t} \, \mathbf{n}(b)), \qquad a, b \in [0,1]. \tag{34}$$

This defines an operation $\mathbf{s}$ dual to the t-norm $\mathbf{t}$. However, note that this needs not necessarily be a t-conorm.

Lemma 9. *Let $\mathbf{n}$ be an involutive negation.*

  (a) If $\mathbf{t}$ is a t-norm then the operation $\mathbf{s}$ defined in (34) is a t-conorm.
  (b) If $\mathbf{s}$ is a t-conorm then the operation $\mathbf{t}$ defined by

$$a \, \mathbf{t} \, b = \mathbf{n}(\mathbf{n}(a) \, \mathbf{s} \, \mathbf{n}(b)), \qquad a, b \in [0,1] \tag{35}$$

  is a t-norm.

5   Fuzzy logic in narrow sense

5.1   Graded formal logical systems

We start with the theory of graded formal logical systems, which is based on the graded approach principle. The grades are supposed to be taken from the complete residuated lattice

$$\mathcal{L} = \langle L, \vee, \wedge, \otimes, \to, \mathbf{0}, \mathbf{1} \rangle. \tag{36}$$

At the same time, (36) is the set of truth values using which the semantics of formulas in fuzzy logic is defined. This two-way view makes us finally possible to formulate generalization of the Gödel completeness theorem. Let us note that in comparison with the other kinds of many-valued logic, there are no designated truth values, i.e. all the truth values are equal in their importance.

An important principle in fuzzy logic is the *maximality principle* according to which, if the same object is assigned more truth values then its final truth assignment is equal to the *maximum (supremum) of all of them*.

Syntax. First, let us consider some formal language J and a set F_J of well formed formulas, which will be specified below. Let us only note that the language J is supposed to contain the logical constants $\mathbf{a}$ for all the truth values $a \in L$. The specific symbols $\bot, \top$ denote logical constants for $\mathbf{0}, \mathbf{1} \in L$, respectively. Furthermore, we will introduce the concept of evaluated formula.

Definition 10. An evaluated formula is a couple

$$a/A \tag{37}$$

where $A \in F_J$ is a formula and $a \in L$ is its syntactic evaluation.

Definition 11. An n-ary inference rule r in the graded logical system is a scheme

$$r : \frac{a_1/A_1, \ldots, a_n/A_n}{r^{evl}(a_1, \ldots, a_n)/r^{syn}(A_1, \ldots, A_n)}, \tag{38}$$

using which the evaluated formulas $a_i/A_i, \ldots, a_n/A_n$ are assigned the evaluated formula $r^{evl}(a_1, \ldots, a_n)/r^{syn}(A_1, \ldots, A_n)$. The syntactic operation r^{syn} is a partial n-ary operation on F_J and the evaluation operation r^{evl} is an n-ary lower semicontinuous operation on L (i.e. it preserves arbitrary suprema in all variables).

Definition 12. An evaluated formal proof of a formula A from a fuzzy set $X \subseteq F_J$ is a finite sequence of evaluated formulas

$$w := a_0/A_0, \ a_1/A_1, \ldots, \ a_n/A_n \tag{39}$$

such that $A_n := A$ and for each $i \leq n$, either there exists an n-ary inference rule r such that

$$a_i/A_i := r^{evl}(a_{i_1}, \ldots, a_{i_n})/r^{syn}(A_{i_1}, \ldots, A_{i_n}), \quad i_1, \ldots, i_n < i,$$

or

$$a_i/A_i := X(A_i)/A_i.$$

The evaluation a_n of the last member in (39) is the *value* of the evaluated proof w. We will usually write the value of the proof w as $\mathrm{Val}(w)$.

Axioms in the graded formal systems are sets of evaluated formulas. Since the evaluations can be interpreted as membership degrees in the fuzzy set, the axioms are at the same time *fuzzy sets of formulas*. Similarly as in the classical case, we distinguish logical and special axioms, namely $\mathrm{LAx}, \mathrm{SAx} \subseteq F_J$. The we may introduce the following definition.

Definition 13. A formal fuzzy theory T in the language J of FLn is a triple

$$T = \langle \mathrm{LAx}, \mathrm{SAx}, R \rangle$$

where $\mathrm{LAx} \subseteq F_J$ is a fuzzy set of logical axioms, $\mathrm{SAx} \subseteq F_J$ is a fuzzy set of special axioms, and R is a set of sound inference rules.

Given a fuzzy theory T. We denote its language by $J(T)$. If w is a proof in T then its value is denoted by $\mathrm{Val}_T(w)$.

Definition 14. Let T be a fuzzy theory and $A \in F_J$ a formula. We say that the formula A is a theorem in the degree a (or provable in the degree a) in the fuzzy theory T if

$$a = \bigvee \{\mathrm{Val}(w) \mid w \text{ is a proof of } A \text{ from } \mathrm{LAx} \cup \mathrm{SAx}\}. \qquad (40)$$

In this case, we will write $T \vdash_a A$.

Note that unlike the classical syntax where finding one finite proof is sufficient, the situation here is not so convenient. In general, a sequence (possibly infinite) of finite proofs must be considered, because maximum of their values does not need to exist, and so, we cannot confine ourselves to one proof only.

Semantics. A *truth valuation* of formulas from F_J is a function $\mathcal{D} : F_J \longrightarrow L$ being a homomorphism with respect to the connectives (see below). If $\mathbf{a}$ is a logical constant for some $a \in L$ then $\mathcal{D}(\mathbf{a}) = a$.

The truth valuation $\mathcal{D}$ is a model of the fuzzy theory T, $\mathcal{D} \models T$, if $\mathrm{SAx}(A) \leq \mathcal{D}(A)$ holds for all formulas $A \in F_J$.

Definition 15. Let T be a fuzzy theory. we say that A is true in the degree a in the fuzzy theory T if

$$a = \bigwedge \{\mathcal{D}(A) \mid \mathcal{D} \models T\} \qquad (41)$$

and write $T \models_a A$.

Because all the truth values are equal in their importance, we may generalize the concept of tautology as follows.

Definition 16. We say that a formula A is an a-tautology (tautology in the degree a) if

$$a = \bigwedge \{\mathcal{D}(A) \mid \mathcal{D} \text{ is a truth valuation}\} \qquad (42)$$

and write $\models_a A$. If $a = 1$ then we write simply $\models A$ and call A a tautology.

Example 14. Let the set of truth values be the Lukasiewicz algebra $\mathcal{L}_L$. Then the formula of the form $A \vee \neg A$ is a 0.5-tautology. Indeed, in any truth valuation $\mathcal{D}$ we have $\mathcal{D}(A \vee \neg A) = \mathcal{D}(A) \vee \neg \mathcal{D}(A) \geq 0.5$.

Soundness and completeness. Recall that the classical syntactic and semantic consequence operations are two different characterizations of true formulas. An important achievement of classical logic is the theorem stating that, even though each characterization is based on entirely different assumptions, both of them lead to the same result.

We say that the graded formal logical system is *sound* if

$$T \vdash_a A \quad \text{and} \quad T \models_b A \quad \text{implies} \quad a \leq b \qquad (43)$$

holds for every theory T and a formula $A \in F_J$.

The graded formal logical system is *complete* if

$$T \vdash_a A \quad \text{iff} \quad T \models_a A \tag{44}$$

holds for every theory T and a formula $A \in F_J$. Note that due to the definition, we always have $T \vdash_a A$ and $T \models_b A$ for some $a, b \in L$, possibly equal to $\mathbf{0}$.

The completeness of the graded formal logical systems with implication, however, can be obtained only in a very special case, which can be seen from the theorem below.

Theorem 3. *Given a graded formal logical system with the implication connective $\Rightarrow$ based on a complete residuated lattice (36). If the interpretation $\rightarrow$ of $\Rightarrow$ does not fulfil the equations*

$$\bigvee_{i \in I} (a \rightarrow b_i) = a \rightarrow \left(\bigvee_{i \in I} b_i \right) \tag{45}$$

$$\bigvee_{i \in I} (a_i \rightarrow b) = \left(\bigwedge_{i \in I} a_i \right) \rightarrow b \tag{46}$$

$$\bigwedge_{i \in I} (a \rightarrow b_i) = a \rightarrow \left(\bigwedge_{i \in I} b_i \right) \tag{47}$$

$$\bigwedge_{i \in I} (a_i \rightarrow b) = \left(\bigvee_{i \in I} a_i \right) \rightarrow b \tag{48}$$

for arbitrary subset of L then such a system cannot be complete.

It follows from this theorem and the properties of t-norms that if we suppose the support of the structure of truth values to be the interval $[0, 1]$ then the implication operation must be continuous and thus, the corresponding residuated lattice must be isomorphic with the Łukasiewicz algebra.

In the sequel, we will give detailed specific definition of the syntax and semantics of fuzzy logic.

5.2   Syntax of predicate first-order fuzzy logic

The *language J* of predicate fuzzy logic consists of:

(i) Countable set of object variables $x, y, \ldots$.
(ii) Finite or countable set of object constants $\mathbf{u}_1, \mathbf{u}_2, \ldots$.
(iii) Finite or countable set of n-ary functional symbols $f, g, \ldots$.
(iv) Nonempty finite or countable set of n-ary predicate symbols $P, Q, \ldots$.
(v) A set of logical constants $\{\mathbf{a} \mid a \in L\}$.

(vi) Symbol for the logical binary connective of implication $\Rightarrow$.
(vii) Symbol for the general quantifier $\forall$.
(viii) Various kinds of brackets as auxiliary symbols.

Terms are symbols denoting objects.

Definition 17. (i) A variable x or constant $\mathbf{u}$ is a (atomic) term.
  (ii) Let f be an n-ary functional symbol and $t_1,\ldots,t_n$ terms. Then the expression $f(t_1,\ldots,t_n)$ is a term.

Definition 18. (i) A logical constant $\mathbf{a}$ is a (atomic) formula.
  (ii) Let P be an n-ary predicate symbol and $t_1,\ldots,t_n$ be terms. Then the expression $P(t_1,\ldots,t_n)$ is a (atomic) formula.
  (iii) If A,B are formulas then $A\Rightarrow B$ is a formula.
  (iv) If x is a variable and A a formula then $(\forall x)A$ is a formula.
  (v) All formulas are formed by finite application of the previous four items.

A set of all terms of the language J is denoted by M_J and a set of all formulas by F_J. Furthermore, we will often use the symbol M_V to denote the set of all closed terms of the language J. The concepts of the scope of the quantifier, sets $FV(t), FV(A)$ of free variables, substitutible terms, open and closed formula are the same as in the classical logic.

We will furthermore work with the following abbreviations of formulas.

$$\neg A := A \Rightarrow \bot \qquad \text{(negation)}$$
$$A \vee B := (B \Rightarrow A) \Rightarrow A \qquad \text{(disjunction)}$$
$$A \wedge B := \neg((B \Rightarrow A) \Rightarrow \neg B) \qquad \text{(conjunction)}$$
$$A \,\&\, B := \neg(A \Rightarrow \neg B) \qquad \text{(Łukasiewicz conjunction)}$$
$$A \nabla B := \neg(\neg A \,\&\, \neg B) \qquad \text{(Łukasiewicz disjunction)}$$
$$A \Leftrightarrow B := (A \Rightarrow B) \wedge (B \Rightarrow A) \qquad \text{(equivalence)}$$
$$A^n := \underbrace{A \,\&\, A \,\&\, \cdots \,\&\, A}_{n-\text{times}} \qquad \text{(n-fold conjunction)}$$

$$nA := \underbrace{A \nabla A \nabla \cdots \nabla A}_{n-\text{times}} \qquad \text{(n-fold disjunction)}$$

$$(\exists x)A := \neg(\forall x)\neg A \qquad \text{(existential quantifier)}$$

Inference rules.

  (i) The inference rule of *modus ponens* is the scheme

$$r_{MP} : \frac{a/A,\; b/A \Rightarrow B}{a \otimes b/B}. \tag{49}$$

(ii) The inference rule of *generalization* is the scheme

$$r_G : \frac{a/A}{a/(\forall x)A}.$$

(50)

(iii) The inference rule of *logical constant introduction* is the scheme

$$r_{LC} : \frac{a/A}{a \rightarrow a/\mathbf{a} \Rightarrow A}$$

(51)

(note that the evaluation operation of r_{LC} is $r_{LC}^{evl}(x) = a \rightarrow x$).

Given an evaluated formula a/A, the rule r_{LC} enables to represent the evaluation a, which otherwise comes from the outside and does not belong to the language J, inside the latter. Consequently, we may identify the evaluated formula with the formula $\mathbf{a} \Rightarrow A$.

Logical axioms.

Definition 19. Given the following schemes of formulas.

(R1) $A \Rightarrow (B \Rightarrow A)$.
(R2) $(A \Rightarrow B) \Rightarrow ((B \Rightarrow C) \Rightarrow (A \Rightarrow C))$.
(R3) $(\neg B \Rightarrow \neg A) \Rightarrow (A \Rightarrow B)$.
(R4) $((A \Rightarrow B) \Rightarrow B) \Rightarrow ((B \Rightarrow A) \Rightarrow A)$.
(B1) $(\mathbf{a} \Rightarrow \mathbf{b}) \Leftrightarrow \overline{(a \rightarrow b)}$
    where $\mathbf{a} \rightarrow \mathbf{b}$ denotes the logical constant (atomic formula) for the truth value $a \rightarrow b$ when a and b are given.
(T1) $(\forall x)A \Rightarrow A_x[t]$
    for any substitutible term t.
(T2) $(\forall x)(A \Rightarrow B) \Rightarrow (A \Rightarrow (\forall x)B)$
    provided that x is not free in A.

The fuzzy set LAx of logical axioms is specified as follows:

$$\text{LAx}(A) = \begin{cases} a \in L & \text{if } A := \mathbf{a}, \\ 1 & \text{if } A \text{ has some of the forms (R1)–(R4), (T1)–(T3),} \\ 0 & \text{otherwise.} \end{cases}$$

5.3 Semantics of predicate first-order fuzzy logic

Definition 20. A structure for the language J of FLn is

$$\mathcal{D} = \langle D, P_D, \ldots, f_D, \ldots, u_1, \ldots \rangle$$

where D is a set, $P_D \subseteq D$, ... are n-ary fuzzy relations assigned to each n-ary predicate symbol $P, \ldots$, and f_D are ordinary (crisp) n-ary functions on D assigned to each n-ary functional symbol f. Finally, $u_1, \ldots \in D$ are designated elements which are assigned to each constant $\mathbf{u}_1, \ldots$ of the language J, respectively.

A *truth valuation* of formulas in FLn is realized by means of their interpretation in some structure $\mathcal{D}$ in a way analogous to classical logic. For the given structure, it is a function $F_J \longrightarrow \mathcal{L}$ defined below.

Let $\mathcal{D}$ be a structure for the language J. We extend J to the language $J(\mathcal{D})$ by new constants being names for all the elements from D, i.e.constants will be denoted by the corresponding bold-face letter, namely $J(\mathcal{D}) = J \cup \{\mathbf{d} \mid d \in D\}$.

Definition 21.   (i) Interpretation of closed terms:

$$\mathcal{D}(\mathbf{u}_i) = u_i, \quad \mathbf{u}_i \in J, u_i \in D,$$
$$\mathcal{D}(\mathbf{d}) = d, \quad d \in D,$$
$$\mathcal{D}(f(t_1, \ldots, t_n)) = f_D(\mathcal{D}(t_1), \ldots, \mathcal{D}(t_n)).$$

(ii) Interpretation of closed formulas: let $t_1, \ldots, t_n$ be closed terms. Then

$$\mathcal{D}(\mathbf{a}) = a, \quad \text{for all} \quad a \in L,$$
$$\mathcal{D}(P(t_1, \ldots, t_n)) = P_D(\mathcal{D}(t_1), \ldots, \mathcal{D}(t_n)),$$
$$\mathcal{D}(A \Rightarrow B) = \mathcal{D}(A) \rightarrow \mathcal{D}(B),$$
$$\mathcal{D}((\forall x)A) = \bigwedge \{\mathcal{D}(A_x[\mathbf{d}]) \mid d \in D\},$$

and in the case of the derived connectives,

$$\mathcal{D}(A \wedge B) = \mathcal{D}(A) \wedge \mathcal{D}(B),$$
$$\mathcal{D}(A \,\&\, B) = \mathcal{D}(A) \otimes \mathcal{D}(B),$$
$$\mathcal{D}(A \vee B) = \mathcal{D}(A) \vee \mathcal{D}(B),$$
$$\mathcal{D}(A \nabla B) = \mathcal{D}(A) \oplus \mathcal{D}(B),$$
$$\mathcal{D}(A \leftrightarrow B) = \mathcal{D}(A) \leftrightarrow \mathcal{D}(B),$$
$$\mathcal{D}((\exists x)A) = \bigvee \{\mathcal{D}(A_x[\mathbf{d}]) \mid d \in D\}.$$

The interpretation of open formulas is taken as infimum of the truth values of all the closed formulas obtained over all assignments of elements taken from the support to their free variables.

Tautologies play central role in every logical system. The following lemma gives simple rules how to verify tautologies in fuzzy logic.

Lemma 10. *Let A, B be formulas in the language J.*

(a) $\models A \Rightarrow B$ *iff* $\mathcal{D}(A) \leq \mathcal{D}(B)$ *holds in every structure $\mathcal{D}$ for the language J.*

(b) $\models A \Leftrightarrow B$ *iff* $\mathcal{D}(A) = \mathcal{D}(B)$ *holds in every structure $\mathcal{D}$ for the language J.*

5.4   Provability in fuzzy theories

Recall that due to our assumption about equal importance of all the truth degrees, every formula is a theorem (provable) in some degree, possibly equal to **0**. In general, $T \vdash_a A$ means only a supremum of the the values of all proofs of A. In special case, however, there may exist a proof with the value a. In this case we will say that A is *effectively provable*.

Lemma 11. *The following are schemes of effectively provable formal propositional tautologies in T, i.e. for each theorem there exists a proof with the degree equal to $\mathbf{1}$.*

(P1) $T \vdash A \Rightarrow A$,

(P2) $T \vdash A \Rightarrow \top$,

(P3) $T \vdash \neg\neg A \Rightarrow A$, $T \vdash A \Rightarrow \neg\neg A$,

(P4) $T \vdash (A \wedge B) \Rightarrow A$, $T \vdash (A \wedge B) \Rightarrow B$,

(P5) $T \vdash (A \,\&\, B) \Rightarrow A$, $T \vdash (A \,\&\, B) \Rightarrow B$,

(P6) $T \vdash A \Rightarrow (A \vee B)$, $T \vdash B \Rightarrow (A \vee B)$,

(P7) $T \vdash (A \,\&\, \neg A) \Rightarrow B$,

(P8) $T \vdash (A \,\&\, B) \Leftrightarrow (B \,\&\, A)$,

(P9) $T \vdash ((A \,\&\, B) \Rightarrow C) \Leftrightarrow (A \Rightarrow (B \Rightarrow C))$,

(P10) $T \vdash \neg(A \Rightarrow B) \Leftrightarrow (A \,\&\, \neg B)$,

(P11) $T \vdash (A \Rightarrow B) \Leftrightarrow (\neg A \nabla B)$,

(P12) $T \vdash (A \Rightarrow B) \vee (B \Rightarrow A)$,

(P13) $T \vdash (A \wedge B) \Leftrightarrow \neg(\neg A \vee \neg B)$,

(P14) $T \vdash (A \Rightarrow (B \Rightarrow C)) \Leftrightarrow (B \Rightarrow (A \Rightarrow C))$,

(P15) $T \vdash (A^m \Rightarrow (B \Rightarrow C)) \Rightarrow ((A^n \Rightarrow B) \Rightarrow (A^{m+n} \Rightarrow C))$, $m, n \in \mathbb{N}^+$,

Lemma 12. *Let T be a fuzzy predicate calculus. The following are schemes of the effectively provable formal predicate theorems in T, i.e. for each theorem there exists a proof with the degree equal to $\mathbf{1}$.*

(Q1) $T \vdash (\forall x)(A \Rightarrow B) \Rightarrow ((\forall x)A \Rightarrow (\forall x)B)$,

(Q2) $T \vdash (\forall x)(A \,\&\, B) \Leftrightarrow ((\forall x)A \,\&\, B)$ *where x is not free in B,*

(Q3) $T \vdash (\forall x)(A \Rightarrow B) \Leftrightarrow (A \Rightarrow (\forall x)B)$ *where x is not free in A,*

(Q4) $T \vdash (\forall x)(A \Rightarrow B) \Leftrightarrow ((\exists x)A \Rightarrow B)$ *where x is not free in B,*

(Q5) $T \vdash (\exists x)(A \Rightarrow B) \Leftrightarrow (A \Rightarrow (\exists x)B)$ *where x is not free in A,*

(Q6) $T \vdash (\exists x)(A \Rightarrow B) \Leftrightarrow ((\forall x)A \Rightarrow B)$ *where x is not free in B,*

(Q7) $T \vdash (\forall x)A \Leftrightarrow \neg(\exists x)\neg A,$
(Q8) $T \vdash A_x[t] \Rightarrow (\exists x)A$ *for every term* $t.$

Lemma 13. *(a) To every proof* w_A *with* $\mathrm{Val}(A) = a$ *there is a proof* $w_{b \Rightarrow A}$
  with the value $\mathrm{Val}(w_{b \Rightarrow A}) = b \to a.$
  (b) In every fuzzy theory, $T \vdash_a A$ *implies* $T \vdash \mathbf{a} \Rightarrow A$ *for all formulas* A
  and $a \in L.$

5.5   Some important theorems of fuzzy logic

Due to the following theorem, we may derive formulas whose syntactic evaluation is at most as big as their semantic interpretation.

Theorem 4 (validity theorem). *Let* T *be a fuzzy theory. If* $T \vdash_a A$ *and* $T \models_b A$ *then* $a \leq b$ *holds for every formula* $A.$

As a special case, it follows from the definition that $T \vdash_a \mathbf{a}$ and $T \models_a \mathbf{a}$ holds for every logical constant $\mathbf{a}$, $a \in L$, in every fuzzy theory $T.$

The following theorem demonstrates that we may confine ourselves only to closed formulas.

Theorem 5 (closure theorem). *Let* T *be a fuzzy theory. Let* $A \in F_{J(T)}$ *and* A' *be its closure. Then* $T \vdash_a A$ *iff* $T \vdash_a A'.$

Theorem 6. *Let* T *be a consistent fuzzy theory and* $T \vdash_a A$ *and* $T \vdash_b B.$

  (a) $T \vdash_c A \Rightarrow B$ *implies* $c \leq a \to b.$
  (b) $T \vdash_c A \& B$ *implies* $c \geq a \otimes b.$
  (c) If $T \vdash_a (\forall x)A$ *then* $a \leq \bigwedge\{b \mid T \vdash_b A_x[t], t \in M_J(T)\}.$
  (d) If $T \vdash_c A \wedge B$, $T \vdash_a A$ *and* $T \vdash_b B$ *then* $c = a \wedge b.$

Characterization of consistent theories.

Definition 22. A fuzzy theory T is contradictory if there is a formula A and proofs w_A and $w_{\neg A}$ of A and $\neg A$, respectively, such that

$$\mathrm{Val}_T(w_A) \otimes \mathrm{Val}_T(w_{\neg A}) > 0. \tag{52}$$

It is consistent in the opposite case.

Obviously, if T is contradictory then $T \vdash_a A$, $T \vdash_b \neg A$ and $a \otimes b > 0.$

The following theorem states that a contradictory fuzzy theory collapses into a degenerated theory just as in classical logic.

Theorem 7 (Contradiction). *A fuzzy theory* T *is contradictory iff* $T \vdash A$
holds for every formula $A \in F_{J(T)}.$

One might think that Definition 22 is too strong and therefore, we may modify (52) by replacing $\otimes$ by the infimum $\wedge$.

Theorem 8. *A fuzzy theory T is consistent iff $\vee\{a \mid T \vdash_a A \wedge \neg A \text{ and } A \in \mathcal{F}_{\mathcal{L}(T)}\} \leq \frac{1}{2}$.*

In other words, if there is a formula A such that $T \vdash_a A \wedge \neg A$ where $a > \frac{1}{2}$ then T is again contradictory in the sense of Definition 22.

The concept of the consistency threshold of a fuzzy theory is an interesting possibility specific for the fuzzy logic, which, on the other hand, has no sense in classical logic.

Definition 23. Let T be a consistent theory and A a formula. Then an element $\neg a$ is a consistency threshold for A in T if $T' = T \cup \{\, b/\neg A\}$ is contradictory for all $b > \neg a$ and consistent otherwise.

Lemma 14. *Let T be a consistent fuzzy theory and $T \vdash_a A$ where A is a closed formula. Let $T' = T \cup \{\, b/\neg A\}$. Then T' is consistent iff $b \leq \neg a$.*

Definition 24. A language J' is an extension of the language J if $J \subseteq J'$. A fuzzy theory T' is an extension of the fuzzy theory T if $J(T) \subseteq J'(T')$ and $T \vdash_a A$ and $T' \vdash_b A$ implies $a \leq b$ for every formula $A \in F_{J(T)}$.

Extension of fuzzy theories. Occasionally, SAx may be extended by some fuzzy set of formulas $\Gamma \subseteq F_J$. Then by $T \cup \Gamma$ we understand an extended fuzzy theory
$$T \cup \Gamma = \langle \text{LAx}, \text{SAx} \cup \Gamma, R \rangle.$$

The extension T' is a conservative extension of T if $T' \vdash_b A$ and $T \vdash_a A$ implies $a = b$ for every formula $A \in F_{J(T)}$. The extension T' is a simple extension of T if $J'(T') = J(T)$.

The following theorem is a generalization of the classical important deduction theorem.

Theorem 9 (deduction theorem). *Let T be a fuzzy theory, A be a closed formula and $T' = T \cup \{1/A\}$. Then to every formula $B \in F_{J(T)}$ there is an n such that*
$$T \vdash_a A^n \Rightarrow B \quad \text{iff} \quad T' \vdash_a B.$$

Completeness theorem. In this section, the analogy of the first form of the completeness theorem is presented. It is a nice generalization of the classical completeness theorem since it gives the provability and truth degrees of a formula into an exact balance, thus concluding the proclaim that these concepts nontrivially generalize the classical concepts of provability and validity of a formula. By the analogy with classical logic, there are two forms of the completeness theorem. It is surprizing that the formulation of the second form of the completeness theorem is exactly the same as in classical logic. However, its proof is not a trivial analogy.

Theorem 10 (Completeness theorem II). *A fuzzy theory T is consistent iff it has a model.*

Theorem 11 (Completeness theorem I).

$$T \vdash_a A \quad \text{iff} \quad T \models_a A$$

holds for every formula $A \in F_J$ and every consistent fuzzy theory T.

Sorites fuzzy theories. The completeness theorem is a deep result with many consequences. In this subsection, we will demonstrate its application to the sorites paradox discussed above.

Theorem 12. *Let T be a fuzzy theory in which all Peano axioms are accepted in the degree 1. Furthermore, let $1 \geq \varepsilon > 0$ and $\mathbb{HR} \notin J(T)$ be a new predicate. Then the fuzzy theory*

$$T^+ = T \cup \{ \, 1/\mathbb{HR}(0), \, 1 - \varepsilon/(\forall x)(\mathbb{HR}(x) \Rightarrow \mathbb{HR}(x+1)), \, 1/(\exists x)\neg\mathbb{HR}(x) \}$$

is a conservative extension of T.

6   Functional systems in fuzzy logic theories

This section relates to what is known as many-valued functional systems, i.e. systems of functions, whose domain and range are given by the same set of truth values. Functional systems relate to logical theories in such a way that formulas of the latter correspond to elements of the former.

The following are special problems to be considered:

(i) characterization of the functional systems related to propositional and predicate calculi,

(ii) construction of special formulas, which generalize Boolean disjunctive and conjunctive normal forms,

(iii) investigation of what can be approximately described by fuzzy logic normal forms in both calculi,

(iv) investigation of the quality of an approximation, which can be achieved by the special choice of a defuzzification.

6.1   Fuzzy logic functions and normal forms

The following definition generalizes the notion of a 2-valued Boolean function.

Definition 25. Let L be a support of a structure of truth values. A function $f(x_1, \ldots, x_n)$, $n \geq 0$, defined on the set L^n and taking values from L is called an L-valued fuzzy logic function (or, shortly, an L-valued FL-function).

The set of all the L-valued FL-functions will be denoted by P_L. In the case that $L = \{0, 1\}$, $P_{\{0,1\}}$ is a set of Boolean functions.

In what follows we restrict ourselves to $[0, 1]$-valued FL-functions and the structure of truth values will be assumed to be a Lukasiewicz MV-algebra. Thus, we drop the attribute "$[0, 1]$-valued".

It is known that each Boolean function can be represented by a certain logical formula of the classical propositional calculus. It can be shown that this is not true for any L-valued FL-function when $L = [0, 1]$ even if the set of constants is extended to $L = [0, 1]$. Precisely, due to McNaughton (1951), Mundici (1994), Perfilieva, Tonis (1995) it has been proved that $[0, 1]$-valued FL-function can be represented by a formula of propositional calculus iff it is a piecewise linear with integer coefficients.

Moreover, the special interest in fuzzy logic is paid to some distinguished formulas, which are linguistically expressed by "IF-THEN" rules (cf. Sections 2 and 7). Many applications are based on described in such a way models. Thus, it is imprecisely supposed that a certain class of dependences can be described by those rules appropriately. We will show that this presupposition is not an exaggeration. To do this, we translate "IF-THEN" rules into fuzzy logic formulas representable either in two forms, generalizing well known Boolean disjunctive and conjunctive normal forms. Then we investigate what kind of dependences (functions) can be represented by these generalized formulas. We will do it both in propopositional and predicate calculi as well.

Normal forms in fuzzy propositional calculus. First, we recall the disjunctive and conjunctive normal forms for Boolean functions and explain our way of generalization. Let $f(x_1, \ldots, x_n)$, $n \geq 1$, be a Boolean function, i.e. its domain and range is the set $\{0, 1\}$. To represent the propositional variable x and its negation, we will write

$$x^\sigma = \begin{cases} x' & \text{if } \sigma = 0, \\ x & \text{if } \sigma = 1. \end{cases}$$

It is easy to see that $x^\sigma = 1$ iff $x = \sigma$. Then $f(x_1, \ldots, x_n)$ can be represented by the disjunctive normal form

$$f(x_1, \ldots, x_n) = \bigvee_{f(\sigma_1, \ldots, \sigma_n)=1} (x_1^{\sigma_1} \wedge \cdots \wedge x_n^{\sigma_n}) =$$

$$= \bigvee_{(\sigma_1, \ldots, \sigma_n)} (x_1^{\sigma_1} \wedge \cdots \wedge x_n^{\sigma_n} \wedge f(\sigma_1, \ldots, \sigma_n))$$

provided that $f \not\equiv 0$, and by the conjunctive normal form

$$f(x_1, \ldots, x_n) = \bigwedge_{f(\tau_1, \ldots, \tau_n)=0} (x_1^{\bar{\tau}_1} \vee \cdots \vee x_n^{\bar{\tau}_n}) =$$

$$= \bigwedge_{(\tau_1, \ldots, \tau_n)} (x_1^{\bar{\tau}_1} \vee \cdots \vee x_n^{\bar{\tau}_n} \vee f(\tau_1, \ldots, \tau_n))$$

provided that $f \not\equiv 1$.

A special interest will be payed to the right-hand sides of both expressions since they are prototypes of the suggested generalization. Analyzing them we see that they contain two parts, namely the characterization (positive or negative) of the set $\{(\sigma_1, \ldots \sigma_n)\} \subset \{0, 1\}^n$ or $\{(\tau_1, \ldots \tau_n)\} \subset \{0, 1\}^n$ respectively, and the value of the represented function defined on it.

Below, we suggest a generalization of the disjunctive and conjunctive normal forms for FL-functions defined on $[0, 1]$ over Lukasiewicz MV-algebra of operations $\mathcal{L}_L$. The induced MV-algebra of functions $\mathcal{P}_{[0,1]}$ is then based on the set $P_{[0,1]} = \{f(x_1, \ldots, x_n) \mid f : [0, 1]^n \longrightarrow [0, 1], \quad n \geq 0\}$ so that

$$\mathcal{P}_{[0,1]} = \langle P_{[0,1]}, \oplus, \otimes, \neg, 0, 1 \rangle$$

and each operation is defined pointwise. The constants $0, 1$ are the respective constant FL-functions.

We will show how the generalized normal forms DNF and CNF look like and moreover, how the functions represented by them can be characterized. We will follow the explained construction of Boolean forms consisting in two parts and will preserve also two parts in each elementary conjunction or disjunction. The first is the characterization (positive or negative respectively) of a certain set and the second is the average value of a certain function on it.

Let m be a positive integer. Define the interval functions

$$I_k^m(x) = \begin{cases} 1, & \text{if } \frac{k}{m} \leq x < \frac{(k+1)}{m}, \\ 0, & \text{otherwise} \end{cases}$$

where $0 \leq k \leq m - 1$, and

$$I_m^m(x) = \begin{cases} 1, & \text{if } x = 1, \\ 0, & \text{otherwise.} \end{cases}$$

As can be seen, each function I_k^m, $0 \le k \le m-1$, is a characteristic function of the subset $\{x \mid \frac{k}{m} \le x < \frac{(k+1)}{m}\}$. Hence, both this function and its negation will be used in the first parts of the respective normal forms.

Definition 26. Let $f(x_1, \ldots, x_n)$ be an FL-function and m be a positive integer. The formula over $\mathcal{P}_{[0,1]}$

$$\mathrm{DNF}(x_1, \ldots, x_n) = \bigvee_{k_1=0}^{m} \cdots \bigvee_{k_n=0}^{m} (I_{k_1}^m(x_1) \otimes \cdots \otimes I_{k_n}^m(x_n) \otimes f(\frac{k_1}{m}, \ldots, \frac{k_n}{m}))$$

is called the MV-disjunctive normal form for $f(x_1, \ldots, x_n)$ and

$$\mathrm{CNF}(x_1, \ldots, x_n) = \bigwedge_{k_1=0}^{m} \cdots \bigwedge_{k_n=0}^{m} (\neg I_{k_1}^m(x_1) \oplus \cdots \oplus \neg I_{k_n}^m(x_n) \oplus f(\frac{k_1}{m}, \ldots, \frac{k_n}{m}))$$

is called the MV-conjunctive normal form for $f(x_1, \ldots, x_n)$.

Let us apply the identity $\neg a \oplus b = a \to b$ to the expression of CNF and thus derive the following equivalent form

$$\mathrm{CNF}(x_1, \ldots, x_n) = \bigwedge_{k_1=0}^{m} \cdots \bigwedge_{k_n=0}^{m} ((I_{k_1}^m(x_1) \otimes \cdots \otimes I_{k_n}^m(x_n)) \to f(\frac{k_1}{m}, \ldots, \frac{k_n}{m})).$$

If the normal forms include one variable then their representation is a simple reduction of certain formulas such that

$$\mathrm{DNF}(x) = \bigvee_{k=0}^{m} \left(I_k^m(x) \otimes f\left(\frac{k}{m}\right) \right),$$

$$\mathrm{CNF}(x) = \bigwedge_{k=0}^{m} \left(\neg I_k^m(x) \oplus f\left(\frac{k}{m}\right) \right).$$

Both these forms represent the same function

$$f_{\mathrm{DNF}}(x) = f_{\mathrm{CNF}}(x) = \begin{cases} f(\frac{k}{m}), & \text{if } \frac{k}{m} \le x < \frac{k+1}{m}, \quad 0 \le k \le m-1, \\ f(1), & \text{if } x = 1. \end{cases}$$

Remark 1. The expressions $\mathrm{DNF}(x_1, \ldots, x_n)$ and $\mathrm{CNF}(x_1, \ldots, x_n)$ are used to denote the corresponding formulas over $\mathcal{P}[0,1]$. When dealing with FL-functions represented by DNF and CNF, the notation $f_{\mathrm{DNF}}(x_1, \ldots, x_n)$ and $f_{\mathrm{CNF}}(x_1, \ldots, x_n)$ will be used respectively.

Generally, the normal form (disjunctive or conjunctive) for the function $f(x_1, \ldots, x_n)$ does not represent the same function as f. We can only prove that there is an effect of approximation. In order to speak about it correctly we will use the notion which plays the role of a distance in MV-algebra. According to the definition of C. C. Chang, the FL-function

$$d(x, y) = (\neg x \otimes y) \oplus (\neg y \otimes x)$$

can be taken as a distance.

Note, that in the case of $L = [0, 1]$, the distance $d(x, y) = |x - y|$ and thus, we can use classical notions of continuity and uniform continuity for our FL-functions.

Definition 27. Let $u \in L$ and $u < 1$. An FL-function $g(x_1, \ldots, x_n)$ is said to u-approximate an FL-function $f(x_1, \ldots, x_n)$, or in other words $g(x_1, \ldots, x_n)$ approximates $f(x_1, \ldots, x_n)$ with the accuracy u, if the inequality

$$d(g(x_1, \ldots, x_n), f(x_1, \ldots, x_n)) \leq u$$

holds for all $(x_1, \ldots, x_n) \in L^n$.

Theorem 13. Let an FL-function $f(x_1, \ldots, x_n)$ be uniformly continuous on L. Then for any $u \in L$, $0 < u < 1$, there exists $v \in L$, $v > 0$, such that each function $f_{\text{DNF}}(x_1, \ldots, x_n)$ or $f_{\text{CNF}}(x_1, \ldots, x_n)$ based on the parameter v, u-approximates $f(x_1, \ldots, x_n)$.

Normal forms in fuzzy predicate calculus. In this paragraph we introduce the disjunctive and conjunctive normal forms and explain approximation theorems for uniformly continuous fuzzy relations. At first, we recall what is understood by relations in connection with fuzzy logic.

Definition 28. Let D be a nonempty set of objects and L be a support of a complete MV-algebra. An n-ary fuzzy relation R on D is a fuzzy set $R \subseteq D^n$, $n \geq 0$, which is given by the membership function $R(x_1, \ldots, x_n)$ defined on the set D^n and taking values from L.

Note that according to this definition, we identify a fuzzy relation with its membership function to be able to work with functions in the sequel.

Definition 29. Let $P_1, \ldots, P_k$ be unary predicate symbols and $E_{i_1 \ldots i_n}$, $1 \leq i_j \leq k$, $1 \leq j \leq n$, $n \geq 1$, be some atomic formulas. The following formulas of fuzzy predicate logic are called the MV-disjunctive normal form

$$\text{DNF}(x_1, \ldots, x_n) = \bigvee_{i_1=1}^{k} \cdots \bigvee_{i_n=1}^{k} (P_{i_1}(x_1) \,\&\, \cdots \,\&\, P_{i_n}(x_n) \,\&\, E_{i_1 \ldots i_n})$$

and the MV-conjunctive normal form

$$\text{CNF}(x_1, \ldots, x_n) = \bigwedge_{i_1=1}^{k} \cdots \bigwedge_{i_n=1}^{k} (\neg P_{i_1}(x_1) \nabla \cdots \nabla \neg P_{i_n}(x_n) \nabla E_{i_1 \ldots i_n}).$$

As in the previous paragraph, the expression CNF can be rewritten into the following one based on implicaton, namely

$$\mathrm{CNF}(x_1,\ldots,x_n) = \bigwedge_{i_1=1}^{k} \cdots \bigwedge_{i_n=1}^{k} (P_{i_1}(x_1) \,\&\cdots\&\, P_{i_n}(x_n) \Rightarrow E_{i_1\ldots i_n}).$$

In the sequel, we will use the shorts DNF and CNF. Contrary to the case of propositional fuzzy logic where we suggested the expressions for the normal forms as algebraic formulas, the latter expressions are formulas of fuzzy predicate logic. In what follows we suppose a support D of a structure $\mathcal{D}$ to be a compact uniform topological space. Let ρ be a pseudometrics, which belongs to the gage of the uniformity on D, i.e. to the family of all pseudometrics which are uniformly continuous on $D \times D$.

Definition 30. An n-ary fuzzy relation $R \subseteq D^n$ is uniformly continuous on D^n if for any ε, $0 < \varepsilon < 1$, there exists a positive number r, such that if $\rho(x_{11}, x_{21}) < r, \ldots, \rho(x_{1n}, x_{2n}) < r$ then

$$|R(x_{11},\ldots,x_{1n}) - R(x_{21},\ldots,x_{2n})| < \varepsilon.$$

Let us stress that this notion coincides with the analogous classical notion for real valued functions defined on a compact uniform space.

Definition 31. A fuzzy relation $S(x_1,\ldots,x_n)$ on D is said to ε-approximate another fuzzy relation $R(x_1,\ldots,x_n)$ on D if

$$|S(x_1,\ldots,x_n) - R(x_1,\ldots,x_n)| \leq \varepsilon$$

is true for all $(x_1,\ldots,x_n) \in D^n$.

Theorem 14. *Let a fuzzy relation $R \subseteq D^n$ be uniformly continuous on D^n. Then for any ε, $0 < \varepsilon < 1$, there are a formula in the form of $\mathrm{DNF}(x_1,\ldots,x_n)$ (or, $\mathrm{CNF}(x_1,\ldots,x_n)$) and a structure $\mathcal{D}$ with the support D such that the fuzzy relation represented by DNF (CNF) w.r.t. $\mathcal{D}$ ε-approximates the fuzzy relation $R(x_1,\ldots,x_n)$.*

6.2 Fuzzy logic formulas as a tool for an approximate description

In the previous subsection we have considered a class of fuzzy relations represented by normal forms (disjunctive or conjunctive) of predicate fuzzy logic and proved that any fuzzy relation with continuous membership function can be approximated by some element from this class. In this subsection, we will focus on analogous problem of approximation of any continuous function on a compact set in the same class.

First, we will apply the above mentioned results. However, this is possible only if we take a function as a particular case of a fuzzy relation with continuous membership function. We will then obtain a non-direct solution of the problem. The second solution will be provided by a direct algorithmic construction of a fuzzy relation represented by a normal form, which approximates the given function.

For simplicity, we will confine ourselves to real valued real functions. Note, that the general case requires to consider many sorted relations, which might lead to unnecessary complexity. Moreover, throughout this subsection when speaking about a formula we will suppose some quantifier-free one of predicate fuzzy logic.

Let us fix some continuous real valued real function $f(x_1, \ldots, x_n)$, $n \geq 1$, defined on a compact set $D^n \in \mathbb{R}^n$. Obviously, $f(x_1, \ldots, x_n)$ is also uniformly continuous on D^n, i.e. for any $\varepsilon > 0$ there exists $\delta > 0$ such that

$$|f(x_{11}, \ldots, x_{1n}) - f(x_{21}, \ldots, x_{2n})| < \varepsilon \tag{53}$$

whenever $|x_{1j} - x_{2j}| < \delta$, $1 \leq j \leq n$, and $(x_{11}, \ldots, x_{1n}), (x_{21}, \ldots, x_{2n}) \in D^n$. Moreover, $f(D^n)$ is a compact set, as well.

Let $\varepsilon > 0$ be an arbitrary number, and let $\delta > 0$ be as described above. Since D is a compact set there exists a finite covering of it by open δ- intervals $I_1, \ldots, I_k$, $k \geq 1$, respectively. It follows that the set D^n is also covered by all the direct products of those intervals. Since f is uniformly continuous, the set $f(D^n)$ will be covered by intervals $f(I_{i_1} \times \cdots \times I_{i_n})$, $1 \leq i_j \leq k$, $1 \leq j \leq n$, whose lengths are not greater than ε. Their centers will be denoted by $y_{i_1 \ldots i_n}$.

For each δ-interval I_i, $1 \leq i \leq k$, choose any closed subinterval $IC_i \subset I_i$ and denote the closure of I_i by CI_i. Furthermore, consider open ε-intervals around the points $y_{i_1 \ldots i_n}$, $1 \leq i_j \leq k$, $1 \leq j \leq n$, and denote them by $H_{i_1 \ldots i_n}$. Observe that $f(I_{i_1} \times \cdots \times I_{i_n}) \subseteq H_{i_1 \ldots i_n}$. Choose any closed subinterval $HC_{i_1 \ldots i_n} \subset H_{i_1 \ldots i_n}$ containing the point $y_{i_1 \ldots i_n}$. By virtue of Urysohn lemma, for each i, $1 \leq i \leq k$, there exists a continuous function $g_i(x)$ on D taking values from $[0, 1]$, which is equal to one on IC_i and equal to zero on $D \setminus I_i$. At the same time, for each n-tuple $(i_1 \ldots i_n)$, $1 \leq i_j \leq k$, $1 \leq j \leq n$, there exists a continuous function $h_{i_1 \ldots i_n}(y)$ on $f(D^n)$ such that it takes values from $[0, 1]$, it is equal to one on $HC_{i_1 \ldots i_n}$ and it is equal to zero on $f(D^n) \setminus H_{i_1 \ldots i_n}$.

Now, for the given function f let us construct an $(n + 1)$- ary fuzzy relation $R_f \subseteq D^n \times f(D^n)$

$$R_f(x_1, \ldots, x_n, y) = \bigvee_{i_1}^{k} \ldots \bigvee_{i_n}^{k} (g_{i_1}(x_1) \cdots g_{i_n}(x_n) \cdot h_{i_1 \ldots i_n}(y)), \tag{54}$$

which approximates f in a sense explained below. Observe, that '·' denotes the ordinary product and thus, the expression (54) is not a formula of fuzzy predicate logic.

Definition 32. Let $f(x_1, \ldots, x_n)$, $n \geq 1$, be a continuous real valued real function defined on a compact set $D^n \in \mathbb{R}^n$ and $Q(x_1, \ldots, x_n, y) \subseteq D^n \times f(D^n)$ be a fuzzy relation. We say that Q ε-approximates f if the condition $Q(x_1, \ldots, x_n, y) > 0$ implies the condition $|y - f(x_1, \ldots, x_n)| \leq \varepsilon$.

Lemma 15. *The fuzzy relation $R_f(x_1, \ldots, x_n, y)$ given by (54) is uniformly continuous on $D^n \times f(D^n)$.*

Lemma 16. *Let $f(x_1, \ldots, x_n)$, $n \geq 1$, be a continuous real valued real function defined on a compact set $D^n \in \mathbb{R}^n$. Furthermore, let $\varepsilon > 0$ be an arbitrary number and $\delta > 0$ be the number, which provides the inequality (53). If $R_f(x_1, \ldots, x_n, y)$ is the fuzzy relation given by (54) with respect to the function f and values ε, δ, then R_f 2ε- approximates f.*

Theorem 15. *Let $f(x_1, \ldots, x_n)$, $n \geq 1$, be a continuous real valued real function defined on a compact set $D^n \in \mathbb{R}^n$, $\varepsilon > 0$ be an arbitrary number and $\delta > 0$ be the number, which provides the inequality (53). Then there exists a fuzzy relation represented by a formula, which 2ε-approximates f.*

The following algorithm gives the precise expression for the fuzzy relation, which ε-approximates f and which can be represented by a formula in the disjunctive normal form. In comparison with the fuzzy relation $R_f(x_1, \ldots, x_n, y)$ given by (54), the one suggested below has not a continuous membership function.

Algorithm 1.

1. Let $f(x_1, \ldots, x_n)$, $n \geq 1$, be a continuous real valued real function defined on a compact set $D^n \in \mathbb{R}^n$ and let $\varepsilon > 0$ be an arbitrary number. Choose $\delta > 0$ such that the inequality (53) holds and find a finite covering of D by open δ-intervals $I_1, \ldots, I_k$. As a result, the set D^n will be also covered by all the direct products of those intervals.
2. Based on the uniform continuity of f on D^n find the corresponding finite covering of $f(D^n)$ by intervals $f(I_{i_1} \times \cdots \times I_{i_n})$, $1 \leq i_j \leq k$, $1 \leq j \leq n$, whose lengths are not greater than ε. Denote their centers by $y_{i_1 \ldots i_n}$ and consider open ε- intervals $H_{i_1 \ldots i_n}$ around the points $y_{i_1 \ldots i_n}$, $1 \leq i_j \leq k$, $1 \leq j \leq n$.
3. Define functions $\bar{g}_1(x), \ldots, \bar{g}_k(x)$ on D and $\bar{h}_{i_1 \ldots i_n}(y)$ on $f(D^n)$, so that

$$\bar{g}_i(x) = \begin{cases} 1, & \text{if } x \in I_i, \\ 0, & \text{otherwise} \end{cases}$$

and

$$\bar{h}_{i_1 \ldots i_n}(y) = \begin{cases} 1, & \text{if } y \in H_{i_1 \ldots i_n} \\ 0, & \text{otherwise}, \end{cases}$$

where $1 \leq i \leq k$, $1 \leq i_j \leq k$, $1 \leq j \leq n$.

4. Form the fuzzy relation

$$\bar{R}_f(x_1,\ldots,x_n,y) = \bigvee_{i_1}^{k} \cdots \bigvee_{i_n}^{k} (\bar{g}_{i_1}(x_1) \otimes \cdots \otimes \bar{g}_{i_n}(x_n) \otimes \bar{h}_{i_1\ldots i_n}(y)). \quad (55)$$

It can be shown that this fuzzy relation ε- approximates f and can be represented by a formula in the disjunctive normal form w.r.t to some structure.

Observe, that the described constructive procedure can be rearranged for the case of the conjunctive normal form.

6.3   Compositional rule of inference and the best approximation

A compositional rule of inference is a procedure, which generally speaking, corresponds a dependent value to a given relation and a certain restriction (fuzzy or crisp) to an independent value(s). Here we use the very special kind of a compositional rule of inference, which is nothing else as substitution. We will use it to extract a normal function from fuzzy relation describing it. This is eventually done by applying the defuzzification procedure.

We can regard the defuzzification as a *generalized fuzzy operation* on a set of elements.

Definition 33. A defuzzification of a non-empty fuzzy set $A \subseteq X$ given by its membership function $A(x) : X \longrightarrow [0,1]$ is a mapping $\Theta : \mathcal{F}(X) \longrightarrow X$ such that

$$A(\Theta(A)) > 0. \quad (56)$$

The set of all the defuzzification operations on $\mathcal{F}(X)$ will be denoted by $\mathrm{DEF}_{\mathcal{F}(X)}$.

Let X be a compact set and $A(x) \not\equiv 0$ be a continuous membership function of a fuzzy set $A \subseteq X$. Put $x^* = \Theta(A)$ where $\Theta \in \mathrm{DEF}_{\mathcal{F}(X)}$. The following specifications of the defuzzification Θ are used most frequently:

$$x^*_{LOM} = \wedge \{x \mid A(x) = \vee_{u \in X} A(u)\}, \quad \text{(Least of Maxima)} \quad (57)$$

$$x^*_{MOM} = \frac{x_L + x_G}{2}, \quad \text{(Mean of Maxima)} \quad (58)$$

where

$$x_L = \wedge\{x \mid A(x) = \vee_{u \in X} A(u)\}, \quad x_G = \vee\{x \mid A(x) = \vee_{u \in X} A(u)\},$$

$$x^*_{COG} = \frac{\int_X xA(x)dx}{\int_X A(x)dx}. \quad \text{(Center of Gravity)} \quad (59)$$

In particular, if $X = \{x_1, \ldots, x_n\}$ then

$$x^*_{COG} = \frac{\sum_{i=1}^{n} x_i A(x_i)}{\sum_{i=1}^{n} A(x_i)}. \tag{60}$$

Remark 2. In the examples of the defuzzification given above the continuity of $A(x)$ on the compact set X guarantees validity of (56). If $A(x)$ is not continuous then the defuzzification may be also realized in accordance with (57)–(59), but the inequality (56) should be verified subsequently. In the sequel, the membership functions of the defuzzified fuzzy sets may not be necessarily continuous.

Now, given a certain defuzzification we can uniquely correspond a function to a fuzzy relation. Denote the latter by $R(x_1, \ldots, x_n, y)$ and suppose that it is defined on a set $X_1 \times \ldots \times X_n \times Y$. Choose a defuzzification $\Theta \in \text{DEF}_{\mathcal{F}(Y)}$ and define the function $f_{R,\Theta}(x_1, \ldots, x_n) : X_1 \times \ldots \times X_n \longrightarrow Y$ by putting

$$f_{R,\Theta}(x_1, \ldots, x_n) = \Theta(R(x_1, \ldots, x_n, y)). \tag{61}$$

The function $f_{R,\Theta}$ defined by (61) is said *to be adjoined to a fuzzy relation R* by means of the chosen defuzzification Θ.

The following theorem shows the relation between functions f and $f_{R,\Theta}$ where R is a fuzzy relation, which approximates f.

Theorem 16. *Let $f(x_1, \ldots, x_n)$, $n \geq 1$, be a continuous real valued real function defined on a compact set $D^n \in \mathbb{R}^n$ and $\varepsilon > 0$ be an arbitrary number. Furthermore, let a fuzzy relation $R(x_1, \ldots, x_n, y)$ ε-approximate f. Then for any defuzzification $\Theta \in \text{DEF}_{\mathcal{F}(Y)}$ the function $f_{R,\Theta}$ defined by (61) also ε-approximates f as well, i.e. for any $(x_1, \ldots, x_n) \in D^n$*

$$|f(x_1, \ldots, x_n) - f_{R,\Theta}(x_1, \ldots, x_n)| \leq \varepsilon.$$

We will consider the problem of the best approximation w.r.t. the choice of a defuzzification and in accordance with a certain criterion. To simplify the description, we will confine ourselves to the case of functions with one variable.

Definition 34. Let $f(x)$ be a continuous real valued function defined on the closed interval $[a, b]$ and let G be a set of approximation, which consists of continuous real valued functions defined also on $[a, b]$. Moreover, let d determine a distance on the set of all continuous real valued functions defined on $[a, b]$. We say that $g^*(x) \in G$ is the best approximation from G to f if the condition

$$d(f, g^*) \leq d(f, g) \tag{62}$$

holds for all $g(x) \in G$.

In the sequel, we will consider a set of approximation consisting of a set of real valued real functions, which are adjoined to some special fuzzy relations represented by formulas by means of some defuzzification operation. We will be interested in finding the best approximation to the given continuous real valued real function $f(x)$ in a sense close to the above defined one.

On the account of results from the previous subsection we may regard those special fuzzy relations represented by normal forms. In particular, they are given by

$$\text{DNF}(x,y) = \bigvee_{i=1}^{k} (P_i(x) \,\&\, Q_i(y)) \tag{63}$$

and

$$\text{CNF}(x,y) = \bigwedge_{i=1}^{k} (P_i(x) \Rightarrow Q_i(y)) \tag{64}$$

where P_i and Q_i, $1 \leq i \leq k$, are unary predicate symbols.

Let $f(x)$ be a continuous real valued function defined on the closed interval $[a,b]$ so that $f : [a,b] \longrightarrow [c,d]$. Choose a structure $\mathcal{D}$ (follow Algorithm 1 in the previous subsection including the notation) sufficient for the interpretation of the normal forms (63) and (64) and such that the fuzzy relations $R_{\text{DNF}}(x,y)$ and $R_{\text{CNF}}(x,y)$ represented by these normal forms w.r.t. $\mathcal{D}$ are equal to $R(x,y)$ where

$$R(x,y) = \begin{cases} 1, & \text{if } (\exists i)(1 \leq i \leq k \quad \text{and} \quad x \in I_i, y \in \bar{f}[I_i]), \\ 0, & \text{otherwise.} \end{cases} \tag{65}$$

Define the set of approximation $\mathcal{G}_f$ for f as the set of all functions adjoined to $R(x,y)$ given by (65) by means of the defuzzifications $\Theta \in \text{DEF}_{\mathcal{F}(D)}$

$$\mathcal{G}_f = \{g(x) \mid g(x) = f_{R,\Theta}(x) = \Theta(R(x,y)), \ \Theta \in \text{DEF}_{\mathcal{F}(D)}\}. \tag{66}$$

In what follows we will fix the structure $\mathcal{D}$, which will lead to $R(x,y)$ and $\mathcal{G}_f$ and preserve all the intermediate notation.

Theorem 17. *Let $f(x) : [a,b] \longrightarrow [c,d]$ be a continuous function and $\mathcal{G}_f$ be the set of approximation for f given by (66). Furthermore, let $g_{MOM}(x) \in \mathcal{G}_f$ be the function specified by the defuzzification (58). Then for each closed interval $CI_i = [x_i, x_{i+1}]$, $1 \leq i \leq k$, the continuous restriction $g_{MOM}|CI_i(x)$ is the best approximation from $\mathcal{G}_{f,i}$ to $f|CI_i(x)$ with respect to the distance d given by*

$$d(u(x), v(x)) = \max_{CI_i} |u(x) - v(x)|.$$

6.4   Representation of continuous functions and its complexity

We are interested here in the precise representation of continuous functions
by fuzzy relations in such a way that given a function find the fuzzy relation,
to which it can be adjoined by means of some defuzzification. In addition,
the fuzzy relation is supposed to be represented by one of the normal forms.

After a thorough analysis of the methodology of approximation described in
the previous subsection we deduce that the problem of representation can
be solved if a correspondence between function $f(x)$ and the fuzzy relation
$R(x,y)$ adjoined to it, is more rigorous. We will suppose that the latter is
built up in such a way that

$$y = f(x) \quad \text{implies} \quad R(x,y) = 1. \tag{67}$$

Moreover, the defuzzification is supposed to be given by (57). As above, we
will work with real valued real functions with one variable.

Till now, both kinds of normal forms have been interchangable. In this section, we will see that the conjunctive normal form is more suitable for the
solution of our problem. The justification of this follows from the subsequent
lemma where the functions adjoined to fuzzy relations represented by DNF
are shown to be very specific.

Lemma 17. *Let $f(x)$ be a real valued function defined on $[a,b]$, which is
adjoined to the fuzzy relation $R(x,y) \subseteq [a,b] \times f([a,b])$ by means of the
defuzzification (57) so that (67) holds. Suppose that $R(x,y)$ is expressed in
the disjunctive normal form*

$$R(x,y) = \bigvee_{i=1}^{k} (P_i(x) \otimes Q_i(y))$$

*where $P_i(x), Q_i(y)$, $1 \leq i \leq k$, are membership functions of normal fuzzy
sets defined on intervals $[a,b]$ and $f([a,b])$, respectively. Moreover, suppose
that those membership functions are continuous and their a-cuts are closed
intervals. Then $f(x)$ is a piecewise constant with finite number of steps (some
steps may degenerate into points).*

Now, we will show that the same assumptions as in the previous lemma except
the choice of the normal form lead to the wider class of real valued functions,
which can be expressed.

Lemma 18. *Let $f(x)$ be a continuous and strictly monotonous real valued
function defined on $[a,b]$. Then there exists a fuzzy relation $R(x,y)$ defined
on $[a,b] \times [f(a), f(b)]$ and expressed in the conjunctive normal form*

$$R(x,y) = \bigwedge_{i=1}^{2} (P_i(x) \to Q_i(y)) \tag{68}$$

so that (67) holds, and $f(x)$ is the function adjoined to $R(x, y)$ by means of the defuzzification (57).

PROOF: Let $f(x)$ fulfil the assumptions. To obtain the proof, it is sufficient to construct the membership functions $P_i(x), Q_i(y)$, $i = 1, 2$, defined on $[a, b]$ and $[f(a), f(b)]$, respectively and then verify the conclusions.

Since $f(x)$ is continuous and monotonous on $[a, b]$ it defines a one-to-one correspondence between $[a, b]$ and $[f(a), f(b)]$. Therefore, the inverse function $f^{-1}(y)$ exists.

Now, for the definiteness, let us assume that $f(x)$ is monotonously increasing. Put the above mentioned membership functions as follows:

$$P_1(x) = 1 - \frac{x - a}{b - a}, \qquad P_2(x) = \frac{x - a}{b - a}, \quad x \in [a, b],$$

$$Q_1(y) = 1 - \frac{f^{-1}(y) - a}{b - a}, \quad Q_2(y) = \frac{f^{-1}(y) - a}{b - a}, \quad y \in [f(a), f(b)].$$

Let $x^0 \in [a, b]$ be an arbitrary element. We will show that

$$f(x^0) = \wedge\{y \mid R(x^0, y) = 1\} \quad \text{and} \quad R(x^0, f(x^0)) = 1.$$

Indeed,

$$P_1(x^0) \to Q_1(y) = 1 \quad \text{iff} \quad P_1(x^0) \leq Q_1(y)$$

$$\text{iff} \quad 1 - \frac{x^0 - a}{b - a} \leq 1 - \frac{f^{-1}(y) - a}{b - a} \quad \text{iff} \quad y \leq f(x^0),$$

$$P_2(x^0) \to Q_2(y) = 1 \quad \text{iff} \quad P_2(x^0) \leq Q_2(y) \quad \text{iff} \quad \frac{x^0 - a}{b - a} \leq \frac{f^{-1}(y) - a}{b - a}$$

$$\text{iff} \quad f(x^0) \leq y.$$

Thus,

$$R(x^0, y) = 1 \quad \text{iff} \quad f(x^0) = y$$

what we wanted to prove.   □

The same technique can be applied for the representation of the more general class of real valued functions which are constituted of strictly monotonous restrictions. To be precise, we introduce the following definition.

Definition 35. By a piecewise monotonous real valued function $f(x)$ defined on $[a, b]$ we mean a function, for which there exists a finite partition of $[a, b]$ into subintervals such that the restriction of f to each subinterval from the partition is strictly monotonous.

Theorem 18. *Let $f(x)$ be a continuous and piecewise monotonous real valued function defined on $[a, b]$. Then there exist a number N, and a fuzzy* *relation $R(x, y)$ defined on $[a, b] \times f([a, b])$ and expressed in the conjunctive normal form*

$$R(x, y) = \bigwedge_{i=1}^{N} \bigwedge_{j=1}^{2} (P_{ij}(x) \rightarrow Q_{ij}(y)) \tag{69}$$

such that (67) holds and $f(x)$ is the function adjoined to $R(x, y)$ by means of the defuzzification (57).

7   Fuzzy logic in broader sense

FLn should be a theoretical basis standing beyond methods and techniques of fuzzy logic. The mediator between the latter and FLn is natural language. Note that most of the applications of fuzzy logic are based on the generalized modus ponens elaborated in fuzzy logic, which is a model of one of the fundamental principles of human reasoning. Since the success of applications of fuzzy logic is convincing, it is important to elaborate good formalization of natural language and to employ it in the model of the commonsense human reasoning. The *fuzzy logic in broader sense* is a contribution to this task. Of course, the formalization concerns only a small fragment of natural language, namely that which is used in evaluation of the behaviour of dynamic systems and various decision situations.

7.1   Partial formalization of natural language

Evaluating and simple conditional linguistic syntagms. The considered fragment of natural language consists of a set S of linguistic syntagms specified below. They are constructed from the following classes of words.

  (i) *Nouns*, which mostly will be taken as primitives without studying their structure and formally represented simply by variables denoting objects.
  (ii) *Adjectives* where we will confine to few proper ones, mostly forming an evaluation linguistic trichotomy of the type "small", "medium" and "big". These words can be taken as prototypical for other similar linguistic trichotomies, such as "weak, medium strong, strong", "clever, average, dull", etc.
  (iii) *Linguistic hedges*, which form a special kind of the so called intensifying adverbs, such as "very, slightly, roughly", etc. We distinguish linguistic hedges with narrowing effect, such as "very, significantly", etc. and those with extending effect, such as "roughly, more or less", etc.
  (iv) *Connectives* "and", "or" and the *conditional clause* "if ... then ...".

Further class to be added in the future are linguistic quantifiers. For the time being, however, we will not consider them.

A specific is the class of *evaluating syntagms*, i.e., linguistic expressions which characterize a position on a bounded ordered scale (usually an interval of some numbers (real, rational, natural)). These are syntagms such as "small, medium, big, very small, roughly large, extremely high", etc.

The symbol $\langle \ldots \rangle$ used below denotes a metavariable for the kind of a word given inside the angle brackets.

Definition 36. An evaluating syntagm is one of the following linguistic expressions:

(i) Atomic evaluating syntagm, which is any of the following:
- An adjective "small", "medium", or "big",
- A fuzzy quantity "approximately z", which is a linguistic expression characterizing some quantity z from an ordered set.

(ii) Negative evaluating syntagm, which is an expression

not $\langle$atomic evaluating syntagm$\rangle$.

(iii) Simple evaluating syntagm, which is an expression

$\langle$linguistic hedge$\rangle\langle$atomic evaluating syntagm$\rangle$.

(iv) Compound evaluating syntagms constructed as follows: if A, B are evaluating syntagms then 'A and B', 'A or B' are compound evaluating syntagms.

Example 15. Atomic evaluating syntagms are *small, medium, big*. Fuzzy quantities are, e.g. *twenty five, the value z*, etc. Simple evaluating syntagms are *very small, more or less medium, roughly big, about twenty five, approximately z*, etc. Compound evaluating syntagms are *roughly small or medium, quite roughly medium and not big*, etc.

The "fuzzy quantity" is a linguistic characterization of some quantity — usually a number. This means that every linguistic characterization of a number is understood imprecisely. We will take the form "approximately z" as canonical.

It is clear that atomic evaluating syntagms usually form pairs of antonyms, i.e. the pairs

"nominal adjective — antonym".

Of course, there are a lot of other examples, e.g. "young — old", "ugly — nice", "stupid — clever", etc.

Definition 37. Let A be an evaluating syntagm. Then the syntagm

$$\langle \text{noun} \rangle \text{ is } A \tag{70}$$

is an evaluating predication. If A is a simple evaluating syntagm then (70) is a a simple evaluating predication. Compound evaluating predications are formed from other ones using the connectives "and, or".

Example 16. Evaluating predications are, e.g. *"temperature is very high"* (here "high" is taken instead of "big"), *"pressure is not small"*, *"frequency is small or medium"*, *"cost is not small and not big"*, *"income is roughly three million"*, etc. Compound evaluating predication is, e.g. *"temperature is high and pressure is very high"*.

The following definition characterizes the set of linguistic expressions which will be employed in FLb.

Definition 38. The set S of syntagms consists of evaluating syntagms, evaluating predications and the conditional clause

$$\text{IF } A \text{ THEN } B \tag{71}$$

where A, B are evaluating linguistic predications.

7.2   Intension and extension in fuzzy logic

In the logical analysis of natural language, three concepts play a distinguished role, namely intension, extension and possible world.

A possible world is a "way of things it could be in order for the expression to be true", or a "particular state of affairs" which determine truth of a sentence.

Intension of a linguistic syntagm, sentence, or of a concept, can be identified with the property denoted it. Intension may lead to different truth values in various possible worlds but it is invariant with respect to them.

Extension is a set of elements determined by an intension, which fall into the meaning of a syntagm in a given possible world. Thus, it depends on the particular context of use.

Definition 39. Let $A(x_1, \ldots, x_n) \in F_J$ be a formula with the free variables $x_1, \ldots, x_n$ of the respective sorts $\iota_1, \ldots, \iota_n$. Then the set

$$\mathbf{A}_{\langle x_1, \ldots, x_n \rangle} = \left\{ a_{t_1, \ldots, t_n} / A_{x_1, \ldots, x_n}[t_1, \ldots, t_n] \,\middle|\, t_1 \in M_{\iota_1}, \ldots, t_n \in M_{\iota_n} \right\} \tag{72}$$

of evaluated formulas being closed instances of $A(x_1, \ldots, x_n)$ is called a multiformula.

Obviously, the multiformula **A** can be at the same time viewed as a *fuzzy set* $\underset{\sim}{A} \subseteq F_J$ of *closed instances* of the formula $A(x_1, \ldots, x_n)$.

Recall that the syntagms $\mathcal{A} \in \mathcal{S}$ are names of properties of objects. The latter can be represented in a formal language of logic using formulas $A \in F_J$. At the same time, vagueness of syntagms must be taken into account and thus, a simple assignment of A to $\mathcal{A}$ is not sufficient. The following definition offers a solution.

Definition 40. Let $\mathcal{A} \in \mathcal{S}$ be a syntagm and $A(x_1, \ldots, x_n) \in F_J$ be a formula assigned to it. Then the intension of $\mathcal{A}$ is the multiformula

$$\mathrm{Int}(\mathcal{A}) = \mathbf{A}_{\langle x_1, \ldots, x_n \rangle}.$$

A structure $\mathcal{D}$ for J (possibly with some specific properties) is a possible world and the extension of $\mathcal{A}$ in a given possible world $\mathcal{D}$ the fuzzy relation

$$\mathrm{Ext}_{\mathcal{D}}(\mathcal{A}) =$$
$$= \left\{ \mathcal{D}(A_{x_1, \ldots, x_n}[t_1, \ldots, t_n]) / \langle \mathcal{D}(t_1), \ldots, \mathcal{D}(t_n) \rangle \, \middle| \, t_i \in M_{\iota_i}, i = 1, \ldots, n \right\}$$

where $\mathcal{D}(t_1) \in D_1, \ldots, \mathcal{D}(t_n) \in D_n$ are interpretations of the terms $t_1, \ldots, t_n$, respectively in the structure (possible world) $\mathcal{D}$.

It is clear that one intension $\mathrm{Int}(\mathcal{A})$ may lead to (infinitely) many extensions $\mathrm{Ext}_{\mathcal{D}}(\mathcal{A})$ with possibly different truth evaluations.

Example 17. Let $\mathcal{A} := Young$. Let $M_J = \{t_0, \ldots, t_{100}, \ldots\}$ be a set of terms representing years.

We choose a formula $Y(x) \in F_J$ and assign it to the syntagm *Young*. Then the following multiformula is the intension of *Young*:

$$\mathrm{Int}(Young) = \mathbf{Y}_{\langle x \rangle} = \{ \, ^{1}/Y_x[t_0], \ldots, \, ^{1}/Y_x[t_{20}], \ldots, \, ^{0.6}/Y_x[t_{30}], \ldots, \\ ^{0.2}/Y_x[t_{45}], \ldots, \, ^{0}/Y_x[t_{60}] \}. \quad (73)$$

Let $\mathcal{D}$ be a structure with $D \subset \mathbb{N}$. Then a possible extension of *Young* can be, for example, the following.

$$\mathrm{Ext}_{\mathcal{D}}(Young) = \{ \, ^{1}/1, \ldots, \, ^{1}/20, \ldots, \, ^{0.6}/30, \ldots, \, ^{0.2}/45, \ldots, \, ^{0}/60 \} \quad (74)$$

where $\mathcal{D}(t_0) = 1, \ldots, \mathcal{D}(t_{20}) = 20, \ldots, \mathcal{D}(t_{30}) = 30, \ldots, \mathcal{D}(t_{45}) = 45, \ldots, \mathcal{D}(t_{60}) = 60$ are interpretations of the terms from M_J when representing age of people. It is clear that another extension can be obtained from this intension when representing ages of trees, dogs, or whatever else.

Recall in these examples that if $^{a}/A$ is an evaluated formula then only the inequality $\mathcal{D}(A) \geq a$ should be fulfilled.

Let us also remark that from the mathematical point of view, given a fuzzy theory T (in our case, determined by some multiformula(s)), it may have very different models with strange properties. However, for us these may not be interesting as extensions of the meaning. Hence, we will limit the term "extension" to models with some specific properties. It is not so far clear, which properties should really be imposed except for one — with respect to our discussion in Section 2 we will suppose that all fuzzy sets and fuzzy relations assigned to the predicate symbols are continuous (in the sense of some topology; in the case of uncountable subset of real numbers, we suppose the ordinary topology).

7.3   The meaning of simple evaluating syntagms

There is a strong parallel between the concepts of finite numbers and the small ones. The finite numbers can be represented by a horizon running from 0 to large numbers, where it "disappears". Also the meaning of the other two members of the evaluation linguistic trichotomy, i.e. "medium" and "big", can be based on the same ideas and so, we can get a unified theory.

Our formalization of them is based on the concept of "horizon" on the basis of the concept of the Sorites fuzzy theory discussed in Section 5. Note that there is a tendency of people to classify three positions on any ordered scale, namely "the leftmost", "the rightmost", and "in the middle". The horizon of *small* lays "somewhere to the right from the leftmost position". Small numbers are numbers which everybody easily understands and is able to verify that they are small. On the other hand, there is no last small number, i.e. small numbers form a class but not set and we may encounter only the horizon of small numbers running "somewhere to big ones".

Thus, small numbers behave similarly as finite ones. In the same way, the horizon of *big* lays "somewhere to the left from the rightmost position" and there is no first big number. In the case of *medium*, the horizon is spread both to the left as well as to the right from the middle — cf. Figure 1.

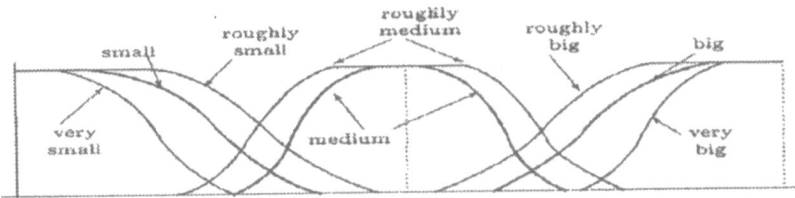

Fig. 1. Typical membership functions of simple evaluating syntagms.

To formalize the meaning of simple evaluating syntagms in FLb we will consider the following laguage of many-sorted FLn:

$$J = \{\leq, L, \Theta = \{\blacktriangleleft_\alpha \mid \alpha \in \xi\}, \{t_{\iota,a} \mid a \in [0,1], \iota = 1, \ldots, \vartheta\}, \eta\}$$

where L is an untyped unary predicate, η is a unary functional symbol, which will represent an order-reversing automorphism and $\blacktriangleleft_\alpha$, $\alpha \in \xi$ are unary connectives interpreted by logically fitting (i.e. Lipschitz continuous — cf. [13]) functions. These will be assigned to linguistic hedges ("very, roughly, significantly", etc.). Recall that L. A. Zadeh proposed to model their meaning using certain transformations of the fuzzy sets, which represent the meaning of the words with linguistic hedges standing before them. From our point of view, the main principle of the model of hedges consists in *shifting* of the horizon. This is achieved by application of certain unary connective. In general, we speak about the linguistic hedges with *narrowing effect* (very, highly, etc.) and those with *widening effect* (more or less, roughly, etc.). The effect of both kinds of hedges when applied to the three atomic syntagms "small", "medium", and "big" is depicted on Figure 1.

We will suppose a more subtle structure of the set of connectives Θ, namely

$$\Theta = \{\blacktriangleleft_{-k}, \ldots, \blacktriangleleft_{-1}, \blacktriangleleft_0, \blacktriangleleft_1, \ldots, \blacktriangleleft_p\}.$$

The central hedge is $\blacktriangleleft_0$. Then the hedges $\blacktriangleleft_{-k}, \ldots, \blacktriangleleft_{-1}$ will be called the *hedges with narrowing effect* and $\blacktriangleleft_1, \ldots, \blacktriangleleft_p$ the *hedges with widening effect*.

Furthermore, we will introduce the following fuzzy set of special axioms SAx:

(i) The classical axioms for crisp binary linear ordering $\leq$ in the degree 1.

(ii)

$$1/(\forall x)(t_{\iota,0} \leq x)\,\&\,(x \leq t_{\iota,1}),$$
$$\left\{ 1/t_{\iota,a} \leq t_{\iota,b} \mid a \leq b,\ a,b \in [0,1] \right\}.$$

(iii) The classical axioms to determine the function assigned to η to be the order reversing automorphism in the degree 1 and, furthermore,

$$1/(\forall x)(x = \eta(\eta(x))),$$
$$1/\eta(t_{\iota,c}) = t_{\iota,c}.$$

(iv) The following axioms specifying the left horizon predicate L

$$1/L(t_{\iota,0}),$$
$$1/\neg L(t_{\iota,c}),$$
$$\left\{ (1 - \tfrac{b-a}{c})/(L_x[t_{\iota,a}] \Rightarrow L_x[t_{\iota,b}]) \mid a,b \in [0,1] \right\}.$$

(v) The axioms characterizing properties of the linguistic hedges

$$\{ \, 1/(\triangleleft_\alpha(A) \Rightarrow \triangleleft_\beta(A)) \mid \alpha \le \beta, \ \alpha, \beta \in \xi \},$$
$$\{ \, 1/(\triangleleft_\alpha(\neg A) \Rightarrow \neg \triangleleft_\alpha(A)) \mid \alpha \in \xi \}.$$

Let us finally introduce the following inference rule

$$\frac{a/A}{\triangleleft_\alpha(a)/\triangleleft_\alpha(A)}, \quad \alpha \in \xi.$$

The operation $\triangleleft_\alpha$ on $[0,1]$ is supposed to be Lispchitz continuous and to fulfil

$$\triangleleft_\alpha(x) = 0 \text{ for } x \in [0, a], \qquad a < 1$$
$$\triangleleft_\alpha(x) = 1 \text{ for } x \in [b, 1], \qquad b > 0$$

so that $\triangleleft_\alpha(x)$ is strictly increasing on $[a, b]$.

Theorem 19 (Completeness). *Let J be a language extended by the logically fitting unary connectives $\triangleleft_\alpha \in \theta$ and the above inference rule. Then for every fuzzy theory T and a formula A, $T \vdash_a A$ iff $T \models_a A$.*

Definition 41.

(a) Basic nominal formula

$$W_\alpha(x) := \triangleleft_\alpha(L(x)).$$

(b) Antonym of (a)

$$AntW_\alpha(x) := W_\alpha(\eta(x)) := \triangleleft_\alpha(L(\eta(x))).$$

(c) Middle member

$$MW_\alpha(x) := \triangleleft_\alpha(\neg W_\alpha(x) \wedge \neg AntW_\alpha(x)).$$

The following lemma can be proved.

Lemma 19. *Let $T \vdash_{e>0} W_\alpha(t_a) \,\&\, AntW_\alpha(t_a)$, $a \in [0,1]$. Then T is contradictory.*

Construction of intension of evaluating predications. Since in the syntax of our logic, we have no means how to specify the objects concretely, $\langle$noun$\rangle$ is assigned a variable x (cf. also Example 17 where $\mathbf{Y}_{\langle x \rangle}$ can be understood as the intension of the evaluating predication " noun is young" where noun may be "man" as well as "dog").

Definition 42.

(i) Let $\mathcal{C} := $ '⟨noun⟩ is $\mathcal{A}$' be an evaluating predication, x be a variable assigned to ⟨noun⟩ and the intension of $\mathcal{A}$ be $\mathbf{A}_{\langle x \rangle}$. Then the intension of $\mathcal{C}$ is

$$\mathrm{Int}(\mathcal{C}) =:= \mathbf{A}_{\langle x \rangle}.$$

Using this definition, we may construct also intension of of the compound evaluating predication when interpreting the connectives "and" and "or" using special logical connectives interpreted by t-norm or t-conorm, respectively.

(ii) Let $\mathcal{C} := $ '⟨linguistic hedge⟩$\mathcal{A}$' be a simple evaluating syntagm where the ⟨linguistic hedge⟩ is assigned the hedge $\vartriangleleft_\alpha \in \Theta \setminus \{\vartriangleleft_0\}$ and the intension $\mathbf{A}$ of its atomic evaluating syntagm $\mathcal{A}$ is one of those given in Definition 41. Then the intension of $\mathcal{C}$ is $\mathbf{C}$ obtained from $\mathbf{A}$ by replacing the hedge $\vartriangleleft_0$ by $\vartriangleleft_\alpha$.

(iii) The intension of the conditional clause $\mathcal{C} := $ ' IF $\mathcal{A}$ THEN $\mathcal{B}$' is

$$\mathrm{Int}(\mathcal{C}) = \mathbf{A}_{\langle x \rangle} \Rightarrow \mathbf{B}_{\langle y \rangle} = \{\, a_t \to b_s / A_x[t] \Rightarrow B_y[s] \big| \, t \in M_{\iota_1}, s \in M_{\iota_2} \} \tag{75}$$

where $\mathrm{Int}(\mathcal{A}) = \mathbf{A}_{\langle x \rangle}$ and $\mathrm{Int}(\mathcal{B}) = \mathbf{B}_{\langle y \rangle}$ are intensions of the corresponding linguistic predications $\mathcal{A}, \mathcal{B}$.

Based on the previous definition, we can more specifically define the intensions of the atomic evaluating syntagms as follows:

Definition 43.

(i) Nominal syntagm $\mathcal{W}$

$$\mathrm{Int}(\mathcal{W}) := \mathbf{W}_{\alpha,\langle x \rangle} = \{\, \vartriangleleft_\alpha(0 \vee \tfrac{c-a}{c}) / \vartriangleleft_\alpha(L_x[t_a]) \mid a \in [0,1]\}.$$

(ii) Antonym $Ant\mathcal{W}$

$$\mathrm{Int}(Ant\mathcal{W}) := \mathbf{Ant W}_{\alpha,\langle x \rangle} = \{\, \vartriangleleft_\alpha(0 \vee \tfrac{a-c}{1-c}) / \vartriangleleft_\alpha(L_x[\eta(t_a)]) \mid a \in [0,1]\}.$$

(iii) Middle syntagm $\mathcal{M}$

$$\mathrm{Int}(\mathcal{M}) := \mathbf{W}_{\alpha,\langle x \rangle} = \{\, \vartriangleleft_\alpha(1 \wedge \tfrac{a}{c} \wedge \tfrac{1-a}{1-c}) / \vartriangleleft_\alpha(\neg L_x[t_a] \wedge \neg L_x[\eta(t_a)]) \mid \\ a \in [0,1]\}.$$

Linguistic variable. We have arrived at the slightly modified concept of the linguistic variable in comparison with the definition originally given by L. A. Zadeh (see [16]).

The *linguistic variable* is given by a set $T(\mathcal{X}) \subset \mathcal{S}$ of evaluating predications (70) *containing the same* ⟨noun⟩, *which is the name $\mathcal{X}$ of the linguistic variable.* The syntactic rule is a rule according to which the evaluating syntagms are constructed (e.g. given by our Definition 37). The significant difference lays in the semantical part since the original definition is purely extensional considering some universe U. Hence, we may define the linguistic variable as follows.

Definition 44. The linguistic variable is a tuple

$$\langle \mathcal{X} := \langle \text{noun} \rangle, T(\mathcal{X}), G, M_\iota, \mathcal{M}, \mathcal{P} \rangle$$

where $T(\mathcal{X}) \subset \mathcal{S}$, G, M_ι have been defined above, $\mathcal{M}$ is a semantical rule assigning to each evaluating predication $\mathcal{A} \in T(\mathcal{X})$ its intension

$$\text{Int}(\mathcal{A}) = \{ a_t / A_x[t] \mid t \in M_\iota \}$$

and

$$\mathcal{P} = \{ \mathcal{D} \mid \mathcal{D} \text{ is a possible world} \}.$$

The possible world is understood to be a specific structure $\mathcal{D}$ for the language J (cf. our discussion above).

7.4 Generalized inference

Fuzzy IF-THEN rules and the linguistic description. An outstanding role in fuzzy logic is played by the conditional clauses (71), which are implications characterized using natural language and used for description of some dynamic process or decision situation. A set of these statements is called the *linguistic description.*

Definition 45. Let $\mathcal{A}_j, \mathcal{B}_j \in \mathcal{S}$, $j = 1, \ldots, m$ be evaluating linguistic predications. Then the linguistic description in FLb is either a finite set $\mathcal{LD}^I$ or a finite set $\mathcal{LD}^A$ of linguistic statements as follows.

(i) $\mathcal{LD}^I = \{ \mathcal{R}_1^I, \ldots, \mathcal{R}_m^I \}$ where

$$\mathcal{R}_j^I = \text{ IF } \mathcal{A}_j \text{ THEN } \mathcal{B}_j, \tag{76}$$

$j = 1, \ldots, m$, are conditional clauses with the intension (75).

(ii) $\mathcal{LD}^A = \{\mathcal{R}_1^A, \dots, \mathcal{R}_m^A\}$ where

$$\mathcal{R}_j^A = \mathcal{A}_j \text{ and } \mathcal{B}_j, \tag{77}$$

$j = 1, \dots, m$, are compound evaluating predications with the intension constructed using Definition 42.

The linguistic description defines a theory of FLb.

Inference based on linguistic implications. We will now consider a simple linguistic description $\mathcal{LD}^I$ consisting of m IF-THEN rules and an evaluating predication $\mathcal{A}'_k$ occurring in the antecedent of some rule from $\mathcal{LD}^I$. Furthermore we will work with the concept of a theory $\mathcal{T}$ of FLb whose special axioms are given by a set of multiformulas. It is clear that this determines a fuzzy theory T, which will be called the *adjoint fuzzy theory*.

Theorem 20. *Let*

$$\mathcal{T} = \{\mathcal{A}'_k, \mathcal{LD}^I\},$$

be a theory of FLb where $\mathcal{LD}^I$ is a simple linguistic description and $\mathcal{A}'_k$ the evaluating predication described above. Then we may derive a conclusion $\mathcal{B}'_k$ with the intension

$$\mathbf{B}'_{k,\langle y \rangle} = \left\{ b'_{k,s} = \bigvee_{t \in M_{\iota_1}} (a'_{k,t} \otimes c_{k,ts}) / B_{k,y}[s] \,\middle|\, s \in M_{\iota_2} \right\} \tag{78}$$

such that all $b'_{k,s}$ for $s \in M_{\iota_2}$ in the multiformula $\mathbf{B}'_{k,\langle y \rangle}$ are maximal.

This theorem explicitly states the following: if we interpret IF-THEN rules as logical implications formed of the simple evaluating predications then the deduction based on them leads to the conclusion, which is the best possible one — in the sense of maximization of the provability degrees within the fuzzy theory ajoint to them.

In fuzzy control, a certain precise value u_0 is usually at disposal, which is obtained by measuring of some characteristics of the controlled process. From our point of view, this value is an interpretation of some closed term t_0 in the language J. Given a linguistic description $\mathcal{LD}^I$, a question now arises whether it is possible to cope with it in such a way that Theorem 20 can be applied. To find an answer, let us consider the following example.

Example 18. Given a linguistic description consisting of two rules

$$\mathcal{R}_1 := \text{ IF } x \text{ is small THEN } y \text{ is big}$$
$$\mathcal{R}_2 := \text{ IF } x \text{ is big THEN } y \text{ is small} .$$

These rules are interpreted by the respective multiformulas $\mathbf{Sm}_{\langle x \rangle} \Rightarrow \mathbf{Bi}_{\langle y \rangle}$ and $\mathbf{Bi}_{\langle x \rangle} \Rightarrow \mathbf{Sm}_{\langle y \rangle}$.

Let us now consider a model $\mathcal{D}_0$ with the support $U = V = [0, 1]$. Onee may agree that small values should lay around 0.3 (or smaller) and big ones around 0.7 (or bigger). Of course, given the measured value, say $x = 0.3$, we expect the result being "big" due to the rule $\mathcal{R}_1$. Similarly, for $x = 0.75$ we expect the result being "small" due to the rule $\mathcal{R}_2$. With respect to our theory, this means that the value 0.3 is an interpretation of some term t_0 in $\mathcal{D}_0$, i.e. $\mathcal{D}_0(t_0) = 0.3$ (and the like for 0.7). Then we have a specific intension of the form $\mathbf{Sm}'_{(x)} = \{\, 1/Sm_x[t_0]\}$ and when applying Theorem 20, we expect the result to be a multiformula $\mathbf{Bi}'_{(y)}$.

The term t_0 considered in this example can be interpreted as a *typical example of the evaluating predication* "⟨noun⟩ is small". In general, given an evaluating predication $\mathcal{A}$ with the intension $\mathbf{A}_{(x)}$, we derive a specific syntagm $Ex(\mathcal{A}) :=$ "typical example of $\mathcal{A}$" with the intension

$$Ex(\mathbf{A}_{(x)}) = \{\, 1/A_x[t_0]\}\,. \tag{79}$$

Determination of the term t_0, however, is out of logic. There are algorithms for solution of this task, implemented, e.g. in the software system LFLC [12]. The described situation is mathematically characterized in the following corollary of Theorem 20.

Corollary 2. *Let*

$$\mathcal{T} = \{Ex(\mathcal{A}_k), \mathcal{LD}^I\}$$

be a theory of FLb where $\mathcal{LD}^I$ is a simple linguistic description and $Ex(\mathcal{A}_k)$ is the syntagm "typical example of $\mathcal{A}_k$" with the intension (79). Then we may derive a conclusion B'_k with the intension

$$\mathbf{B}'_{k,(y)} = \left\{\, b'_{k,s} = c_{k,t_0\,s}/B_{k,y}[s] \,\middle|\, s \in M_{\iota_2} \right\} \tag{80}$$

and all $b'_{k,s}$ for $s \in M_{\iota_2}$ in the multiformula $\mathbf{B}'_{k,(y)}$ are maximal.

Since the intension of k-th rule of $\mathcal{LD}^I$ has the form $\mathbf{A}_{k,(x)} \Rightarrow \mathbf{B}_{k,(y)}$, the $b'_{k,s}$ are equal to $b'_{k,s} = a_{k,t_0} \to b_{k,s}$.

Theorem 21. *Let*

$$\mathcal{T} = \{\mathcal{A}', \mathcal{LD}^I\}$$

be a theory of FLb and T be its adjoint fuzzy theory. Let $\mathcal{LD}^I$ contain rules of the form

$$\text{IF } \mathcal{A}_j \text{ THEN } \mathcal{W}_j$$
$$\text{IF } \mathcal{A}_j \text{ THEN } Ant\,\mathcal{W}_j.$$

Then the fuzzy theory T is contradictory.

This theorem explicitly states that IF-THEN rules which are in contradiction lead to a contradictory fuzzy theory — a degenerated fuzzy theory equal to the set of all the well-formed formulas F_J.

Fuzzy interpolation of an imprecisely given function. First, we will consider the following classical situation. Let f be a function to be interpolated. Our knowledge of f is limited only to m of its points (d_j, e_j), $j = 1, \ldots, m$. Furthermore, we know that if x is "near to d_j" then y is "near to e_j", $j = 1, \ldots, m$. We are looking for some function g which interpolates f, i.e. it coincides with f in the points d_j, $g(d_j) = f(d_j)$, $j = 1, \ldots, m$ and approximates it for $x \neq d_j$. A question arises, how g can be specified. In the frame of fuzzy logic, the interpolation problem can be formulated when replacing the precise equality by the fuzzy one. We will speak about *fuzzy interpolation.* Then the function f is characterized imprecisely using fuzzy quantities in the points $d_1, \ldots, d_m$ according to the table

$$f : \frac{x \| \text{approximately } d_1 | \cdots | \text{approximately } d_m}{y \| \text{approximately } e_1 | \cdots | \text{approximately } e_m.} \qquad (81)$$

The following theorem gives an answer to the question, how the fuzzy interpolation can be formalized.

Theorem 22. *Let $\mathbf{d}_j \in M_{\iota_1}$, $\mathbf{e}_j \in M_{\iota_2}$ be terms (corresponding to d_j, e_j) in the language J and let the fuzzy quantities "approximately d_j", "approximately e_j" have the intensions $\mathbf{Q}[\mathbf{d}_j]_{\langle x \rangle}$ and $\mathbf{Q}[\mathbf{e}_j]_{\langle y \rangle}$, respectively, $j = 1, \ldots, m$. Furthermore, we suppose that the fuzzy quantities "approximately d_j" and "approximately d_k" are disjoint for each $j \neq k$. Then there is a simple linguistic description*

$$\mathcal{LD}^A = \{ \text{"approximately } d_j\text{" and "approximately } e_j\text{"} \mid j = 1, \ldots, m \}, \quad (82)$$

which determines the theory $\mathcal{T} = \mathcal{LD}^A$ of FLb such that the adjoint fuzzy theory T is consistent.

Furthermore, we introduce a new binary predicate symbol $F \notin J(T)$ and extend the fuzzy theory T into

$$T' = T \cup \left\{ 1/(F(x,y) \Leftrightarrow \bigvee_{j=1}^{m} ((x \doteq t_j) \wedge (y \doteq s_j))) \right\}.$$

Then to every term t we can derive in T' the multiformula

$$\mathbf{C}_{\langle y \rangle, t} = \left\{ \left. \textstyle\bigvee_{j=1}^{m} (a_{j,t} \wedge b_{j,s})/F_{xy}[t,s] \right| s \in M_{\iota_2} \right\}, \qquad (83)$$

which is a fuzzy quantity $\mathbf{C}_{\langle y \rangle, \mathbf{d}_j} = \mathbf{Q}[\mathbf{e}_j]_{\langle y \rangle}$ for each $j = 1, \ldots, m$.

8 Conclusion

In this chapter we have presented some of the main concepts of fuzzy logic both in narrow as well as in broader sense. Let us remark that one of the main goals of fuzzy logic is to fulfil its specific agenda, which is the bunch of tasks leading to methods and means, which tolerate vagueness and impreciseness. The tasks include the model of the meaning of some part of natural language in connection with the concept of linguistic variable, the theory of approximate or interpolative reasoning, fuzzy IF-THEN rules, linguistic quantification, defuzzification, and others. The reader may see that fuzzy logic is now a well developed non-trivial theory, in which a lot of work on this agenda has already been done thus demonstrating that the formalism is capable to fulfil it. On the other hand, a lot of problems are still before us. For example, we have not included linguistic quantifiers since their theory is still not developed in fuzzy logic, though very important. Except for linguistic motivation, there is also a lot of purely mathematical problems to be solved, such as more thorough development of fuzzy model theory, and others. For some more details see, e.g. the books [7,8,10,13].

References

1. Black , M. (1937). "Vagueness: An Exercise in Logical Analysis," Philosophy of Science 4, 427–455. Reprinted in Int. J. of General Systems **17**(1990), 107–128.
2. Butnariu, D. and Klement, E. P. (1993). *Triangular Norm-based Measures and Games with Fuzzy Coalitions*. Dordrecht: Kluwer.
3. Chang, C. C. and Keisler, H. J. (1973). *Model Theory*, Amsterdam: North-Holland.
4. Cohen, P. M. (1965), *Universal algebra*, New York: Harper & Row.
5. van Dalen, D. (1994). *Logic and Structure*, Berlin: Springer.
6. Gottwald, S. (1993).*Fuzzy Sets and Fuzzy Logic*. Wiesbaden: Vieweg.
7. Gottwald, S.: **A Treatise on Many-Valued Logics**. Research Studies Press Ltd., Baldock, Herfordshire, UK (to appear)
8. Hájek, P. (1998). *Metamathematics of fuzzy logic*. Dordrecht: Kluwer.
9. Klir, G.J. and Yuan, B. (1995). *Fuzzy Sets and Fuzzy Logic: Theory and Applications*. New York: Prentice-Hall.
10. Mundici, D., Cignoli, R. and D'Ottaviano, I.M.L. (2000). *Algebraic foundations of many-valued Reasoning*. Dordrech: Kluwer.
11. Novák, V. (1992). *The Alternative Mathematical Model of Linguistic Semantics and Pragmatics*. New York: Plenum.
12. Novák, V. (1995). "Linguistically Oriented Fuzzy Logic Controller and Its Design," Int. J. of Approximate Reasoning 1995, **12**, 263–277.
13. Novák, V., Perfilieva, I. and J. Močkoř (1999). *Mathematical Principles of Fuzzy Logic*. Boston: Kluwer.
14. Vopěnka, P. (1979). *Mathematics In the Alternative Set Theory*. Leipzig: Teubner.
15. Zadeh, L. A. (1965). "Fuzzy Sets," Inf. Control. **8**, 338–353.
16. Zadeh, L.A. (1975). "The concept of a linguistic variable and its application to approximate reasoning I, II, III," Inf. Sci., **8**, 199–257, 301–357; **9**, 43–80.

Fuzzy Systems and Data Mining

Witold Pedrycz[1][2]

[1] Department of Electrical and Computer Engineering
   University of Alberta, Edmonton, Canada
[2] Systems Research Institute
   Polish Academy of Sciences, 01-447 Warsaw, Poland
   pedrycz@ee.ualberta.ca

Abstract. Data mining and fuzzy systems share an important common feature that is information granulation. Information granules, and fuzzy sets exploited in the setting of this study, are used to reveal stable, transparent and meaningful patterns in databases. While there exists panoply of various forms of patterns, we focus on associations and rules as the two commonly encountered constructs that exist both in data mining and fuzzy systems. Associations are modeled in the language of fuzzy relations and are direction-free concepts meaning that they are not concerned as to the question "what implies what". Rules, on the other hand, are direction - based constructs with clearly delineated cause and effect (condition and conclusion). Moreover, it is shown that associations and rules are tied together: associations may entail rules but no other way around. We discuss the role of information granularity in determining consistency of the rules and analyze an impact that linguistic quantification of fuzzy sets has on the consistency of the individual rules. An idea of rule growing is also discussed.

Keywords data mining, fuzzy systems, information granularity, granulation, associations, rules, fuzzy sets, associations versus rules, directionality, attributes

1 Introduction

Fuzzy systems (models) have been around from the very inception of fuzzy sets. The paradigm of rule-based modeling permeates the area. The quality of any fuzzy model (that may involve a family of criteria such as accuracy, interpretability and alike) depends on the size of information granules - fuzzy sets being used as building blocks of any fuzzy model. The leitmotiv of fuzzy models - capturing and describing relationships at the linguistic level and design user-oriented structures is shared by data mining. Data mining in databases tries to make sense of raw data by revealing meaningful and easily interpretable relationships (see [1,3,4,8,18,21,24]). The domain of data mining is highly heterogeneous embracing a number of well-established information technologies including statistical pattern recognition, neural networks, machine learning, knowledge-based systems, etc [2,6,9–11,14,22,23,25,28]. The synergistic character of data mining is definitely one of its dominant and visible features that make this pursuit to emerge as a new area of research and applications. The ultimate goal of data mining is revealing patterns that are easy to perceive, interpret, and manipulate. One may ask, in turn, what makes necessary to develop such patterns. To address this essential question, we should revisit what makes humans so superb at perceiving, understanding, and acting in complex situations and yet so limited in basic arithmetic operations, manipulating numbers, etc. The cornerstone of human cognition is the concept of information granules and information granulation. Information granules help us cope with an abundance of detailed numeric data. Numbers are important, yet humans tend to produce abstractions that are more tangible and easy to deal with. Abstractions manifest themselves in the form of information granules – entities that encapsulate a collection of fine grain entities (in particular, numbers) into a single construct thus making them indistinguishable. The level of detail retained depends on the size of the information granules and is directly implied by the problem at hand. Information granulation makes all data mining pursuits more user-oriented, and allows the user to become more proactive in the overall process.

Granulation and information granules are found in many different frameworks, such as set-based environments and their generalizations (fuzzy sets, rough sets, shadowed sets, random sets, etc.) [15,13,19,20,29–31,17,16], as well as probabilistic frameworks (subsequently leading to probabilistic information granules). In this study, we concentrate on the use of fuzzy sets regarded as a conceptual environment of information granulation. Nevertheless, this environment of data mining along with the ensuing methodology is valid for some other scenarios of information granulation.

The study is organized as follows. We start with a concept of granulation of information, discuss its role in data mining and elaborate on the role of fuzzy sets as a useful vehicle of information granulation. Associations, as generic entities of data mining, are studied in Section 3. They are contrasted with rules as direction-based patterns. A way of converting associations to rules along with the characterization of the validity of the obtained rules is covered in Section 5 and 6. The quality of the rules can be enhanced: this can be accomplished by a linguistic modification of the fuzzy sets (Section 7) or through a process of growing rules (Section 8). Conclusions are covered in Section 9.

2   Granulation of information

In this section, we concentrate on the essence of information granulation and the role of information granules in a spectrum of perception processes carried out by humans. Then our focus moves into the construction of information granules in the setting of fuzzy sets.

2.1   Prerequisites: the role of information granulation in data mining processes

The essence of information granulation lies in a conceptual transformation in which a vast amount of numbers is condensed into a small number of and meaningful entities - information granules [16,29,31]. Information granules are abstract, synthesized, and user-oriented entitities. They are easily comprehended, memorized and used as building blocks helpful when perceiving more complex concepts. Information granules are manifestations of abstraction. The level of such abstraction depends upon the objective of a perception process carried out by humans. This, in turn, is implied by the goal of data mining and a level of the related decision - making processes. Strategic, long-term decision processes invoke the use of coarse and more stable information granules. Short-term decision-making processes involving immediate actions require another look at the same database that requires fine grain information granules. In this way, the size of information granule becomes crucial to the successful process of data mining. So far, we have not defined information granules and their "size" (granularity) in any formal fashion. As a matter of fact, such a definition has to be linked with the formal framework in which such information granules are constructed (one should emphasize that fuzzy sets form one among possible formal environments of information granulation). This means that when using, for instance, fuzzy sets, the size of the granules needs to be expressed in terms of the language of fuzzy set theory. Another way of expressing granularity should hold for probabilistic granules. Nevertheless, on the intuitive side, we can envision the general relationship: the larger the number of elements embraced by the information granule, the lower the granularity of such construct. And conversely, the lower the number of elements in the information granule, the higher its granularity. We can also use the term specificity as being the inverse to the notion of granularity.

2.2   Information granulation with the aid of fuzzy sets

In this study, we concentrate on the use of fuzzy sets as a vehicle of information granulation. There are three main ways in which information granules - fuzzy sets or fuzzy relations can be constructed

> User-oriented. It is a user or designer of the system who completely identifies the form of the information granules. For instance, they could be a priori defined as a series of triangular fuzzy numbers. Moreover, the number of these terms as well as their parameters is totally specified in advance.
> Algorithmic approach to information granulation. In this case, information granules are determined as a result of optimization of a certain performance index (objective function). Clustering algorithms are representative examples of such algorithms of unsupervised learning that lead to

the formation of information granules. Quite commonly, the granules are fuzzy sets (or fuzzy relations) when using FCM and similar algorithms or sets (or relations) when dealing with the methods such as ISODATA [5][12]

A combination of these two. The methods that fall under this category are a hybrid of user-based and algorithmic driven methods. For instance, some parameters of the information granulation process can be set up by the user while the detailed parameters of the information granules can be determined (or refined) through some optimization mechanism during the second phase. The influence of the user versus the influence of data varies from case to case

One should become aware of the advantages and potential drawbacks of the two first methods (the third one is a compromise between the user and data-driven methods and as such may reduce the disadvantages associated with its components). The user-based approach, even quite appealing and commonly used, may not reflect the specificity of the problem (and, more importantly, the data to be granulated). There could be a serious danger of forming fuzzy sets not conveying any experimental evidence. In other words, we may end up with a fuzzy set whose existence could hardly be justified in light of the currently available data. The issue of the experimental legitimization of fuzzy sets along with some algorithmic investigations has been studied in detail in [15]. On the other hand, the algorithmic-based approach might not be able to reflect the semantics of the problem. Essentially, the membership functions are built as constructs minimizing a given performance index. This index itself may not capture the semantics of the information granules derived in this fashion. Moreover, the data-driven information granulation may be computationally intensive, especially when dealing with large sets of multidimensional data (that are common to many tasks of data mining). This may eventually hamper the usage of clustering as a highly viable and strongly recommended option in data mining.

Bearing in mind the computational facet of data mining, we consider a process of granulation that takes place for each variable (attribute) separately. There are several advantages to follow this approach. First, the computational aspect of data mining pursuits is addressed. Second, there is no need for any prior normalization of the data that could eventually result in an extra distortion of relationships within the database; this phenomenon has been well known in statistical pattern recognition [10]. The drawback of not capturing the relationships between the variables can be considered minor in comparison to the advantages of this approach. As far as the size of information granules is concerned, a σ-count (cardinality) is a viable option (at least for normal fuzzy sets).

3   Building associations in databases

Once the information granules have been constructed for each variable in the database separately, they need to be combined. The aggregated (composite) granule

granule A_i and granule B_j

is a fuzzy relation defined as a Cartesian product of the corresponding coordinates

$$A_{ij} = A_i \times B_j$$

The membership function is defined by taking the and-combination (aggregation) of the contributing membership functions, namely

$$(A_i(x)tB_j(y))$$

with t denoting the triangular norm. The definition easily extends to any number of the information granules defined in the corresponding universes of discourse, say

$$A_{ij...l} = A_i \times B_j \times \ldots \times Z_l$$

that is

$$A_{ij...l}(x) = A_{ij...l}(x_1, x_2,, x_n) = A_I(x_1)tB_j(x_2)\ldots t \ldots Z_l(x_n)$$

with n being the number of the attributes (variables) encountered in the problem. Assume that for each coordinate (variable, attribute), we have constructed c granules. With the fixed number of attributes equal to n, we come up with c^n different Cartesian products. Only a certain fraction of these combinations would be legitimate in light of the experimental data available in the database. Each Cartesian combination of the information granules as outlined above is quantified by computing its σ-count. The information granules are then ranked according to the values of their σ-counts. We can form an agenda $\mathbf{D}$ of the most significant Cartesian products that is those ones characterized by the highest values of the σ-counts. The construction of the agenda of size p can be characterized as follows - cycle through all Cartesian products and retain those characterized by the highest values of σ-counts With an increasing number of attributes existing in databases, there is an explosion in the number of possible combinations. Say, for $c = 7$ (which could be a fairly typical value of the number of the information granules) and $n = 40$, we end up with all possible combinations of 7^{40} composite granules (Cartesian products). Subsequently, if an exhaustive search is out of question, one may consider various evolutionary techniques as a viable alternative. With the increasing number of information granules defined for each attribute, the likelihood of having strongly supported Cartesian products may be lower. Larger information granules promote Cartesian products described by higher values of the experimental evidence, that is, higher values of the corresponding σ-counts.

4   From associations to rules in databases

The Cartesian products constructed in the previous section, are associations - basic entities that are the tangible results of data mining. The agenda **D** retains the most significant (that is, data legitimate) findings in the database. It captures the most essential dependencies. It is important to stress that associations are *direction free*. They do not commit to any causal link between the variables, or more specifically, between the information granules. In this sense, associations are general constructs. One may even emphasize that these are the most generic entities to be used in mining static relationships in data mining. It is needless to say that the form of the associations, their number as well as the underlying experimental evidence, hinges on the information granules being used across all activities of data mining.

Rules, on the other hand, are *direction-oriented* constructs. They are conditional statements of the form

if condition (s) then action (s)

The form of the rule clearly stipulates the direction of the construct: the values of the conditions stipulate certain actions. The direction makes the construct more detailed and fundamentally distinct from associations. All rules are associations but not all associations are rules. This observation is merely a translation of associations and rules in the language of mathematics: all relations (Cartesian products) are functions (rules) but not the other way around.

In data mining, the issue of distinction between associations and rules plays a primordial role. We may not be certain as to the direction between attributes (what implies what). Therefore it is prudent to proceed with a two-phase design: first, reveal associations and then analyze if some of them could become rules. Note that when dealing with rules, they could be articulated once we decide upon the split of attributes (variables) into inputs and outputs. The task looks quite obvious for modeling physical systems (very few variables with an obvious direction between them). In data mining, though, this could

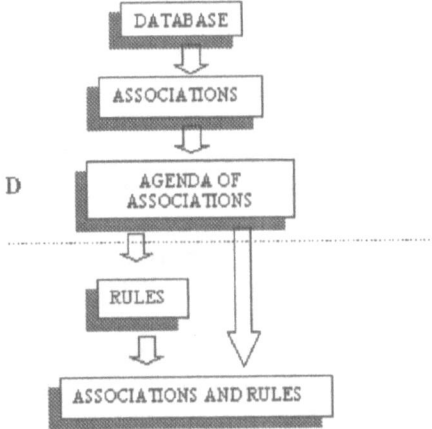

Fig. 1. A two-stage process of data mining forming a collection of associations and rules

be a part of the data mining pursuit. The two-phase process of data mining associations - rules is illustrated in Figure 1.

In the next section, we discuss how to realize the second phase and produce rules. Hopefully, the objective of this pursuit becomes clear: the data mining of rules realized from the very beginning of the process is false and unnecessarily restrictive. This common practice praised and followed by many, could be dangerous from the conceptual point of view: seeing functions where there are only relations is not appropriate.

5   Building rules in data mining processes

Associations are basic constructs (patterns) in a database. The rules are a result of further refinement in the sense of some directionality added to them. The introduced directionality component gives rise to conflicting rules. We study how to measure the level of conflict. It is shown that the conflict requires a formulation of the similarity between information granules. Noticing the conflict, the study concentrates on the way in which its level could be alleviated. Two main ways are anticipated: the first one handles the rules at the structural level meaning that we revisit a collection of variables (attributes) that should come at the condition or conclusion part. The second one is to modify granularity of the fuzzy sets occurring in the rules; we show that these modifications support the enhancement of the consistency level of the rules.

5.1   Expressing similarity between fuzzy sets: an equality index

Consider two fuzzy sets A and B defined in the same finite space $\mathbf{X}$, $card(\mathbf{X}) = \mathbf{n}$. How similar are these two fuzzy sets? There are several ways of quantifying this phenomenon. The one that is closely linked with set theory and subsequently with fuzzy sets is the one that states that sets are equal if the the following inclusions conditions are simultaneously satisfied

$$(A \text{ is } included \text{ in } B) \text{ } and \text{ } (B \text{ is } included \text{ in } A)$$

As we are concerned with fuzzy sets, this statement converts into a fuzzy set construct. The inclusion operation is modeled via an implication operation while the and connective is realized by means of the standard minimum. The degree of equality (matching) of two membership grades $A(x)$ and $B(x)$ is expressed in the form

$$A(x) \equiv B(x) = \text{Min}[(A(x) \Rightarrow B(x)), (B(x) \Rightarrow A(x))]$$

Here the implication is defined as the residuation operation [15]

$$a \Rightarrow b = \text{Sup}\{c \in [0,1] \mid atc \leq b\}, a, b \in [0,1]$$

Next, an overall level of matching is computed by straight averaging over $\mathbf{X}$, namely

$$A \equiv B = \frac{1}{n} \sum_{k=1}^{n} (A(x_k) \equiv B(x_k))$$

When it comes to expressing matching between fuzzy relations, the same definition applies yet the calculations are carried out for each coordinate of the relation independently and the results are combined via the minimum operation. Say, if we have two items $A \times C$ and $B \times D$, with A and B defined in the same space $\mathbf{A}$ and C and D defined in $\mathbf{C}$, the overall matching level is in the form

$$(A \times C) \equiv (B \times D) = \text{Min}(A \equiv B, C \equiv D)$$

5.2   Transforming associations into rules - adding a directionality dimension to the constructs

The starting point of this design is a collection of *associations*. Some of them could be recognized as *potential* rules. For the clarity of presentation, we consider only three variables and the associations therein in the form of the following Cartesian products

$$A_i \times B_i \times C_i; i = 1, 2, .., r.$$

The first step is to split the variables (attributes) into inputs and outputs. For instance, the first two are regarded as inputs, the third one as an output. The potential rules read as follows

$$\text{if } A_i \text{ and } B_i \text{ then } C_i$$

By taking C_i as input and retaining the two others as outputs, we get another potential rule

$$\text{if } C_i \text{ then } (A_i \text{ and } B_i)$$

Obviously, there are far more different arrangements of the variables with respect to their directionality.

In general, when dealing with multivariable associations, numerous potential rules are possible depending on the allocation of the variables. Figure 2 illustrates this by showing the agenda **D** with its entries being identified as inputs or outputs.

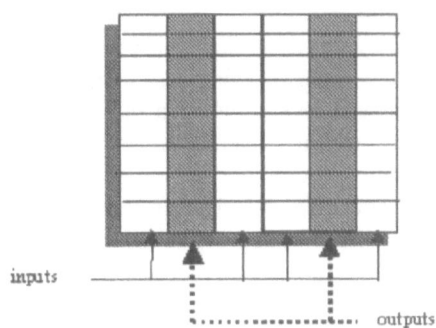

Fig. 2. The agenda of associations in database with the attributes identified as inputs (shadowed entries) and outputs

The detailed implementation of the rules (involving various models of the implication operator and rule aggregation) were studied in numerous volumes under the banner of fuzzy inference, and is not of particular interest here; the reader may refer to [5,10] as two selected points of reference). The crucial point here is how to identify that some associations are rules. This identification occurs with regard to pairs of associations. The underlying principle is straightforward: the rules obtained from the associations are free of conflict. We say that two rules are conflicting if they have quite similar conditions yet they lead to very different (distinct) conclusions. The effect of conflict is illustrated succinctly in Figure 3.

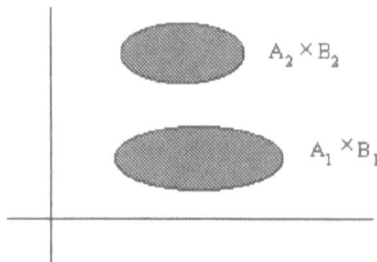

Fig. 3. Two associations and rule induction

Here we have two associations: $A_1 \times B_1$ and $A_2 \times B_2$. When converted into rules of the form *if A_i then B_i* then these rules are in conflict. Noticeably, the same associations when converted into rules *if B_i then A_i* (which exhibit another direction) are not in conflict. To measure the level of conflict, we first have to express similarity (or difference) between two fuzzy relations. There are numerous ways of completing this task. By referring to the literature on fuzzy sets, the reader may encounter a long list of methods. The basic selection criteria would involve efficiency of the method as well as its computational overhead. Moving to the consistency of the rules, we can distinguish the following general situations as to the similarity of the conditions and the conclusions of some potential rules, see Table 1

Table 1. Four general cases of similarity levels between conditions and conclusions of two rules along with their consistency

Conditions	Conclusions	Consistency of rules
similar	different	inconsistent
similar	similar	consistent
different	similar	consistent
different	different	consistent

The above scenarios suggest that a plausible consistency index of the two rules can be based on the fuzzy implication

$$\text{Cons(rule-1, rule-2)} = A_1 \times B_1 \equiv A_1 \times B_1 \Rightarrow C_1 \equiv C_2$$

Linking the above formula with the qualitative analysis shown before, it becomes apparent that the values of the consistency index, Cons(., .) attains higher values for lower values of the observed equality for the condition part and higher values for the observed equality occurring at the conclusion side. In particular, two implications are of interest

$$a \Rightarrow b = \{1 \text{ if } a \leq b \\ =\{b \text{ if } a \gneq b$$

and

$$a \Rightarrow b = \{1 \text{ if } a \leq b \\ =\{\frac{b}{a} \text{ if } a \gneq b$$

The multidimensional form of the consistency reads as

$$\text{Cons (rule-1, rule-2)} = \text{(condition part of -1 condition part of -2)}$$

$$\text{(conclusion part -1 conclusion part -2)}$$

where the condition part and conclusion part involve the Cartesian products of the information granules of the attributes placed in the condition and conclusion parts of the rules.

So far, we have investigated a pair of rules. Obviously, when dealing with a collection of associations (and rules afterwards), we would like to gather a global view as to the consistency of the given rule with regard to the rest of the rules. A systematic way of dealing with the problem is to arrange consistency values into a form of a consistency matrix $\mathbf{C}$ having N rows and N columns (as we are concerned with N associations). The (i, j) th entry of this matrix denotes a level of consistency of these two rules. The matrix is symmetrical with all diagonal entries being equal to 1. As a matter of fact, it is enough to compute the lower half of the matrix. The overall consistency of the ith rule is captured by the average of the entries of the ith column (or row) of $\mathbf{C}$,

$$Cons(i, \mathbf{D}) = \frac{1}{N} \sum_{j=1}^{N} c_{ij}$$

This gives rise to the linear order of the consistency of the rules. This arrangement helps us convert only a portion of the associations into rules while retaining the rest of them as direction-free constructs. What we end up, is a mixture of heterogeneous constructs as illustrated in Figure 4.

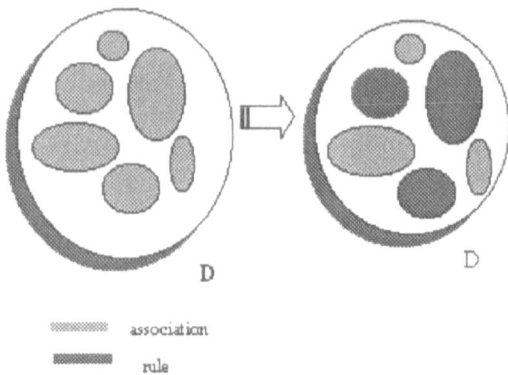

Fig. 4. By selecting highly consistent rules, the result of data mining is a mixture of associations and rules

Obviously, by lowering the threshold level (viz. accepting less consistent rules), more associations can be elevated to the position of rules.

An interesting question arises to the quality of rules, and how a given association could be converted to a rule.

6   Properties of rules induced by associations

How to produce more rules out of associations and make these rules more consistent? There is a lot of flexibility in answering this question. The rules may have a different number of conditions and conclusions. Some attributes can be dropped and not showing up in the rules. To illustrate the point, consider the associations involving four attributes, namely

$$A_i \times B_i \times C_i \times D_i$$

The following are general observations (they come from the interpretation of the consistency index (x) we adhere to)

Increasing the number of attributes in the condition part promotes higher consistency of the rules. That is, the rules

- if A_i and B_i and C_i then D_i

are more consistent than the rules in which the first attribute has been dropped, such as

- if B_i and C_i then D_i

This is easy to see in light of the main properties of the implication operation that is

$$atb \rightarrow c \geq a \rightarrow c, ab, c \in [0,1].$$

A drop in the number of attributes in the condition part contributes to rules that tend to be more *general*, i.e., they apply to a broad spectrum of situations. By adding more conditions, we make rules more *specific* (viz. they are appropriate for a smaller number of cases).The increased generality of the rules comes hand in hand with their elevated level of inconsistency.

Interestingly, the analysis of the consistency of the overall set of rules (say, by determining the sum of all entries of $\mathbf{C}$, say $\Delta = \sum_{i,j} c_{ij}$) brings us to the examination of the relevance of the attributes: if dropping a certain attribute from the condition part does not reduce the values of Δ then the attribute may be regarded as irrelevant. The more evident reduction in Δ linked with the elimination of the given attribute, the more essential this attribute is. This corresponds to the well-known problem of feature selection in pattern recognition [10]. The difference here lies in the fact that the discriminatory properties of a given attribute are quantified not for the attribute itself but a manifestation of this property is determined for the assumed level of granularity (that is the number of fuzzy sets defined there). In other words, if the granularity of the attribute has been changed (say, by increasing the number of information granules therein), it may happen that its discriminatory properties could be affected as well.

> By removing attributes from the conclusion part (for a fixed number of attributes in the condition part), the rules become more consistent. Again, by following the definition of inconsistency, this tendency becomes evident as we have
>
> $$atb \Rightarrow ctd \leq a \Rightarrow c$$
>
> The finding concurs with our intuition: by removing more attributes, the conclusions tend to become less "disjoint" thus reducing the level of potential inconsistency. In principle, the rule becomes less specific (viz. it supports more general conclusions).

One should stress that the above analysis is carried out for the fixed associations. In particular, we have not affected the granularity of the original information granules. The size of information granules may substantially affect the consistency of the rules.

7   Linguistic modifications of information granules in the rules

The consistency of the rules can be enhanced by affecting the granularity of the fuzzy sets occuring in the rules. Instead of the rule

$$\text{if } A \text{ then } B$$

we consider the rule involving the linguistically modified fuzzy sets

$$\text{if } \tau(A) \text{ then } \mu(B)$$

with τ and μ being the linguistic modifiers [15] of A and B such as *very*, *more or less*, etc. The general model takes on the power form $\tau(A) = A^p$ or in terms of the membership functions $\tau(A)(x) = A^p(x)$. When $p \neq 1$ we are dealing with a concentration operation. The values of p lower than 1 give rise to the dilution effect and produce the fuzzy set with lower granularity. Using the notation

$$\tau(A) = A^p \quad \mu(B) = B^r$$

several observations can be drawn as to the character of the rules with the linguistically modified fuzzy sets. The p-r plane, Figure 5, splits into four quadrants

(1) Ω 1: in this region we encounter rules that are more general than the original one; subsequently one may anticipate higher likelihood of conflict
(2) Ω 3: the rules are more specific (less general) as more specific condition allows us to derive less detailed conclusion. The likelihood of conflict is reduced.
(3) Ω 2 and Ω 4: the level of generality is similar to the original rule. The only difference is that the condition and conclusion are affected in the same way and some monotonicity property is preserved.

The rules in $\Omega 2$ have more detailed condition and this specificity translates to the increased granularity of the conclusion. For instance, we may have a rule *if very(A) then very(B)*. The converse situation happens in $\Omega 4$. The rule *if more or less(A) then more or less(B)* is located in this region. In particular, if the support of A is equal to $\mathbf{X}$ and p approaches zero, the rule transforms into an unconditional statement: "if *anything* then $\mu(B)$"

The linguistic modifications of the fuzzy sets affect the rules and impact their level of conflict yet the changes promoted by the power-like modifications of the membership functions are not very radical as the membership functions retain their original supports.

8   Growing rules through manipulation of information granules

The idea of granularity of information is essential to the evaluation of rules. Especially, by manipulating the level of information granules occurring in

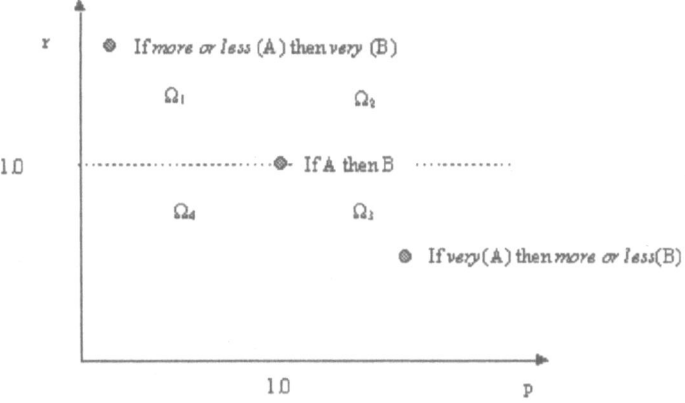

Fig. 5. The p-r plane of linguistically modified rules

the rule, we gain a better insight into its performance. Especially, by manipulating this level, we can enhance the quality of the rule. This leads us to a
new idea of rule growing (rather than constructing in a usual way). Let us
concentrate on the simple rule linking A and B *if A then B* where A and
B are two fuzzy sets (information granules) defined in the two corresponding spaces A and B. The two popular measures expressing rules relevance
concern confidence and support. Both of them are given in the language of
probabilities. The first one is defined as a ratio of two probabilities

$$\text{conf} = \frac{Prob(A \times B)}{Prob(A)}$$

Intuitively, if A and B are unrelated, then we anticipate that the relevance
of the rule is low; this is implied by the low values of the probability of the
Cartesian product standing in the nominator of the confidence measure. The
support of the rule is expressed in the form

$$\text{Support} = Prob(A \times B)$$

In the case of fuzzy sets, the computations involve probabilities of fuzzy
events. Then we derive

$$Prob(A) = \frac{1}{N} \sum_{k=1}^{N} A(x(k))$$

$$Prob(A \times B) = \frac{1}{N} \sum_{k=1}^{N} A(x(k))B(y(k))$$

where $x(k)$ and $y(k)$ are the elements occurring in the database.

The quality of the rule is measured by its confidence and support: both of them should be high to enhance the quality of the rule. Note that that when the size of information granules A and B increases, the support of the rule increases. This may not be necessarily true for the confidence level. Nevertheless, we may expect a certain form of monotonic behavior of the confidence and support measure while changing the size of the information granules A and B. Taking a weighted sum of the two measures

$$\text{confidence} + \alpha \text{ support}$$

(with α being a weight factor), one can attempt find its maximum by affecting the size of the granules. This gives rise to the notion of growing rules: starting from the specific fuzzy sets A and B, we start expanding them, namely start *growing* the rule, Figure 6.

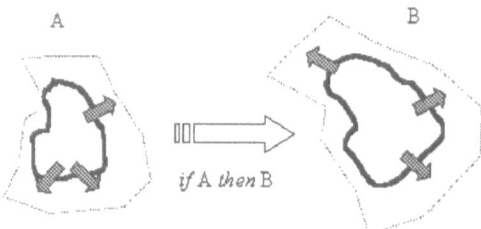

Fig. 6. Growing a rule by varying information granularity of the condition and conclusion fuzzy sets

While the growing the rules occurs individually for each rule, one has to monitor a level of potential conflict. This gives rise to the two-level structure, Figure 7, where at the lower level the growth of the rules takes places and a control of conflict is realized at the global level.

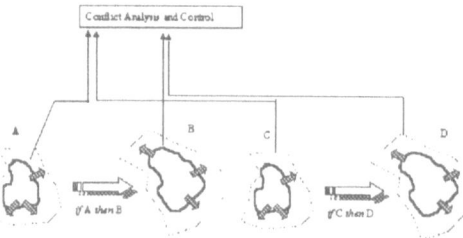

Fig. 7. Growing individual rules and controlling a level of conflict at the global scale of the rulebase

9   Conclusions

We have discussed the idea of information granulation realized with the aid of fuzzy sets, and developed a complete algorithmic framework that helps reveal patterns in databases. The study makes a clear distinction between associations and rules by showing that rules are simply directional constructs that originate from associations. We characterized a collection of rules with respect to its consistency. Various means of the enhancement of the rules were shown including the one involving the linguistic modification of fuzzy sets standing in the rule. We have made a clear point that any prior design commitment to directionality between variables in databases could be too restrictive; the search for associations does make sense while jumping into the formation of rules could be dangerously premature. By the same token, one should beware of exploiting and endorsing standard techniques of model identification and rule-based systems, as the algorithmic skeleton there is too limited and somewhat biased.

Acknowledgment

The support from the Natural Sciences and Engineering Research Council of Canada (NSERC) is gratefully acknowledged.

References

1. R. Agrawal, T. Imielinski, A. Swami, *Database mining: a performance perspective*, IEEE Transactions on Knowledge and Data Engineering, 5, (1993), 914-925.
2. J. Buckley, Y. Hayashi, *Fuzzy neural networks: a survey*, Fuzzy Sets and Systems, 66, (1994), 1-14.
3. K. Cios, W. Pedrycz, R. Swiniarski, *Data Mining Techniques*, Kluwer Academic Publishers, Boston, (1998).
4. J. Chattratichat, *Large scale data mining: challenges and responses* In: Proc. 3rd Int. Conf. on Knowledge Discovery and Data Mining, Newport Beach, CA, August, 14-17, (1997), pp.143-146.
5. B.S. Everitt, *Cluster Analysis*, Heinemann, Berlin, (1974).
6. C.J. Harris, C.G. Moore, M. Brown, *Intelligent Control - Aspects of Fuzzy Logic and Neural Nets*, World Scientific, Singapore, (1993).
7. D.O. Hebb, *The Organization of Behavior: A Neuropsychological Theory*, J. Wiley, N. York, (1949).
8. P.J. Huber, *From large to huge: a statistician's reaction to KDD and DM*, In: Proc. 3rd Int. Conf. on Knowledge Discovery and Data Mining, Newport Beach, CA, August 14-17, (1997), 304-308.
9. J. S. R. Jang, C.T. Sun, E. Mizutani, *Neuro-Fuzzy and Soft Computing*, Prentice Hall, Upper Saddle River, NJ, (1997).
10. A. Kandel, *Fuzzy Mathematical Techniques with Applications*, Addison-Wesley, Reading, MA, (1986).

11. N. Kasabov, *Foundations of Neural Networks, Fuzzy Systems, and Knowledge Engineering*, MIT Press, Cambridge, MA, (1996).
12. L. Kaufman and P.J. Rousseeuw, *Finding Groups in Data*, J. Wiley, New York, (1990).
13. Z. Pawlak, Rough Sets: *Theoretical Aspects of Reasoning about Data*, Kluwer Academic, Dordrecht, (1991).
14. W. Pedrycz, *Computational Intelligence: An Introduction*, CRC Press, Boca Raton, FL, (1997).
15. W. Pedrycz, F. Gomide, *An Introduction to Fuzzy Sets*, Cambridge, MIT Press, Cambridge, MA, (1998).
16. W. Pedrycz, M.H. Smith, *Granular correlation analysis in data mining*, Proc. 18th Int Conf of the North American Fuzzy Information Processing Society (NAFIPS), New York, June 1-12, (1999), 715-719.
17. W. Pedrycz, E. Roventa, *From fuzzy information processing to fuzzy communication channels*, Kybernetes, vol. 28, no.5, (1999), 515-527.
18. W. Pedrycz, *Fuzzy set technology in knowledge discovery*, Fuzzy Sets and Systems, 3, (1998), 279-290.
19. W. Pedrycz, *Shadowed sets: representing and processing fuzzy sets*, IEEE Trans. on Systems, Man, and Cybernetics, part B, 28, (1998), 103-109.
20. Pedrycz, W. Vukovich, G. *Quantification of fuzzy mappings: a relevance of rule-based architectures*, Proc. 18th Int Conf of the North American Fuzzy Information Processing Society (NAFIPS), New York, June 1-12, (1999), 105-109.
21. G. Piatetsky-Shapiro and W. J. Frawley, (eds.) *Knowledge Discovery in Databases*, AAAI Press, Menlo Park, California, (1991).
22. J. R. Quinlan, *Induction of Decision Trees*, Machine Learning 1, 1, 81-106, (1986).
23. J. R. Quinlan, *C4.5: Programs for Machine Learning*, Morgan Kaufmann Publishers, San Mateo, California, (1993).
24. H. Toivonen, *Sampling large databases for association rules* In: Proc. 22nd Int. Conf. on Very Large Databases, (1996), 134-145.
25. L.H. Tsoukalas, R.E. Uhrig, *Fuzzy and Neural Approaches in Engineering* J. Wiley, New York, (1997).
26. R. R. Yager, *Entropy and specificity in a mathematical theory of evidence* Int. J. Gen. Syst., 9, (1983), 249-260.
27. K. Yoda, T. Fukuda, Y. Morimoto, *Computing optimized rectilinear regions for association rules* In: Proc. 3rd Int. Conf. on Knowledge Discovery and Data Mining, Newport Beach, CA, August 14-17, (1997), 96-103.
28. J. Wnek and R. S. Michalski, *Conceptual Transition from Logic to Arithmetic in Concept Learning*, Reports of Machine Learning and Inference Laboratory, MLI 94-7, Center for MLI, George Mason University, December (1994).
29. L. A Zadeh, *Fuzzy sets and information granularity*, In: M.M. Gupta, R.K. Ragade, R.R. Yager, eds., Advances in Fuzzy Set Theory and Applications, North Holland, Amsterdam, (1979), 3-18.
30. L. A. Zadeh, *Fuzzy logic = Computing with words, IEEE Trans. on Fuzzy Systems*, vol. 4, 2, (1996), 103-111.
31. L. A. Zadeh, *Toward a theory of fuzzy information granulation and its centrality in human reasoning and fuzzy logic*, Fuzzy Sets and Systems, 90, (1997), 111-117.

Flexible Querying in Deductive Database

Maria I. Sessa

Dip. Matematica e Informatica - Università di Salerno
via S. Allende, 84081 Baronissi (SA), Italy - mis@unisa.it

Abstract. In this paper we frame the approach proposed in [9] in the field of Deductive Databases. Approximate information can be managed by introducing a similarity relation $\mathcal{R}$ in the set of predicate names and object names of the language. The notion of fuzzy least Herbrand model is also introduced.

1 Introduction and previous works

A Database Management System (DBMS) is a software tool that allows to store, gain access and modify large quantities of data. In standard DBMS an implicit double assumption is made: the information to be managed are either accurately known or totally unknown and user needs are modelled by Boolean predicates which allow the qualification of the elements or sets of elements.

It seems natural to extend the functionality of DBMSs so that they can take under consideration the notion of imprecision/uncertainty which may apply to data stored and/or to the queries addressed to the system. Theory of fuzzy sets provided useful tools to the extension of DBMS in this direction. Tree main topics in the research activity can be identified:

- the possibility of dealing with incompletely known data; in particular, it is usefully exploited the use of formal tools which allow to represent imperfect information when the uncertainty pervading data is not of stochastic nature, but rather subjective.
- the adjunction of the capability of performing flexible querying, i.e., the introduction in user queries of preferences at the level of both atomic conditions and their combination.
- the handling of integrity constraints involving gradual properties.

The use of fuzzy set theory has been studied in many Database models (as: relational, network, object oriented, etc.) (see [4] and [12] for a survey). Indeed, fuzzy sets are especially appropriate to model preferences and approximate information. Due to its sound mathematical foundations, the relational framework has been frequently chosen as a basis to define extensions with a clear semantics. Some approaches introduce the notion of fuzzy relation whose tuples are weighted. Then, the extension of the relational algebra and the implementation of flexible queries have been developed with several techniques. In particular, the notion of similarity has been exploited to enrich the relation model [5]. The idea is to consider that some domains can be provided with similarity (or proximity) relations describing some semantic relationship (such as interchangeability) between values.

A particular approach to the design of DBMS is given by the *Deductive Database*. On one hand, Deductive Databases are an attempt to adapt PROLOG, which has a "small-data" view of the word, to a "large-data" word. On the other, since relation are naturally thought of as the value of a logical predicate, it is easy to see the Deductive Database as an advanced form of relational system.

In this paper we propose the introduction of approximate representation and inference capabilities in Deductive Databases systems by exploiting the extension of the Logic Programming paradigm proposed in [9]. As well as in [5], the basic feature of this extension is the use of similarity relations in some domains of symbols (alphabet). By using an abstract interpretation technique, in [9] a declarative semantics is provided which allows to highlight the meaning of the proposed extension. The notion of fuzzy least Herbrand model is also provided.

2   Similarity-based Deductive Database Systems

Deductive Database are DBMS whose query language and storage structure are based on a logical model of data. The basic idea is that a Deductive Database divides its information into two categories:

1. *Data*, or facts, normally represented by a predicate with constant arguments (*ground atom*). For example, the fact *parent(joe, sue)*, means that Sue is a parent of Joe. Here, *parent* is the name of the predicate, and this predicate is represented *extensionally*, that is, by storing in the database a relation of all the true tuples for this predicate. Thus, (*joe, sue*) would be one of the tuples in the stored relation.
2. *Rules*, or program, normally written in PROLOG-style notation as
   $$p : - q_1, ..., q_n.$$
   This rule is read declaratively as "q_1 and q_2 and ... and q_n implies p". each of p (the *head*) and the q_is (the *subgoals* of the *body*) are *atomic formulae* (*literals*), consisting of a predicate applied to *terms*, which are either constants, variables, or functions applied to terms. Programs in which therms are either constants or variables are often referred to as *Datalog* programs. The data are referred to as the *Extensional Database* (*EDB*), and the rules as *Intensional Database* (*IDB*).

As an Example, let us consider the following program

$anc(X, Y) : - par(X, Y)$

$anc(X, Y) : - par(X, Z), anc(Z, Y)$

and let the query be

$anc(john, Y)?$

Assume that a Database contains a parenthood relation *par*. Then the program defines a derived relation describing ancestors, and the query asks for the ancestors of *john*. A well known strategy for evaluating answers for query to logic programs is the *bottom-up* strategy. It computes the least fixpoint semantic, i.e., the least Herbrand model of the given program. While the strategy is reasonably efficient when the query does not contain instantiated variables, the previous example shows that it is very inefficient when bindings for some variables are given in the query. The reason is that it computes the complete *anc* relation and then applies selection to it. Thus, all ancestors are computed, even though only the ancestors of *john* are needed. A *top-down* strategy (as used by PROLOG) may do much better by computing only the ancestors of *john*. However, it is worth to stress that, for positive programs without function symbols, the bottom-up evaluation process always terminates. On the contrary, the PROLOG's depth-first strategy can lead to infinite loops even in this case.

The optimization of the bottom-up evaluation technique has been an attractive research area. Several query answering methods have been proposed in literature and a number of projects have lead to implemented systems (see [13] for a survey). In particular, many techniques exploit transformation strategies that can be applied to an arbitrary program and a query to produce a program that is equivalent to the given one with respect to the query, and that uses the bindings in the query to direct the computation. It improves the efficiency of the computation since only the facts that are "relevant" to the given query are considered. These transformation are in many cases generalizations of the Magic Sets and Counting techniques [2] which in their initial formulation were of limited applicability.

Approximate representation and inference capabilities can be naturally introduced in Deductive Databases systems by exploiting the extension of the Logic Programming paradigm proposed in [9]. Following [18] and [3], the basic idea is that approximate reasoning can be performed on the basis of "analogy" or "similarity". Thus, we consider reasonings that may be approximated by allowing the antecedent clauses of a rule to match its premises only approximately. As an example, we can consider the following inference scheme

- $output(x) : - rel_1(x, a)$

- $rel_2(b, c)$

- rel_1 "is similar to" rel_2 for me
- c "is similar to" a for me

$output(b)$.

Obviously, the similarity is a graded notion [17], so the degree at which we can admit the conclusion "$output(b)$" depends on the degree of similarity between the predicates "rel_1" and "rel_2" and between the constants "a" and "c". An important feature of such a kind of inference is the fact that the similarity is defined between symbols in the alphabet of the language, i.e. it is exploited at a syntactic-level. This makes this approach different from the usual one where the similarity is defined in the set of the interpretations (semantic-level) [14], [7], [8]. More resemblance can be found with the similarity-based approach proposed in the framework of relational databases [5], where the membership value of a tuple belonging to a response relation to a query is computed on the basis of a similarity defined over the domain.

In [9] this approach has been introduced in the paradigm of Logic Programming without function symbols, in order to obtain approximate reasoning capabilities. In [15] this methodology has been extended to the case of first order languages. This approach can be directly framed in the context of Deductive Databases. The basic idea is that the exact matching between different entities is relaxed by considering a similarity relation between constants and between predicate names in the language of a function free logic program P. Then, we can extend the given Deductive Databas P by adding new rules and facts which are similar at least with a fixed value $\lambda \in (0, 1]$ to the given ones. The obtained set of facts and rules, named *extended Deductive Database*, allows to enhance the resolution process with approximate inference capabilities.

A well known property states that a min-transitive similarity can be described "level by level" by a family of classical equivalence relations. Any equivalence relation of this family, named *cut of level* λ and denoted with $\equiv_{\mathcal{R},\lambda}$, is obtained by considering as equivalent two elements which have a similarity value greater than a fixed value $\lambda \in [0, 1]$. Then, in [9] an alternative transformation technique of function free logic programs is also defined by exploiting the abstraction provided by the relation $\equiv_{\mathcal{R},\lambda}$. This transformation can be applied to Deductive Databases as well. It allows to construct a set of rules and facts, named *abstract Deductive Database*, on a different alphabet.

The equivalence between these two inference processes can be proved by using an abstract interpretation technique. Thus, without changing the standard strategies for evaluating answers for a query (as: Magic Sets, counting algorithm, factoring optimization, etc.), both these transformations can be exploited to manage approximate information in a Deductive Database system. Following this approach, the fixed level λ, which allows to construct the extended and abstract set of rules and facts, provides a measure of the approximation of the obtained answers. The notion of fuzzy least Herband model introduced in [9] provides the semantics of the proposed fuzzy extension of the Deductive Database model.

In the sequel we recall the main features of these transformation methodologies. In particular, since similarity relation plays a central rule in this approach, some insights on this notion are provided in the next Sextions 3 and 4.

3   Similarity relation

In [9] approximate reasoning capabilities are introduced in the Logic Programming paradigm by exploiting the notion of similarity, which is a mathematical tool that provides a way to manage alternative instances of an entity that can be considered "equal" with a given degree [17]. In this section we recall some classical results concerning this notion [10] that will be exploited in the sequel. At first, we synthetically give some well known notions concerning closure operators and equivalence relations

Definition 1. Let $\mathcal{U}$ be a set, $\mathcal{P}(\mathcal{U})$ its power set and $\preceq$ a partial ordering in $\mathcal{P}(\mathcal{U})$. An operator $H : \mathcal{P}(\mathcal{U}) \to \mathcal{P}(\mathcal{U})$ is called a *closure* (resp. *reductive*) operator if the following properties hold:

i) $X \preceq H(X)$ (*resp.* $H(X) \preceq X$)

ii) $H(H(X)) = H(X)$

iii) $X \preceq Y \Longrightarrow H(X) \preceq H(Y)$.

In particular, we are interested to the following construction of a closure operator associated with an equivalence relation. This construction is a crucial tool in our approach

Proposition 1. *Let* $\equiv$ *be an equivalence relation on a set* S. *Then, the operator* $H_\equiv : \mathcal{P}(S) \to \mathcal{P}(S)$ *such that for any* $X \subseteq \mathcal{P}(S)$

$$H_\equiv(X) = \{x' \in S \mid \exists x \in X : \quad x' \equiv x)\}$$

is a closure operator.

The following Example shows an application of the previous construction which provides a classical closure operator

Example 1.

Let $f : S \longrightarrow T$ be a function. Then it is well known that the relation *Kernel of* f, denoted with $Kern_f$, and defined as follows:

$$Kern_f \;\; = \;\; \{(x,y) \in S \times S \quad | \quad f(x) = f(y)\}$$

is an equivalence relation. Moreover, the operator H_{Kern_f} defined by setting for any $X \subseteq \mathcal{P}(S)$

$$H_{Kern_f}(X) = \{x' \in S \quad | \quad \exists x \in X : \quad f(x') = f(x)\}$$

is a closure operator.

Now we summarize some results concerning similarity relation. In this notion we esploit the binary operation $\wedge : [0,1] \times [0,1] \to [0,1]$ such that $x \wedge y$ denotes the *minimum* between the two elements $x, y \in [0,1]$. This is a necessary assumption for the validity of the results obtained in the proposed approach.

Moreover, let us recall that, given a set $\mathcal{U}$, a *fuzzy subset of* $\mathcal{U}$ is any map $s : \mathcal{U} \to [0,1]$. In particular, we call *fuzzy relation in* $\mathcal{U}$ a fuzzy subset $\mathcal{R} : \mathcal{U} \times \mathcal{U} \to [0,1]$.

Definition 2. A *similarity* on a domain $\mathcal{U}$ is a fuzzy relation $\mathcal{R} : \mathcal{U} \times \mathcal{U} \to [0,1]$ in $\mathcal{U}$ such that the following properties hold

i) $\mathcal{R}(x,x) = 1$ for any $x \in \mathcal{U}$ (reflexivity)

ii) $\mathcal{R}(x,y) = \mathcal{R}(y,x)$ for any $x, y \in \mathcal{U}$ (symmetry)

iii) $\mathcal{R}(x,z) \succeq \mathcal{R}(x,y) \wedge \mathcal{R}(y,z)$ for any $x, y, z \in \mathcal{U}$ (transitivity).

Similarity relations are strictly related with equivalence relations and, then, to closure operators as stated by the following results

Definition 3. Let $\mathcal{U}$ be a domain and $\mathcal{R} : \mathcal{U} \times \mathcal{U} \to [0,1]$ a fuzzy relation in $\mathcal{U}$. Then, for any $\lambda \in [0,1]$, the relation $\equiv_{\mathcal{R},\lambda}$ in $\mathcal{U}$ defined as

$$x \equiv_{\mathcal{R},\lambda} y \quad \Longleftrightarrow \quad \mathcal{R}(x,y) \succeq \lambda$$

is an equivalence relation named *cut of level* λ (in short λ-*cut*) of $\mathcal{R}$.

The notion of $\lambda - cut$ allows us to define a similarity relation by means of a suitable family of equivalence relations according to the following result that can be easily proven

Proposition 2. *Let $\mathcal{R}$ be a similarity in a domain $\mathcal{U}$ and, for any $\lambda \in [0,1]$ let $\equiv_{\mathcal{R},\lambda}$ be the $\lambda - cut$ of $\mathcal{R}$. Then, $\{\equiv_{\mathcal{R},\lambda}\}_{\lambda \in [0,1]}$ is a family of equivalence relations such that,*

i) for any μ and λ in $[0,1]$, $\lambda \preceq \mu \quad \Rightarrow \quad \equiv_{\mathcal{R},\lambda} \supseteq \equiv_{\mathcal{R},\mu}$

ii) for any μ in $[0,1]$, $\bigcap_{\lambda \preceq \mu} \equiv_{\mathcal{R},\lambda} = \equiv_{\mathcal{R},\mu}$.

Conversely, let $\{\equiv_\lambda\}_{\lambda \in [0,1]}$ be a family of equivalence relations satisfying conditions i) and ii). Then the relation $\mathcal{R}$ defined by setting

$$\mathcal{R}(x,y) = Sup\{\lambda \in [0,1] \ | \ x \equiv_\lambda y\}$$

is a similarity whose λ-cuts are equal to the relations $\equiv_\lambda$ belonging to the given family.

The λ-cut can be considered as a generalization of the equality and it is a crucial notion in our approach. Indeed, the relation $\equiv_{\mathcal{R},\lambda}$ will be the formal tool that in the sequel we exploit in order to formalize the idea that two constant symbols can be considered equal with a fixed approximation level $\lambda \in [0,1]$. Such a level provides a measure of the allowed approximation in order to avoid failure of matching between constant symbols.

Finally, let us note that in accordance with Proposition 1, the operator H_λ : $\mathcal{P}(\mathcal{U}) \to \mathcal{P}(\mathcal{U})$ such that for any $X \in \mathcal{P}(\mathcal{U})$

$$H_\lambda(X) = \{z \in \mathcal{U} \ | \ \exists x \in X : \ x \equiv_{\mathcal{R},\lambda} y\} = \{z \in \mathcal{U} \ | \ \exists x \in X : \ \mathcal{R}(z,x) \succeq \lambda\}$$

is a closure operator.

In the sequel, given $a \in \mathcal{U}$ the simplified notation $H_\lambda(a)$ will be exploited instead of $H_\lambda(\{a\})$.

Example 2.

Let U be a set of animals denoted with

$$U = \{M, \ B, \ G, \ E, \ P, \ H, \ S, \ T, \ C, \ W, \ D\}$$

where these letters stand for

$$M = man, \ B = bear, \ G = gorilla, \ E = eagle, P = pigeon, \ H = hawk,$$

$$S = shark, \ T = tiger, C = cat, \ W = wolf, \ D = dog.$$

We can define a similarity $\mathcal{R}$ between elements in U by providing for any $x, y \in U$ the similarity value as follows:

$$\mathcal{R}(x,y) = \mathcal{R}(y,x)$$

$$\mathcal{R}(x,y) = 1 \quad if \ x = y$$

$$\mathcal{R}(D,W) = \mathcal{R}(E,H) = .8$$

$$\mathcal{R}(M,G) = \mathcal{R}(C,T) = .5$$

$$\mathcal{R}(M,B) = \mathcal{R}(B,G) = \mathcal{R}(E,P) = \mathcal{R}(H,P) = \mathcal{R}(D,C) = \mathcal{R}(D,T) =$$
$$= \mathcal{R}(W,T) = \mathcal{R}(W,C) = .2$$

$$\mathcal{R}(x,y) = 0 \quad otherwise.$$

As an example, by considering the cut relation $\equiv_{\mathcal{R},.2}$ of level $\lambda = .2$, the equivalence class of the element W is $\{W,D,C,T\}$, whereas by considering the cut relation $\equiv_{\mathcal{R},.7}$ of level $\lambda = .7$, the equivalence class of the element W is $\{W, D\}$. According to the definition, it is $H_{.2}(\{W, B\}) = \{W, D, C, T, B, M, G\}$ and $H_{.7}(\{W, B\}) = \{W, D, B\}$.

4 Interactive procedures to define a similarity relation

We concern with similarity defined in a finite domain $\mathcal{U}$. It implies that we have a discrete and ordered set of possible similarity values $\lambda_i \in [0, 1]$, with i belonging to a finite set I of indexes. Then, the family $\{\equiv_{\mathcal{R}, \lambda}\}_{\lambda \in [0,1]}$ in Proposition 2 is given by $\{\equiv_{\mathcal{R}, \lambda_i}\}_{i \in I}$.

It is not easy to define a similarity in a set of elements by providing the similarity values $\mathcal{R}(x, y)$ for any $x, y \in U$. Indeed, the transitivity constraint can produce side effects which contradict the given similarity values. According with Proposition 2, a level-by-level construction of the family $\{\equiv_{\lambda_i}\}_{0 \preceq i \preceq n}$ of $\lambda_i - cuts$ allows to overcome this problem.

The following algorithm is an interactive step-by-step procedure which in output gives the quotient sets of the family of $\lambda_i - cuts$. These partitions are defined according to guided choices, interactively provided by the user, which express the subjective evaluation of the similarity between elements.

Algorithm 1. BOTTOM-UP ALGORITHM.

INPUT: a domain $U = \{a_1, ..., a_m\}$ and the ordered set $\Lambda = \{\lambda_0, ..., \lambda_n\}$, of similarity levels, with $0 = \lambda_0 \prec \lambda_1 \prec ... \prec \lambda_n = 1$

OUTPUT: the quotient sets $Q_i = \{C_1^i, ..., C_{k_i}^i\}$, $0 \preceq i \preceq n$, associated with the λ–cuts in the family $\{\equiv_{\lambda_i}\}_{0 \preceq i \preceq n}$.

set $k_0 = 1$, $C_{k_0}^0 = U$ and $Q_0 = \{C_{k_0}^0\}$;

for $i = 1, ..., n$

    for $j = 1, ..., k_{i-1}$

        if C_j^{i-1} is a singleton;

            then continue;

      else

            get in input from the user $C_{j_1}^i, ..., C_{j_h}^i$ such that:

            $C_j^{i-1} = \cup C_{j_r}^i$, $1 \preceq r \preceq h$, and

            $C_p^i \cap C_q^i$ =empty, for any $p, q \in \{j_1, ..., j_h\}$;

            set $Q_i = (Q_{i-1} - C_j^{i-1}) \cup \{C_{j_1}^i, ..., C_{j_h}^i\}$;

    end for

end for

end

The previous algorithm starts from the trivial partition $Q_0 = U/_{\equiv_0} = \{U\}$ corresponding in the intended similarity to the cut of level 0, where all the element are in the same equivalence class. At any step a refinement of the previous partition is constructed. In particular, at any iterative step, the user provides the splitting of an equivalence class C_j^{i-1} in the quotient set Q_{i-1} of $\equiv_{\lambda_{i-1}}$, on the basis of a subjective evaluation of a more significant similarity-level $\lambda_i \succ \lambda_{i-1}$ between elements in this class. Let us stress that, in general, we can allow a similarity value 1 also between different elements.

The following Example shows an application of the Algorithm 1

Example 3.

Let $\Lambda = \{0, .4, .6, .8, 1\}$ be an ordered set of similarity values and U a set of geometric figures denoted with $U = \{F_3, F_4, F_5, F_6, F_8, C\}$ where these letters stand for

$F_3 = triangle \quad F_4 = square \quad F_5 = pentagon$

$F_6 = hexagon \quad F_8 = octagon \quad C = circle.$

The following steps can be obtained by exploiting the Algorithm 1

- $Q_0 = U/_{\equiv_{.0}} = \{\{F_3, F_4, F_5, F_6, F_8, C\}\}$

- $Q_1 = U/_{\equiv_{.4}} = \{\{F_3\}, \{F_4, F_5, F_6, F_8, C\}\}$

- $Q_2 = U/_{\equiv_{.6}} = \{\{F_3\}, \{F_4, F_5, F_6\}, \{F_8, C\}\}$

- $Q_3 = U/_{\equiv_{.8}} = \{\{F_3\}, \{F_4\}, \{F_5, F_6\}, \{F_8\}, \{C\}\}$

- $Q_4 = U/_{\equiv_1} = \{\{F_3\}, \{F_4\}, \{F_5\}, \{F_6\}, \{F_8\}, \{C\}\}$.

The following table shows the corresponding Similarity values between pairs of elements

	F_3	F_4	F_5	F_6	F_8	C
F_3	1	0	0	0	0	0
F_4	0	1	.6	.6	.4	.4
F_5	0	.6	1	.8	.4	.4
F_6	0	.6	.8	1	.4	.4
F_8	0	.4	.4	.4	1	.6
C	0	.4	.4	.4	.6	1

Let us stress that in the bottom-up Algorithm 1 the user provides the definition of the new classes $C_{j_1}^i, \ldots, C_{j_h}^i$ on the basis of subjective choices. If for the current level of similarity λ_i the user thinks that there are not significant difference with respect to the previous level λ_{i-1}, the resulting quotient set Q_i will be equal to Q_{i-1}.

A symmetric procedure can be also defined in a Top-Down way starting from the partition corresponding to the greatest similarity value 1. In this case, at any step a classe in the new partition is constructed by considering union of classes belonging to the previous partition. Let us stress that also in this algorithm we allow the definition of a similarity value 1 between different elements.

Algorithm 2. TOP-DOWN ALGORITHM.

INPUT: a domain $U = \{a_1, ..., a_m\}$ and the ordered set $\Lambda = \{\lambda_0, ..., \lambda_n\}$, of similarity levels, with $0 = \lambda_0 \prec \lambda_1 \prec ... \prec \lambda_n = 1$

OUTPUT: the quotient sets $Q_i = \{C_1^i, ..., C_{k_i}^i\}$, $0 \preceq i \preceq n$, associated with the λ−cuts in the family $\{\equiv_{\lambda_i}\}_{0 \preceq i \preceq n}$.

Set $Q_n = \{C_1^n, ..., C_{k_n}^n\}$, $1 \preceq k_n \preceq m$, *such that:*

with $U = \cup C_r^n$, $1 \preceq r \preceq k_n$, and

$C_p^n \cap C_q^n =$empty, $p, q \in \{1, ..., k_n\}$;

for $i = n, ..., 2$

    if $k_i = 1$

        then set $Q_{i-1} = Q_i$

    else

        get in input from the user $C_1^{i-1}, ..., C_{k_{i-1}}^{i-1}$ *such that:*

        $C_j^{i-1} = \cup C_r^i$, $r \in I_i$, $1 \preceq j \preceq k_{i-1}$, $I_i \subseteq \{1, ..., k_i\}$,

        and $C_p^{i-1} \cap C_q^{i-1} =$empty, $p, q \in \{1, ..., k_{i-1}\}$;

        set $Q_{i-1} = \{C_1^{i-1}, ..., C_{k_{i-1}}^{i-1}\}$;

end for

set $Q_0 = \{U\}$;

end

The following example shows an application of the Algorithm 2:

Example 4.

Let $\mathcal{R}$ be the similarity considered in the Example 2. An equivalent representation of $\mathcal{R}$ can be given by generating the quotient sets of the $\lambda_i - cuts$

in the family $\{\equiv_{\mathcal{R},\lambda_i}\}_{0 \preceq i \preceq 4}$, corresponding to the different similarity levels $\{\lambda_0, \lambda_1, \lambda_2, \lambda_3, \lambda_4\} = \{0, .2, .5, .8, 1\}$. The following steps can be obtained by exploiting the Algorithm 2:

$$U/\equiv_{\mathcal{R},1} = \{\{M\},\{B\},\{G\},\{E\},\{P\},\{H\},\{S\},\{T\},\{C\},\{W\},\{D\}\},$$

$$U/\equiv_{\mathcal{R},.8} = \{\{M\},\{B\},\{G\},\{E,H\},\{P\},\{S\},\{T\},\{C\},\{W,D\}\},$$

$$U/\equiv_{\mathcal{R},.5} = \{\{M,G\},\{B\},\{E,H\},\{P\},\{S\},\{T,C\},\{W,C\}\},$$

$$U/\equiv_{\mathcal{R},.2} = \{\{M,B,G\},\{E,P,H\},\{S\},\{T,C,W,D\}\}$$

$$U/\equiv_{\mathcal{R},0} = \{\{M,\ B,\ G,\ E,\ P,\ H,\ S,\ T,\ C,\ W,\ D\}\} = \{U\}$$

For any x, y in U, the similarity value $\mathcal{R}(x,y)$ can be obtained by considering the maximum level λ_i, $i \in \{0, 1, 2, 3, 4\}$, such that the elements x and y belongs to a same equivalence class in $\equiv_{\mathcal{R},\lambda_i}$.

When in Definition 2 the T-norm $\wedge$ is the minimum, a simple characterization of the similarity is given by the Proposition 3, where the following Definition is exploited [15]:

Definition 4. We say that the *min-triangular* property is satisfied by a fuzzy relation $\mathcal{R} : U \times U \longrightarrow [0,1]$, if for any x, y, z in U the similarity values $\mathcal{R}(x,y)$, $\mathcal{R}(x,z)$, $\mathcal{R}(y,z)$ are such that two are equal and the third one is greater or equal than the others.

Proposition 3. *Let $\mathcal{R}$ be a fuzzy relation in a domain U satisfying conditions i) and ii) in Definition 2. Then,*

$\mathcal{R}$ *is a similarity* $\iff$ $\mathcal{R}$ *satisfies the min-triangular property.*

5   Logic programming and abstract interpretations

Let us recall that the declarative paradigm of Logic Programming is a refutation system which allows to compute logic consequence of a conjunction of Horn clauses named *logic program* [1], [11]. Let $\mathcal{L}$ be a first order language and P a logic program in $\mathcal{L}$. We denote with $B_{\mathcal{L}}$ the set of ground atomic formulae in $\mathcal{L}$, i.e. the Herbrand base of $\mathcal{L}$. A well known results states that the least Herbrand model of P is given by:

$$M_P = \{L \in B_P \mid P \models L\}.$$

Let us denote with T_P the immediate consequence operator $T_P : \mathcal{P}(B_{\mathcal{L}}) \to \mathcal{P}(B_{\mathcal{L}})$ defined by

$$T_P(X) = \{a \mid a \longleftarrow a_1, ..., a_n \in \Gamma(P) \text{ and } a_i \in X, \ n \succeq i \succeq 1\}$$

where $\Gamma(P)$ denotes the set of all ground instances of clauses in P. The application of Tarski's fixpoint theorem yields an alternative characterization of the semantics of the logic program P. Indeed, it can be proved that

$$M_P = lfp(T_P) = \bigcup_{n \succeq 0} T_P^n(empty)$$

with $T_P^n(empty) \subseteq T_P^{n+1}(empty)$, $\forall n \succeq 0$.

In [15] the general definition of *structural translation* of a first order language has been introduced. Exploiting this notion, the generalization to first order languages, of the transformation techniques introduced in [9] for function free logic programs, is given.

Let us consider two first order languages $\mathcal{L}$ and $\mathcal{L}'$ where the sets of function symbols and of predicate symbols are denoted with F and R in $\mathcal{L}$, and with F' and R' in $\mathcal{L}'$. The set of variable is denoted with V. By introducing the following definition, we characterize a class of functions which allow to translate a language into another one by preserving both the predicate/function arities and the structure of the formulae. Here and in the sequel, the notation "/" is exploited when a proposition can be given for the elements both on the left and on the right of "/".

Definition 5. Given two first order language $\mathcal{L}$ and $\mathcal{L}'$, we call *structural translation* of $\mathcal{L}$ in $\mathcal{L}'$ any function $\Delta : F \cup R \longrightarrow F' \cup R'$ such that for any function/predicate symbol $s \in F \cup R$ with arity n,

$$\Delta(s) = s'$$

with $s' \in F' \cup R'$ function/predicate symbol of same arity n.

Recursively, we define the extension $\Delta : \mathcal{L} \longrightarrow \mathcal{L}'$ by setting

- for any variable $x \in V$,

$$\Delta(x) = x$$

- for any $t_1, ..., t_n$ terms in $\mathcal{L}$, and f function/predicate symbol in $F \cup R$,

$$\Delta(f(t_1, ..., t_n)) = \Delta(f)(\Delta(t_1), ..., \Delta(t_n))$$

- $\Delta(\square) = \square$, with $\square$ empty clause

- for any A, B formulae in $\mathcal{L}$ and $x \in V$

$$\Delta(A \wedge B) = \Delta(A) \wedge \Delta(B), \quad \Delta(A \vee B) = \Delta(A) \vee \Delta(B),$$

$$\Delta(\neg A) = \neg\Delta(A), \quad \Delta(\forall x A) = \forall x \Delta(A), \quad \Delta(\exists x A) = \exists x \Delta(A).$$

It is easy to see that both a structural translation Δ and the equivalence relation $Kern_\Delta$ preserve the structure of a formula C. In particular, the structural translation of a clause in $\mathcal{L}$ is a clause in $\mathcal{L}'$. Analogously, if a formula C' is equivalent to a clause C modulo $Kern_\Delta$, then C' is a clause, too.

As a consequence, recalling that $\Gamma(P)$ is the set of all ground instances of clauses in P, we can state the following proposition:

Proposition 4. *Given two first order language $\mathcal{L}$ and $\mathcal{L}'$ and a structural translation Δ of $\mathcal{L}$ in $\mathcal{L}'$. Then, for any logic program P on $\mathcal{L}$, both $\Delta(\Gamma(P))$ and $H_{Kern_\Delta}(\Gamma(P))$ are logic programs on $\mathcal{L}'$.*

Now, we restrict our attention to structural translation of logic programs. Let us denote with $B_\mathcal{L}$ and $B_{\mathcal{L}'}$ the Herbrand bases of $\mathcal{L}$ and $\mathcal{L}'$, and with $T_{H_{Kern_\Delta}(\Gamma(P))}$ and $T_{\Delta(\Gamma(P))}$ the immediate consequence operator of the programs $H_{Kern_\Delta}(\Gamma(P))$ and $\Delta(\Gamma(P))$, respectively. We can consider the logic program $\Delta(\Gamma(P))$ as the representation of an abstraction process related to symbols in the alphabet of P. By denoting with $M_{H_{Kern_\Delta}(\Gamma(P))}$ and $M_{\Delta(\Gamma(P))}$ the least Herbrand models of the logic programs $H_{Kern_\Delta}(\Gamma(P))$ and $\Delta(\Gamma(P))$, respectively, the following result allows us to clarify the relation which links the semantics of these programs.

We briefly recall that abstract interpretation is a theory of semantics approximation widely exploited to prove properties of programs written in any programming language [6]. Roughly speaking, the correspondence between concrete and abstract properties is established by a pair of functions which is a Galois connection formalizing the loss of information. The notion of approximation is modelled by the *abstraction function* α that, for any concrete property $p^\flat \in P^\flat$ provides the best approximation $\alpha(p^\flat)$ in the abstract domain $P^\sharp$. The semantic of the abstract properties is given by the *concretization function* γ that, for any abstract description $p^\sharp \in P^\sharp$, provides the corresponding concrete property $\gamma(p^\sharp)$ in the concrete domain $P^\flat$. The formal definition is as follows:

Definition 6. Let $(P^\flat, \preceq^\flat)$ and $(P^\sharp, \preceq^\sharp)$ are posets, a Galois connection is a pair of maps, $\alpha : P^\flat \longrightarrow P^\sharp$ and $\gamma : P^\sharp \longrightarrow P^\flat$, such that for any $p^\sharp \in P^\sharp$ *and* $p^\flat \in P^\flat$,

$$\alpha(p^\flat) \preceq^\sharp p^\sharp \iff p^\flat \preceq^\flat \gamma(p^\sharp).$$

We have a Galois surjetion when the abstraction function α is a surjection. The following proposition states that a basic property for associated operators in a Galois connection holds for the two operators $T_{H_{Kern_\Delta}(\Gamma(P))}$ and $T_{\Delta(\Gamma(P))}$ This property states a link between their fixpoints, denoted respectively with $lfp(T_{H_{Kern_\Delta}(\Gamma(P))})$ and $lfp(T_{\Delta(\Gamma(P))})$, which provide respectively the *concrete semantics* and the *abstract semantics*.

Proposition 5. *Given two first order languages $\mathcal{L}$ and $\mathcal{L}'$ and a structural translation Δ of $\mathcal{L}$ in $\mathcal{L}'$. The pair of functions $\alpha : \mathcal{P}(B_\mathcal{L}) \to \mathcal{P}(B_{\mathcal{L}'})$ and $\gamma : \mathcal{P}(B_{\mathcal{L}'}) \to \mathcal{P}(B_\mathcal{L})$ such that, for any $X \in \mathcal{P}(B_\mathcal{L})$, and $Y \in \mathcal{P}(B_{\mathcal{L}'})$*

$$\alpha(X) = \Delta(X) \qquad \gamma(Y) = \Delta^{-1}(Y),$$

provides a Galois surjection between the complete lattices $(\mathcal{P}(B_{\mathcal{L}}), \subseteq)$ and $(\mathcal{P}(B_{\mathcal{L}'}), \subseteq)$, with $\gamma\alpha = H_{Kern_\Delta}$.

Moreover, the abstrac semantics is optimal with respect to the associate operators $T_{H_{Kern_\Delta}(\Gamma(P))}$ and $T_{\Delta(\Gamma(P))}$, i.e.

$$\alpha(M_{H_{Kern_\Delta}(\Gamma(P))}) = M_{\Delta(\Gamma(P))}.$$

The α-optimality property stated by the previous proposition ensures that the abstract semantics $lfp(T_{\Delta(\Gamma(P))}) = M_{\Delta(\Gamma(P))}$ is a complete and correct approximation of the concrete semantics $lfp(T_{H_{Kern_\Delta}(\Gamma(P))}) = M_{H_{Kern_\Delta}(\Gamma(P))}$. In other words, the abstract operator $T_{\Delta(\Gamma(P))}$ computes inferences which are exhaustive and coherent with respect to inferences computed by the concrete operator $T_{H_{Kern_\Delta}(\Gamma(P))}$. The structural translation Δ provides the abstraction function α. Then, if we are interested to information related to this considered abstraction, we can equivalently perform our computations in the abstract or in the concrete domain.

Since a Deductive Database is also a logic program, all the results recalled in this section can be given for Deductive Databases as well.

6   Extended and abstract deductive database

In [9] an extension of Logic Programming is proposed in order to provide approximate reasoning capabilities. In that paper logic programs on function-free languages are considered, i.e. without function symbols. In [15] this approach has been generalized to logic programs on first order languages, by framing the approach introduced in [9] in the context of a structural translation of languages. We now briefly recall these results and frame them in the context of Deductive Databases.

In the classical case, function and predicate symbols of a first order language $\mathcal{L}$ are crisp elements, i.e., distinct elements represent distinct information and no matching is possible. e relax this constraint and suppose that it is possible to consider different functions, or different predicates with the same arity, as "similar" with a degree expressed by a value in $[0, 1]$. This notion extends the usual identity relation, indeed equals elements have similarity degree 1 and completely different elements have similarity degree 0, and it is well modelled by the definition of similarity relation given in Section 3. More formally, let us denote with

- V the set of *variable* symbols, ordered in a sequence $x_1, x_2, \dots$.

- F the set of *function* symbols

- R the set of *predicate* symbols.

As usual, constants can be considered as functions of arity zero. Let us consider a similarity relation $\mathcal{R}$ in $F \cup R \cup V$ such that $\mathcal{R}(t, t') = 0$ whenever one of the following cases occurs:

- t and t' are not both in R or F or V

- t and t' are predicates in R with different arities

- t and t' are functions in F with different arities

- t and t' are variable and $t \neq t'$.

In other words, $\mathcal{R}$ provides a non-zero similarity value only for function / predicate symbols with the same arity in $F \cup R$, whereas it is the identity relation for variables in V. We can recursively extend this similarity to a fuzzy relation in the set of terms / atoms / clauses in $\mathcal{L}$ that we call *structural extension of* $\mathcal{R}$. Let $s_1, ..., s_n, r_1, ..., r_m$ are terms and f, g function/predicate symbols with arities n and m, respectively. If $n = m$, we set

$$\mathcal{R}(f(s_1, ..., s_n), g(r_1, ..., r_n)) = \mathcal{R}(f, g) \wedge (\textstyle\bigwedge_{i=1}^{n} \mathcal{R}(s_i, r_i)),$$

otherwise, if $n \neq m$ the similarity value is zero. Then, let $C = A_0 \longleftarrow A_1, ..., A_n$ and $C' = A'_0 \longleftarrow A'_1, ..., A'_m$, are two Horn clauses. If $n = m$ we set

$$\mathcal{R}(C, C') = \textstyle\bigwedge_{i=0}^{n} \mathcal{R}(A_i, A'_i)$$

otherwise, if $n \neq m$ we set $\mathcal{R}(C, C') = 0$. In particular, we have that $\mathcal{R}(C, \square) = 0$ for any $C \neq \square$.

The following proposition can be easily proved:

Proposition 6. *The structural extension of a similarity $\mathcal{R}$ to terms / atoms / clauses in $\mathcal{L}$ is a similarity.*

In parallel, according with Proposition 1, the closure operator H_λ associated to the λ-cut of $\mathcal{R}$, with $\lambda \in (0, 1]$, can be defined on sets of terms, atoms and clauses in $\mathcal{L}$, too. In particular, denoted with $\square$ the empty clause, it is $H_\lambda(\square) = \square$.

If P is a Deductive Database, the set

$$H_\lambda(P) = \{C' \in \mathcal{L} \mid \exists C \in \Gamma(P) : \quad \mathcal{R}(C, C') \succeq \lambda\}$$

is constructed by adding to P all the clauses in $\mathcal{L}$ that can be obtained by replacing function and predicate symbols of a clause in P with symbols having similarity degree not minor than λ. As a consequence, also $H_\lambda(P)$ is a Deductive Database, that we name *extended-Deductive Database of level λ*. Thus, a first way to manage the weakening of the equality relation between function/predicate symbols, expressed by the similarity relation $\mathcal{R}$, can be given by considering inferences with respect to $H_\lambda(P)$.

Obviously, if $T_{H_\lambda(P)}$ is the immediate consequence operator of the Deductive Database $H_\lambda(P)$, then the least Herbrand model of $H_\lambda(P)$ is given by

$$M_{H_\lambda(P)} = lfp(T_{H_\lambda(P)}) = \bigcup_{n \geq 0} T^n_{H_\lambda(P)}(empty).$$

An alternative way to manage the information carried on by the similarity introduced between function/predicate symbols in P, can be given by considering as an unique element different symbols which have similarity degree greater or equal to λ. In other word, we can consider the quotient sets $F/_{\equiv_{\mathcal{R},\lambda}}$ and $R/_{\equiv_{\mathcal{R},\lambda}}$ as new sets of function and predicate symbols, respectively. We denote with $[s]$ the equivalence class of a symbol $s \in F \cup R$, and with $\mathcal{L}_\lambda$ the first order language related to the new alphabet. We can introduce a link between the two languages $\mathcal{L}$ and $\mathcal{L}_\lambda$ by exploiting the notion of structural translation given by Definition 5. It leads to a function $\tau_\lambda : \mathcal{L} \longrightarrow \mathcal{L}_\lambda$, named *translation up to* $\equiv_{\mathcal{R},\lambda}$, that associates to a formula G in $\mathcal{L}$, the formula $\tau_\lambda(G)$ in $\mathcal{L}_\lambda$ obtained by replacing predicate and function symbols in G with their equivalence classes in $F/_{\equiv_{\mathcal{R},\lambda}}$ and $R/_{\equiv_{\mathcal{R},\lambda}}$, respectively.

More formally, according to Definition 5 we give the following definition

Definition 7. Given a first order language $\mathcal{L}$, a similarity $\mathcal{R}$ and $\lambda \in (0,1]$, we call *translation up to* $\equiv_{\mathcal{R},\lambda}$ the structural translation $\tau_\lambda : F \cup R \longrightarrow F/_{\equiv_{\mathcal{R},\lambda}} \cup R/_{\equiv_{\mathcal{R},\lambda}}$ defined by setting

$$\tau_\lambda(f) = [f]$$

for any function/predicate symbol $f \in F \cup R$.

Let us consider a Deductive Database P on the language $\mathcal{L}$. The set

$$P_\lambda = \tau_\lambda(\Gamma(P)) = \{K \in \mathcal{L}_\lambda \quad | \quad K = \tau_\lambda(H), \quad H \text{ clause in } \Gamma(P)\}$$

is obtained by replacing function and predicate symbols of a clause in $\Gamma(P)$ with the related equivalence classes modulo $\equiv_{\mathcal{R},\lambda}$. By Proposition 4, also P_λ is a Deductive Database that we name *abstract-Deductive Database of level* λ.

Let $B_{\mathcal{L}_\lambda}$ denote the Herbrand base of $\mathcal{L}_\lambda$, and T_{P_λ} the immediate consequence operator of P_λ, the least Herbrand model of P_λ is given by

$$M_{P_\lambda} = lfp(T_{P_\lambda}) = \bigcup_{n \geq 0} T^n_{P_\lambda}(empty).$$

Then, P_λ could be used to manage similarity-based reasoning as well as $H_\lambda(P)$.

The following result states the equivalence of these two transformation techniques:

Theorem 1. *Let P be a Deductive Database on a first order language $\mathcal{L}$ and $\mathcal{R}$ a structural extension of a similarity in $\mathcal{L}$. Then, the functions $\alpha :$*

$\mathcal{P}(B_{\mathcal{L}}) \to \mathcal{P}(B_{\mathcal{L}_\lambda})$ and $\gamma : \mathcal{P}(B_{\mathcal{L}_\lambda}) \to \mathcal{P}(B_{\mathcal{L}})$ *such that,* $\forall X \in \mathcal{P}(B_{\mathcal{L}})$, *and* $\forall Y \in \mathcal{P}(B_{\mathcal{L}_\lambda})$

$$\alpha(X) = \tau_\lambda(X) \quad ; \quad \gamma(Y) = \tau_\lambda^{-1}(Y)$$

provides a Galois surjection between the complete lattices $(\mathcal{P}(B_{\mathcal{L}}), \subseteq)$ *and* $(\mathcal{P}(B_{\mathcal{L}_\lambda}), \subseteq)$, *such that* $\gamma\alpha = H_\lambda$.

Moreover, the abstract semantics is optimal with respect to the operators $T_{H_\lambda(P)}$ *and* T_{P_λ}, *i.e.*

$$\alpha\left(M_{H_\lambda(P)}\right) = M_{P_\lambda}.$$

The previous theorem states that both the extended and the abstract Deductive Databases can be used for processing information provided by a similarity relation defined in the language of P. It is worth to stress that the introduced similarity generally changes the semantic of the original Deductive Database. Indeed, it allows us to add new clauses to P providing the extended Deductive Database $H_\lambda(P)$. This is the more straight way to implement the approximated inference process based on similarity. On the other hand, by considering the abstract Deductive Database $P_\lambda = \tau_\lambda(\Gamma(P))$, it is possible to express information provided by the similarity relation in a syntectic way exploiting the quotient language $\mathcal{L}_\lambda$. Both these Deductive Databases allow to perform inference by assuming a "tolerance" level $\lambda \in (0, 1]$ in the relaxed matching between different function/predicate symbols.

7   Fuzzy least Herbrand model

In [9] the notion of *fuzzy least Herbrand model* of a program P on a function free language $\mathcal{L}$ with a similarity $\mathcal{R}$ has been introduced. In [15] it has beeen extended to first order languages. This notion is grounded on the observation that an higher values of the fixed similarity level λ corresponds to a lower possibility of inferences in the extended Deductive Database $H_\lambda(P)$. More precisely, if we consider $\lambda, \eta \in (0, 1]$ with $\eta \prec \lambda$, then, by Proposition 2i) and by definition of the closure operator H_η, it follows that $H_\eta(P) \supseteq H_\lambda(P)$. As a consequence, since Logic Programming is a monotone inference system, we have that $M_{H_\eta(P)} \supseteq M_{H_\lambda(P)}$. This observation leads to the following definition, that we frame in the context of Deductive Databases:

Definition 8. Let P be a Deductive Database on a first order language $\mathcal{L}$ with a similarity relation $\mathcal{R}$, and $\left\{M_{H_\lambda(P)}\right\}_{\lambda \in [0,1]}$ the family of Herbrand models defined in Section 6. The *fuzzy least Herbrand model* $M_{(P,\mathcal{R})} : B_{\mathcal{L}} \longrightarrow [0, 1]$ of the Deductive Database P with the similarity $\mathcal{R}$ is defined by setting for any $A \in B_{\mathcal{L}}$

$$M_{P,\mathcal{R}}(A) = Sup\left\{\lambda \quad | \quad A \in M_{H_\lambda(P)}\right\}$$

or, equivalently

$$M_{P,\mathcal{R}}(A) = Sup\{\lambda \mid H_\lambda(P) \models A\}.$$

Intuitively, the degree of membership of an atom A is given by the best "tolerance" level $\lambda \in (0,1]$ which allows to prove A exploiting the corresponding extended Deductive Database $H_\lambda(P)$.

The α-optimality of the Galois connection defined in the previous section, allows to give the following alternative characterization of the fuzzy least Herbrand model $M_{P,\mathcal{R}}$.

Theorem 2. *Let P be a Deductive Database on a first order language $\mathcal{L}$ with a similarity relation $\mathcal{R}$. Then for any $A \in B_{\mathcal{L}}$*

$$M_{P,\mathcal{R}}(A) = Sup\{\lambda \mid \tau_\lambda(A) \in M_{P_\lambda}\}$$

or, equivalently

$$M_{P,\mathcal{R}}(A) = Sup\{\lambda \mid P_\lambda \models \tau_\lambda(A)\}.$$

Thus, we conclude that in order to compute the fuzzy least Herbrand model of a Deductive Database P with a similarity $\mathcal{R}$, according to the previous results we can equivalently perform our computations in the extended or in the abstract domain.

By a computational point of view, we observe that the membership function that provides the fuzzy least Herbrand model can be computed by means of a bounded number of top-down generation procedures of least Herbrand models. Indeed, we recall that the alphabet of the language $\mathcal{L}$ with the similarity relation $\mathcal{R}$ is finite, then $\mathcal{R}$ has a finite set of similarity values $\{\lambda_1, \lambda_2, ..., \lambda_n\}$. By the definition, we have that the set of similarity values of the structural extension of $\mathcal{R}$ to terms/atoms/formulae in $\mathcal{L}$ is included in the set $\{\lambda_1, \lambda_2, ..., \lambda_n\}$. Thus, the family $\{\equiv_{\mathcal{R},\lambda_i}\}_{1 \preceq i \preceq n}$ provides all the different λ−cuts associated to the $\mathcal{R}$, and all the associated closure operators $H_{\lambda_i}, 1 \preceq i \preceq n$. As a consequence, for any $A \in B_{\mathcal{L}}$, the value $M_{P,\mathcal{R}}(A)$, that provides the membership degree of the atom A to the fuzzy least Herbrand model, is obtained in Definition 8 by computing the maximum on the finite set $\{\lambda_1, \lambda_2, ..., \lambda_n\}$.

8   Conclusion and future works

In this paper the similarity-based approach to Logic Programming proposed in [9] and [16] has been framed in the field of Deductive Database systems.

By the procedural point of view, information carried by this extension could be equivalently managed by exploiting standard query answering techniques (as Magic Sets, counting algorithm, factoring optimization, etc.) in the different transformations of the given Deductive Database P which we call, respectively, extended and abstract Deductive Database. Since the extended Deductive Database is obtained by adding new clauses or facts to the given set, and the abstract Deductive Database is constructed on a different language, both these procedures require some time consuming preprocessing steps. Then, modified version of the standard query answering techniques could be studied in order to avoid these preprocessing steps. The aim is to deal directly with rules and facts of the given Deductive Database so that the approximation introduced by the considered similarity does not modify the exact knowledge base.

References

1. Apt, R.K. (1990) Logic Programming, in: J. van Leeuwen (Ed.), *Handbook of Theoretical Computer Science*, vol. B, pp. 492-574, Amsterdam: Elsevier.
2. Bancilhon, F., Maier, D., Sagiv, Y., Ullman, J.D. (1986) Magic Sets and other strange ways to implement logic programs, in: *Proc. ACM Symp. on Principles of Database systems*, pp. 1-15.
3. Biacino, L., Gerla, G., Ying, M. (2000) Approximate reasoning based on similarity, *Math. Logic Quart.*, **46**.
4. Bosc, P., Buckles, B.B., Petry, F.E., Pivert, O. (1999) Fuzzy Databases, in: Bezdek, J.C., Dubois, D., Prade, H. (Ed.s), *Fuzzy Sets in Approximate Reasoning and Information Systems*, pp. 403-468, Boston: Kluwer.
5. Bukles, B., Petry, F. (1982) A fuzzy model for relational databases, *Fuzzy Sets and Systems*, **7**:213-226.
6. Cousot, P., Cousot, R. (1992) Abstract Interpretation and application to Logic Programs, *Journal of Logic Programming*, **13**:103-179.
7. Dubois, D., Esteva, F., Garcia, P., Godo, L. (1997) A logical approach to interpolation based on similarity, *Int. Journal of Approximate Reasoning*, **17**:1-36.
8. Esteva, F., Garcia, P. Godo, L. Rodriguez, R. (1997) A modal account of similarity-based reasoning, *Int. Journal of Approximate Reasoning* **16**:312-344.
9. Gerla, G., Sessa, M.I. (1999) Similarity in Logic Programming in: G. Chen, M. Ying, K.-Y. Cai (Ed.s), *Fuzzy Logic and Soft Computing*, pp. 19-31, Boston: Kluwer.
10. Klawonn, F., Castro, J.L. (1995) Similarity in Fuzzy Reasoning, *Mathware & Soft Computing*, **2**:197-228.
11. Lloyd, J.W. (1987) *Foundations of Logic Programming*, Berlin: Springer.
12. Petry, F.E. (with the collaboration of Bosc P.) (1996) *Fuzzy Databases - Principles and Applications*, Boston: Kluwer.
13. Ramakrishnan, R., Ullman, J.D. (1995) A survey of Deductive Database systems, *Journal of Logic Programming* **23**:125-149.
14. Ruspini, E.H. (1991) On the semantics of fuzzy logic, *Int. Journal of Approximate Reasoning*, **5**: 45-88.
15. Sessa, M.I. (2000) Translation and Similarity-based Logic Programming (to appear on *Soft Computing*).
16. Subrahmanian, V.S. (1986) On the semantics of quantitative logic, in: *Proc. IEEE Symposium on Logic Programming*, pp. 173-182.

17. Zadeh, L.A. (1971) Similarity Relations and Fuzzy Orderings, *Information Sciences*, **3**:177-200.
18. Ying, M.S. (1994) A Logic for Approximated Reasoning, *The Journal of Symbolic Logic*, **59**:830-837.

Neural Networks for Pattern Recognition, Image and Signal Processing

Roberto Tagliaferri

Dip. di Matematica e Informatica, Soft-Computing Lab., Università di Salerno
84081 Baronissi (SA), Italy - robtag@unisa.it
INFM unità di Salerno - 84081 Baronissi (SA), Italy
IIASS "E. R. Caianiello" - Vietri s/m (SA), Italy

Abstract. In this paper Neural Networks are presented in the context of Statistical Pattern Recognition, focusing the attention on all the steps needed to classify and interpolate input data. Standard multi-layer models are briefly illustrated, and then proved to be good instruments for data interpolation and Bayesian classification. Furthermore, Neural Networks are presented in the pre-processing stage, both for input reduction and clustering. Finally, two applications to signal and image processing are summarized to show the potentiality of Neural Network based systems in real world Statistical Pattern Recognition problems.

1 Introduction

Neural Networks (NN's) are a standard methodology in Statistical Pattern Recognition. They are used in noisy environments to approximate unknown functions, to classify input patterns, to model data distributions (time series analysis, system identification, probability density functions (pdf) estimation, etc.), by learning from examples. Although NN's study was born to model the brain, nowadays it is mainly directed to areas concerning Pattern Recognition, Signal and Image Processing, Time Series Prediction, System Identification, Robotics and Control. Some times NN's are used by experts in specific application areas or by researchers not experts in Statistics: the results are ad-hoc methods and an insufficient statistical validation of the experimental results. On the other hand, as we shall see in this paper, another important issue concerning the use of NN's is the data choice and processing: in fact, for example, we need to divide the data set into three independent sets (training, validation and test sets) with the same pdf of the input patterns to statistically validate our experiments and to suitably pre and post process the data to obtain significant results.

I shall refer to real world applications developed by my group in cooperation with Physicians and Astronomers to better illustrate some of the previous arguments.

In the following section the problems concerning data analysis and processing are discussed. In section 3 Multi-layer NN's are summarized. Next, three classical uses of NN's in Statistical Pattern Recognition: classification, interpolation (regression) and preprocessing (Principal Component Analysis (PCA), and clustering) are illustrated. Finally, in section 6 practical uses of NN's in two real world applications are used to illustrate in practice the theoretic concepts.

2   Neural Networks and the data

The first problem we meet when dealing with NN's is that concerning the choice of the data to test and validate the models. In fact, there are two different problems: the first one concerning the repeatability of the experiments, the second one concerning the fact that, in many cases, we receive the data without the possibility of controlling the way they were obtained, as a *status quo*.

In fact, to control the goodness of our approach and models we should compare them with standard models, by using well known data sets. For example, in the field of function approximation the time series generated by the Mackey-Glass differential equations are often used as a good benchmark, while in the field of classification there are many standard data bases in the www machine learning sites (for example, at the repository of machine learning databases and domain theories at University of California at Irvine (http://www.uci.edu:pub/machine-learning.databases or

ftp.ics.uci.edu:pub/machine-learning-databases)).

Moreover, many times specific problems must be solved and the data sets are not standards: in this case the NN's researchers must provide enough experiments with different methodologies to statistically prove the goodness of their approach.

Before starting the experiments, standard statistical operations must be applied to the data to study its distribution and to divide data sets into the classical training, test and validation sets.

The second step consists in the input normalization and coding: if an input variable assumes much more higher values than another one, it really influences the NN's training process more than the other one. Therefore, we must normalize all the input values in the same range. For example, if $\bar{x}_i$ is the mean value of the i–th input variable and σ_i^2 is its variance, then a simple normalization for the n–th input of the i–th variable is given by

$$\tilde{x}_i{}^n = \frac{x_i^n - \bar{x}_i}{\sigma_i^2} \tag{1}$$

The result is that the linearly transformed variables have zero mean and unit standard deviation over the transformed training set.

A bit more sophisticated linear rescaling of the variables is the whitening or sphering [7]: in this case the transformed training set has zero mean and a covariance matrix which is given by the unit matrix.

Let us assume to group the input variables x_i into a vector $\mathbf{x} = (x_1, \ldots, x_d)^T$, with d the input dimension, $\bar{\mathbf{x}}$ the sample mean vector and the covariance matrix with respect to the N data points of the training set given by:

$$\Sigma = \frac{1}{N-1} \sum_{n=1}^{N} (\mathbf{x}^n - \bar{\mathbf{x}})(\mathbf{x}^n - \bar{\mathbf{x}})^T \tag{2}$$

If we introduce the eigenvalue equation for the covariance matrix

$$\Sigma \mathbf{u}_j = \lambda_j \mathbf{u}_j \tag{3}$$

then the vector of the linearly transformed input variables is given by

$$\tilde{\mathbf{x}}^n = \Lambda^{-1/2} \mathbf{U}^T (\mathbf{x}^n - \bar{\mathbf{x}}) \tag{4}$$

where

$$\mathbf{U} = (\mathbf{u}_1, \ldots, \mathbf{u}_d) \tag{5}$$

$$\Lambda = diag(\lambda_1, \ldots, \lambda_d). \tag{6}$$

A different situation can be met when dealing with categorical discrete variables. In this case we do not want to impose an artificial ordering on the data: a good solution consists in using a 1-of-c coding for the input data, i.e. we use c binary inputs, each one assuming value 1 only if the corresponding feature is given in input and 0 otherwise.

Many problems can occur when there are missing data or outliers: in these cases a great attention must be paid and a robust statistical knowhow is needed to avoid trivial mistakes.

Finally, the last problem in pattern recognition we want to present here is the input dimension reduction: there are two possible approaches, the former by using feature selection techniques, the latter reducing the input dimensionality through PCA or similar techniques.

Any procedure for feature selection is based on two components: a criterion to judge whether one subset of features is better then another, and a systematic procedure for searching through candidate subsets of features. The criterion used varies depending on regression or classification problems.

In the former case we could use a sum-of-squares error with linear or non-linear models depending on the computational costs of the experiments while in the latter case we could use the probability of mis-classification or the percent of correct classified patterns.

For what concerns the search procedure, it is clear that in the most of the cases it is inherently untractable to consider all the possible subsets of features. If the used criterion satisfies a monotonicity relation, then it is possible to use the classical accelerated procedure known as *branch and bound* [27]. Alternatively, it is possible to use approximated solutions, for example, by using *sequential search techniques* [7]. A good approach is the *sequential forward selection* where at each step a new feature is added to the set of the selected features on the basis of which of the possible candidates at that stage gives rise to the largest increase in the value of the selection criterion. A better, but more time consuming, approach is the *sequential backward selection*, where at each stage the feature which gives the smallest reduction in the value of the selection criterion is deleted from the set of the selected features.

A standard unsupervised technique used to input dimension reduction is the PCA: the aim of this method is to create a matrix to linearly map vectors $\mathbf{x}^n$ in a d-dimensional space onto vectors $\mathbf{z}^n$ in an M-dimensional space, where $M < d$. It can be proved that if $\boldsymbol{\Sigma}$ is the covariance matrix of the set of vectors $\{\mathbf{x}^n\}$, as defined in Equation (2), then the best choice of the columns of the matrix is that of the eigenvectors of $\boldsymbol{\Sigma}$ corresponding to the M greatest eigenvalues computed by using Equation (3). The subspace spanned by the principal eigenvectors $\mathbf{u}_1, \ldots, \mathbf{u}_M$, $M < L$ is called the PCA subspace (of dimensionality M). The linear dimensionality reduction procedure derived above is discussed at length in Jollife [15]

In some cases linear PCA is not good enough to represent the data or the unsupervised algorithms do not find the best representation of the input data with respect to the classification task we want to perform. In the former case we must use non-linear PCA, in the latter we must use supervised techniques as the auto-associative Multi-layer Perceptrons [33] or the Fisher's linear discriminant [10], [7].

The interested reader can find a more deeper analysis of these arguments in [7].

3   Multi-layer Neural Networks

The most adopted model in the NN's literature is the well known "Multi-layer Perceptron" [33] shown in Fig. 1

It consists of one input layer, one output layer and one or two hidden layers. Each neuron of a given layer is connected to all the neurons of the next layer. Our model is synchronous: at each time every neuron receives as input the

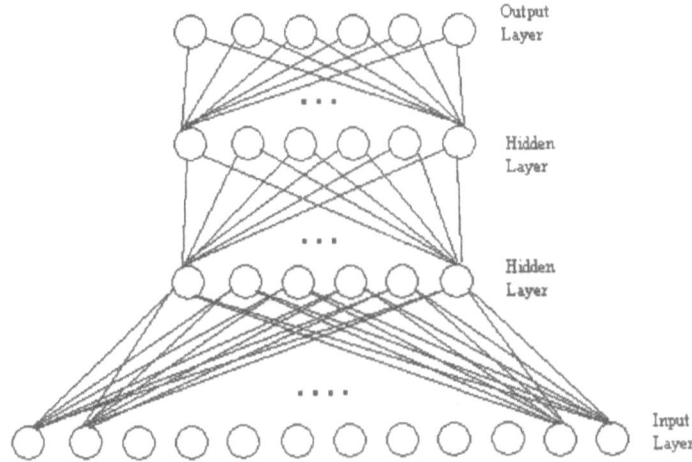

Fig. 1. The architecture of a three layer Multi-layer Perceptron

weighted sum of the input patterns and/or of the other neuron outputs, as shown in the following equations (in the case of a two-layer NN):

$$z_k = f\left[\sum_n w_{kn} x_n - B_k\right] \tag{7}$$

$$y_k = f\left[\sum_n w_{kn} z_n - B_k\right] \tag{8}$$

Where w_{kn} is the weight associated to the link from the n-th input and neuron k in Equation (7) and the weight associated to the link from neuron n to neuron k in Equation (8), z_n is the output of the hidden neuron n, x_n is the n-th input, y_n is the output of the output neuron n, and B_k is the threshold of neuron k, generally called bias. The neuron output is a continuous and derivable function of its net input with values in the $[0, 1]$ range. This function f is non-linear for the hidden units. In many cases we choose the sigmoidal function which looks like:

$$f(x) = 1/(1 + e^{-x}) \tag{9}$$

The training procedure is the so called "back propagation" which works as follows: we give to the input neurons the first pattern and then the net gives its output. If it is not equal to the desired output pattern, we compute the difference (error) between these two values and change the weights in order to minimize it. We repeat these operations for each input pattern until we minimize the mean square error of the system. Given the p-th pattern in input, a classical error function E_p (called sum-of-squares) is:

$$E_p = \frac{1}{2} \sum_j (t_j^p - y_j^p)^2 \qquad (10)$$

where t_j^p is the p-th desired output value and y_j^p is the output of the corresponding neuron. Different error functions and optimized algorithms for error function minimization can be considered [7]. Something about them will be discussed in the next sections

4   Neural Networks for classification and regression

4.1   Bayes' theorem and discriminant functions

Let us suppose to have a classification problems with c classes $C_1, \ldots, C_c$ and input patterns $\mathbf{x}$: the classical Bayes' theorem is applied to compute the posterior probabilities $P(C_k|\mathbf{x})$ combining the prior probabilities $P(C_k)$ and the class-conditional densities $p(\mathbf{x}|C_k)$:

$$P(C_k|\mathbf{x}) = \frac{p(\mathbf{x}|C_k)P(C_k)}{p(\mathbf{x})} \qquad (11)$$

where the unconditional densities $p(\mathbf{x})$ are given by:

$$p(\mathbf{x}) = \sum_{k=1}^{c} p(\mathbf{x}|C_k)P(C_k) \qquad (12)$$

which ensures that the posterior probabilities sum to one.

A pattern classifier provides a rule for assigning each input point $\mathbf{x}$ to one of the c classes. We can therefore regard the input space as being divided up into c decision regions $\Re_1, \ldots, \Re_c$ such that a point falling in region $\Re_k$ is assigned to class C_k. The boundaries between these regions are known as *decision surfaces* or *decision boundaries*.

We can formulate the classification process in terms of a set of *discriminant functions* $y_1(\mathbf{x}), \ldots, y_c(\mathbf{x})$ such that an input vector $\mathbf{x}$ is assigned to class C_k if

$$y_k(\mathbf{x}) > y_j(\mathbf{x}) \qquad \forall j \neq k. \qquad (13)$$

The decision rule for minimizing the probability of mis-classification is obtained simply by choosing

$$y_k(\mathbf{x}) = P(C_k|\mathbf{x}). \tag{14}$$

Since it is only the relative magnitude of the discriminant functions which is important in determining the class, we can replace $y_k(\mathbf{x})$ with any monotonic function of it and the decisions of the classifier will not be affected.

4.2   Neural Networks for interpolation

Multi-layer Perceptrons are good function approximators; moreover another class of Multi-layer NN's, the Radial Basis Function NN's possess the property of *best approximation* [12].

Here we want to show one important result for the interpretation of the outputs of the NN's trained by minimizing a sum-of-squares error function [7]. It is true in the limit the size N of the training data set goes to infinity. In this case, the sum-of-squares error can be written as

$$E = \frac{1}{2}\sum_k \int \{y_k(\mathbf{x};\mathbf{w}) - \langle t_k|\mathbf{x}\rangle\}^2 p(\mathbf{x})d\mathbf{x} + \frac{1}{2}\sum_k \int \{\langle t_k^2|\mathbf{x}\rangle - \langle t_k|\mathbf{x}\rangle^2\}p(\mathbf{x})d\mathbf{x} \tag{15}$$

where

$$\langle t_k|\mathbf{x}\rangle = \int t_k p(t_k|\mathbf{x})dt_k \tag{16}$$

and

$$\langle t_k^2|\mathbf{x}\rangle = \int t_k^2 p(t_k|\mathbf{x})dt_k \tag{17}$$

are conditional averages of the target data. Since the second term in Equation (15) is independent of the NN function mapping $y_k(\mathbf{x})$, it can be neglected in determining the network weights by error minimization. Moreover, the integrand in the first term in Equation (15) is non-negative and therefore the absolute minimum of the error function occurs when the first term vanishes, which corresponds to:

$$y_k(\mathbf{x};\mathbf{w}^*) = \langle t_k|\mathbf{x}\rangle \tag{18}$$

where $\mathbf{w}^*$ is the weight vector at the minimum of the error function. Equation (18) says that the NN mapping is given by the conditional average of the target data, i.e. by the *regression* of t_k conditioned by $\mathbf{x}$.

This result depends on

a) a data set sufficiently large to approximate an infinite data set,
b) a number of adaptive weights of the NN sufficiently large and
c) the fact that the NN is optimized in such a way as to find the appropriate minimum of the cost function.

A consequence of the last point is that we need good algorithms (both in error performance and in computational time) to find the minimum error NN configuration: to this aim many work has been done in the last years by many researchers in the topic of MLP learning optimization, by inserting classical and original optimization techniques in the BP algorithm (see [7] and papers therein quoted) : gradient descent with momentum, conjugate gradients, Newton's methods, the Levenberg-Marquardt algorithm are listed as an example.

Finally, we must say that the sum-of-squares error function is derived [7] from the principle of maximum likelihood by assuming that the distribution of the target data could be described by a Gaussian function with an $\mathbf{x}$-dependent mean, and a single global variance parameter. Although the sum-of-squares error does not require that the distribution of target variables be Gaussian, this error function cannot distinguish between the true distribution, and a Gaussian distribution having the same $\mathbf{x}$-dependent mean and average variance.

4.3 Neural Networks as discriminant functions and Bayesians' classifiers

It is clear that NN's can be used as a general class of discriminant functions, but we are more interested in the fact that the outputs can be used to approximate the posterior probabilities $P(C_k|\mathbf{x})$ when $\mathbf{x}$ is presented in input to the NN [7].

When the experimental data is limited, a NN could still be a good discriminant function even though its estimates of probabilities may be poor. On the other hand regression is usually based on the assumption of Gaussian noise and sum-of-squares error, which are inappropriate for binary data. Only in the limit of an infinite set of data, the NN outputs approximate the posterior probabilities.

A good error function for classification is the *cross-entropy* error function which is derived from the maximum likelihood function:

$$E = -\sum_p \{t^p \ln y^p + (1 - t^p) \ln(1 - y^p)\} \tag{19}$$

where y^p and t^p are the output of the NN and the target corresponding to the p-th input pattern, respectively.

In the case of the two-class problem, the neurons must compute the logistic function to calculate posterior probabilities both for 1-of-c target coding scheme and for continuous variables in the range $(0, 1)$.

In the case of Multi-class problems, the *cross-entropy* error function works equally well, but the activation functions, assuming Gaussian distribution of hidden unit activations, must be the *soft-max* functions:

$$y_k = P(C_k|\mathbf{x}) = \frac{\exp(a_k)}{\sum_{k'} \exp(a_{k'})}. \tag{20}$$

5   Neural Networks for the pre-processing

NN's can be used also in the pre-processing stage to solve different goals. We briefly show two of them, as examples: we shall use them in our applications.

The first is regarding the use of NN's for the PCA, the second for clustering.

5.1   Neural Networks for the PCA

PCA's can be neurally realized in various ways [3], [16], [28], [29], [31] and [34]. The PCA neural network described by us is one layer feedforward neural network which is able to extract the principal components of the stream of input vectors. Typically, Hebbian type learning rules are used, based on the one unit learning algorithm originally proposed by Oja [28]. Many different versions and extensions of this basic algorithm have been proposed during the recent years; see [17], [18], [30] and [34].

The structure of the PCA NN can be summarized as follows [17], [18], [30] and [34]: there is one input layer, and one forward layer of neurons totally connected to the inputs; during the learning phase there are feedback links among neurons, that classify the network structure as either hierarchical or symmetric (see Fig. 2). After the learning phase the network becomes purely feedforward. The hierarchical case leads to the well known GHA algorithm ([18], [34]); in the symmetric case we have the Oja's subspace network [28].

PCA neural algorithms can be derived from optimization problems, such as variance maximization and representation error minimization [17], [18] so obtaining nonlinear algorithms (and relative NN's). These NN's have the same

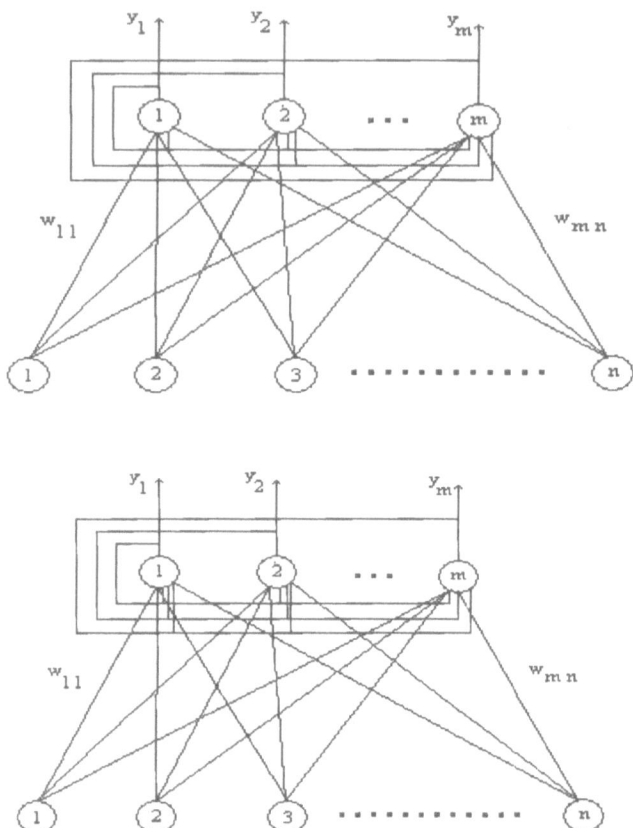

Fig. 2. Hierarchical PCA Neural Net (up) and Symmetric PCA Neural Net (down) architectures

architecture of the linear ones: either hierarchical or symmetric. These learning algorithms can be further classified in: robust PCA algorithms and nonlinear PCA algorithms. We define robust a PCA algorithm when the objective function grows less than quadratically [17] and [18]. The nonlinear learning function appears at selected places only. In nonlinear PCA algorithms all the outputs of the neurons are nonlinear functions of the responses.

Robust PCA algorithms. In the robust generalization of variance maximisation, the objective function $f(z)$ is assumed to be a valid cost function ([17]

and [18]), such as $\ln \cos(z)$ e $|z|$. This leads to the algorithm:

$$\mathbf{w}_i^{(t+1)} = \mathbf{w}_i^{(t)} + \mu^{(t)} g(y_i^{(t)}) \mathbf{e}_i^{(t)}, \tag{21}$$

$$\mathbf{e}_i^{(t)} = \mathbf{x}^{(t)} - \sum_{j=1}^{I(i)} y_j^{(t)} \mathbf{w}_j^t$$

where $\mathbf{w}_i^{(t)}$ is the weight vector associated to the links to neuron i at time t, $\mu^{(t)}$ is the learning rate at time t, and $y_j^{(t)}$ is the output neuron j.

In the hierarchical case we have $I(i) = i$. In the symmetric case $I(i) = M$, the error vector $\mathbf{e}_i^{(t)}$ becomes the same $\mathbf{e}^{(t)}$ for all the neurons, and Equation (21) can be compactly written as:

$$\mathbf{W}^{(t+1)} = \mathbf{W}^{(t)} + \mu^{(t)} \mathbf{e}^{(t)} g(\mathbf{y}^{(t)T}) \tag{22}$$

where $\mathbf{y} = \mathbf{W}^T \mathbf{x}$ is the instantaneous vector of neuron responses. The learning function g, derivative of f, is applied separately to each component of the argument vector.

The robust generalisation of the representation error problem ([17] and [18]), with $f(t) \leq t^2$, leads to the stochastic gradient algorithm :

$$\mathbf{w}_i^{(t+1)} = \mathbf{w}_i^{(t)} + \mu^{(t)} (\mathbf{w}_i^{(t)T} g(\mathbf{e}_i^{(t+1)}) \mathbf{x}^{(t)} + \tag{23}$$
$$+ \mathbf{x}^{(t)T} \mathbf{w}_i^{(t)} g(\mathbf{e}_i^{(t)}))$$

This algorithm can be again considered in both hierarchical and symmetric cases. In the symmetric case $I(i) = M$, the error vector is the same $\mathbf{e}^{(t)}$ for all the weights $\mathbf{w}^{(t)}$. In the hierarchical case $I(i) = i$, Equation (23) gives the robust counterparts of principal eigenvectors $\mathbf{c}_i$.

Approximated Algorithms. The first update term $\mathbf{w}_i^{(t)T} g(\mathbf{e}_i^{(t)}) \mathbf{x}^{(t)}$ in Equation (23) is proportional to the same vector $\mathbf{x}^{(t)}$ for all the weights $\mathbf{w}_i^{(t)}$. Furthermore, we can assume that the error vector $\mathbf{e}^{(t)}$ should be relatively small after the initial convergence. Hence, we can neglect the first term in Equation (23) and this leads to:

$$\mathbf{w}_i^{(t+1)} = \mathbf{w}_i^{(t)} + \mu^{(t)} \mathbf{x}^{(t)T} y_i^{(t)} g(\mathbf{e}_i^{(t)}) \tag{24}$$

Nonlinear PCA Algorithms. Let us consider now the nonlinear extensions of PCA algorithms. We can obtain them in a heuristic way by requiring all neuron outputs to be always nonlinear in the Equation (21) ([17] and [18]). This leads to:

$$\mathbf{w}_i^{(t+1)} = \mathbf{w}_i^{(t)} + \mu^{(t)} g(y_i^{(t)}) \mathbf{b}_i^{(t)}, \tag{25}$$

$$\mathbf{b}_i^{(t)} = \mathbf{x}^{(t)} - \sum_{j=1}^{I(i)} g(y_j^{(t)}) \mathbf{w}_j^{(t)} \quad \forall i = 1, \dots, p$$

5.2   Neural Networks for clustering

Unsupervised NN's partition the input space into clusters and assign to each neuron a weight vector which univocally individuates the template characteristic of one cluster in the input feature space. After the learning phase, all the input patterns are classified.

Unsupervised Neural Networks. Kohonen's ([21] and [22]) Self Organizing Maps (SOM) are composed by one neuron layer structured in a rectangular grid of m neurons. When a pattern $\mathbf{x}$ is presented to the NN, each neuron i receives the input and computes the distance d_i between its weight vector $\mathbf{w}_i$ and $\mathbf{x}$. The neuron which has the minimum d_i is the winner. The adaptation step consists in modifying the weights of the neurons in the following way:

$$\mathbf{w}_j^{(t+1)} = \mathbf{w}_j^{(t)} + \varepsilon^{(t)} h_{\sigma(t)} \left(d(j,k)\right) \left(\mathbf{x} - \mathbf{w}_j^{(t)}\right) \tag{26}$$

where $\varepsilon^{(t)}$ is the learning rate ($0 \le \varepsilon^{(t)} \le 1$) decreasing in time, $d(j,k)$ is the distance in the grid between the j and the k neurons and $h_{\sigma(t)}(x)$ is a unimodal function with variance $\sigma^{(t)}$ decreasing with x.

The Neural-Gas NN is composed by a linear layer of neurons and a modified learning algorithm [24]. It classifies the neurons in an ordered list $(j_1, \dots, j_m)$ accordingly to their distances from the input pattern. The weight adaptation depends on the position $rank(j)$ of the j-th neuron in the list

$$\mathbf{w}_j^{(t+1)} = \mathbf{w}_j^{(t)} + \varepsilon^{(t)} h_{\sigma(t)} \left(rank(j)\right) \left(\mathbf{x} - \mathbf{w}_j^{(t)}\right) \tag{27}$$

and works better than the preceding one: in fact, it is quicker and reaches a lower average distortion value[1].

---

[1] Let $P(\mathbf{x})$ be the pattern probability distribution over the set $V \subseteq \Re^n$ and let $\mathbf{w}_i(\mathbf{x})$ be the weight vector of the neuron which classifies the pattern $\mathbf{x}$. The average distortion is defined as $E = \int P(\mathbf{x}) (\mathbf{x} - \mathbf{w}_i(\mathbf{x}))^2 d\mathbf{x}$

The Growing Cell Structure (GCS) [11] is a NN which is capable to change its structure depending on the data set. Aim of the net is to map the input pattern space into a two-dimensional discrete structure S in such a way that similar patterns are represented by topological neighboring elements. The structure S is a two-dimensional simplex where the vertices are the neurons and the edges attain the topological information. Every modification of the net always maintains the simplex properties. The learning algorithm starts with a simple three node simplex and tries to obtain an optimal network by a controlled growing process: *id est*, for each pattern x of the training set, the winner k and the neighbors weights are adapted as follows:

$$\mathbf{w}_k^{(t+1)} = \mathbf{w}_k^{(t)} + \varepsilon_b \left(\mathbf{x} - \mathbf{w}_k^{(t)}\right); \quad \mathbf{w}_j^{(t+1)} = \mathbf{w}_j^{(t)} + \varepsilon_n \left(\mathbf{x} - \mathbf{w}_j^{(t)}\right) \qquad (28)$$

$\forall j$ connected to k; where ε_b and ε_n are constants which determine the adaptation strength for the winner and for the neighbors, respectively.

The insertion of a new node is made after a fixed number λ of adaptation steps. The new neuron is inserted between the most frequent winner neuron and the more distant of its topological neighbors. The algorithm stops when the network reaches a pre-defined number of elements.

The on-line K-means clustering algorithm [23] is a simpler algorithm which applies the Gradient Descent directly to the average distortion function as follows:

$$\mathbf{w}_j^{(t+1)} = \mathbf{w}_j^{(t)} + \varepsilon^{(t)} \left(\mathbf{x} - \mathbf{w}_j^{(t)}\right) \qquad (29)$$

The main limitation of this technique is that the error function presents many local minima which stop the learning before reaching the optimal configuration.

Finally, the Maximum Entropy NN [32] applies the Gradient Descent to the error function to obtain the adaptation step:

$$\mathbf{w}_j^{(t+1)} = \mathbf{w}_j^{(t)} + \varepsilon^{(t)} \frac{\exp\left(-\beta^{(t)} d_j\right)}{\sum_{k=1}^{m} \exp\left(-\beta^{(t)} d_k\right)} \left(\mathbf{x} - \mathbf{w}_j^{(t)}\right) \qquad (30)$$

where β is the inverse temperature and takes value increasing in time and d_j is the distance between the j-th and the winner neurons.

Hybrid neural nets. Hybrid NN's are composed by a clustering algorithm which makes use of the information derived by one unsupervised single layer NN. After the learning phase of the NN, the clustering algorithm splits the output neurons in a number of subsets which is equal to the number of the desired output classes. Since the aim is to put similar input patterns in the same class and dissimilar input patterns in different classes, a good strategy consists in applying a clustering algorithm directly to the weight vectors of the unsupervised NN.

A non-neural agglomeration clustering algorithm that divides the pattern set (in this case the weights of the neurons) $W = \{\mathbf{w}_1, ..., \mathbf{w}_m\}$ in l clusters (with $l < m$) can be briefly summarized as follows:

1 initially divide W in m clusters $C_1, ..., C_m$ such that $C_j = \{\mathbf{w}_j\}$;
2 compute the distance matrix D with elements $D_{ij} = d(C_i, C_j)$;
3 find the smallest element D_{ij} and unify the clusters C_i and C_j in a new one $C_{ij} = C_i \cup C_j$;
4 if the number of clusters is greater than l then go to step 2, else stop.

Many algorithms quoted in literature [9] differ only in the way in which the distance function is computed. For example:

$$d(C_i, C_j) = \min_{\mathbf{w}_k \in C_i \text{ and } \mathbf{w}_l \in C_j} \|\mathbf{w}_k - \mathbf{w}_l\|$$

(nearest neighbor algorithm);

$$d(C_i, C_j) = \left\| \frac{1}{|C_i|} \sum_{\mathbf{w}_k \in C_i} \mathbf{w}_k - \frac{1}{|C_j|} \sum_{\mathbf{w}_l \in C_j} \mathbf{w}_l \right\|$$

(centroid method);

$$d(C_i, C_j) = \frac{1}{|C_i||C_j|} \sum_{\mathbf{w}_k \in C_i, \, \mathbf{w}_l \in C_j} \|\mathbf{w}_k - \mathbf{w}_l\|$$

(average between groups).

The output of the clustering algorithm will be a labeling of the patterns (in this case neurons) in l different classes.

Unsupervised hierarchical neural nets. Unsupervised hierarchical NN's add one or more unsupervised single layers NN to any unsupervised NN, instead of a clustering algorithm as it happens in hybrid NN's.

In this way, the second layer NN learns from the weights of the first layer NN and clusters the neurons on the basis of a similarity measure or a distance. The iteration of this process to a few layers gives the unsupervised hierarchical NN's.

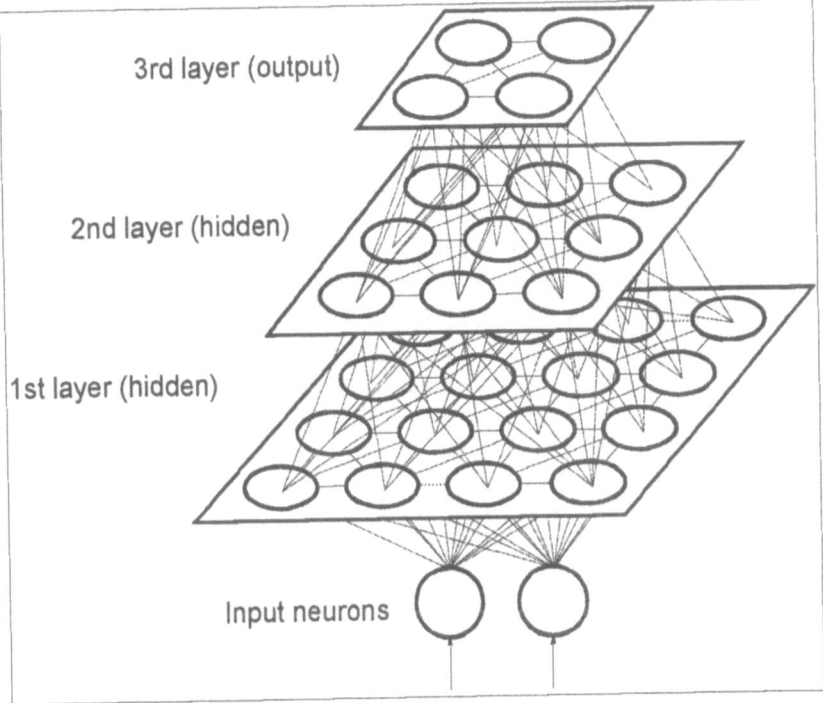

Fig. 3. Example of an unsupervised hierarchical neural network

The number of neurons at each layer decreases from the first to the output layer and, as a consequence, the NN takes the pyramidal aspect shown in Fig. 3.

The NN takes as input a pattern x and then the first layer finds the winner neuron. The second layer takes the first layer winner weight vector as input and finds the second layer winner neuron and so on up to the top layer. The activation value of the output layer neurons is 1 for the winner unit and 0 for all the others. In short: the learning steps of a s layer hierarchical NN with training set X are the following:

- the first layer is trained on the patterns of X with one of the learning algorithms for unsupervised NN's;
- the second layer is trained on the elements of the set X_2 which is composed by the weight vectors of the first layer winner units;
- the process is iterated to the $i-th$ layer NN $(i > 2)$ on the training set which is composed by the weight vectors of the winner neurons of the $(i-1)-th$ layer when presenting X to the first layer NN, X_2 to the second layer and so on.

By varying the learning algorithms we obtain different NN's with different properties and abilities. For instance, by using only SOM's we have a Multi-layer SOM (ML-SOM) [20] where every layer is a two-dimensional grid. We can easily obtain [36] *ML-Neural Gas, ML-Maximum Entropy* or *ML-K means* organized on a hierarchy of linear layers. The ML-GCS has a more complex architecture and has at least 3 units for layer.

By varying the learning algorithms in the different layers, we can take advantage from the properties of each model (for instance, since we cannot have a ML-GCS with 2 output units we can use another NN in the output layer).

A hierarchical NN with a number of output layer neurons equal to the number of the output classes simplifies the expensive post-processing step of labelling the output neurons in classes, without reducing the generalization capacity of the NN.

6   Neural Networks' applications

As examples we presents two systems based on NN's to perform signal and image processing aims: the neural network based estimator system for the estimation of frequencies in unevenly sampled and very noise sequences of data and the NEXT system for the star/galaxy classification in astronomical images.

6.1   The neural network based estimator system

The process for periodicity analysis can be divided in the following steps ([37] and [39]):

- Preprocessing

We remove the harmonics (low frequencies) due to the number of points which compose the signals and than we normalize the data to obtain a zero mean and unitary variance process.

- Neural computing

The fundamental learning parameters are:

1) the initial weight matrix;

2) the number of neurons, that is the number of principal eigenvectors that we need, equal to twice the number of signal periodicities (for real signals);

3) ϵ, i.e. the threshold parameter for convergence;

4) α, the nonlinear learning function parameter;

5) μ, that is the learning rate;

6) the number N of input components.

We initialize the weight matrix **W** assigning the first input patterns values.

Experimental results show that α can be fixed to : 1., 5., 10., 20. Moreover, the learning rate μ can be also decreased during the learning phase, but we fix it between 0.05 e 0.00001.

We use a simple criterion to decide if the neural network has reached the convergence: we calculate the distance between the weight matrix at step $t+1$, $\mathbf{W}^{(t+1)}$, and the matrix at the previous step $\mathbf{W}^{(t)}$, and if this distance is less than a fixed error threshold (ϵ) we stop the learning process. Alternatively, we can use an early stopping criterion to avoid overfitting problems. In this case, we stop the learning when all the estimates of the Multiple Signal Classificator (*MUSIC*) ([19] and [25]) are greater than 0 (see next pages).

We finally have the following general algorithm in which STEP 4 is one of the neural learning algorithms seen above in section 5.1:

1 Initialize the weight vectors $\mathbf{w}_i^{(0)}$ $\forall i = 1, \ldots, p$. Initialize the learning threshold ϵ, the learning rate μ. Reset the pattern counter $t = 0$.
2 Input the t-th pattern $\mathbf{x}^{(t)} = [x_1^{(t)}, \ldots, x_N^{(t)}]$ where N is the number of input components.
3 Calculate the output for each neuron $y_j^{(t)} = \mathbf{w}_j^{(t)T}\mathbf{x}_i^{(t)}$ $\forall i = 1, \ldots, p$.
4 Modify the weights $\mathbf{w}_i^{(t+1)} = \mathbf{w}_i^{(t)} + \mu^{(t)}g(y_i^{(t)})\mathbf{e}_i^{(t)}$ $\forall i = 1, \ldots, p$.
5 Calculate $testnorma = \sqrt{\sum_{j=1}^{p}\sum_{i=1}^{N}(w_{ij}^{(t+1)} - w_{ij}^{(t)})^2}$.
6 Convergence test: if ($testnorma < \epsilon$) or the Music estimates > 0 then goto **STEP 8**.
7 $k = k + 1$. Goto **STEP 2**.
8 End.

- Frequency estimator

We exploit the frequency estimator *MUSIC*. It takes as input the weight matrix columns after the learning. The estimated signal frequencies are obtained as the peak locations of the following function ([19] and [25]):

$$P_{MUSIC} = \frac{1}{M - \sum_{i=1}^{M}|\mathbf{e}_f^H\mathbf{w}_i|^2} \tag{31}$$

where $\mathbf{w}_i$ is the i-th weight vector after learning, and $\mathbf{e}_f^H$ is the pure sinusoidal vector: $\mathbf{e}_f^H = [1, e_f^{j2\pi f}, \ldots, e_f^{j2\pi f(L-1)}]^H$.

When f is the frequency of the i-th sinusoidal component, $f = f_i$, we have $\mathbf{e} = \mathbf{e}_i$ and $P_{MUSIC} \to \infty$. In practice we have a peak near and in correspondence of the component frequency. Estimates are related to the highest peaks. The output neurons are twice the number of frequencies we have to estimate.

To reduce the convergence time and to obtain a better estimate, it is better to have only 2 output neurons each time. In this way we estimate the first frequency. Then we eliminate this frequency from the signal by using IIR and Butterworth filtering [25]. We apply this procedure to the remaining signal, so obtaining the other frequencies, one at each step.

In the case of unevenly sampled data, we use an extension of Music to directly include unevenly sampled data without using the interpolation pre-processing step of the previous algorithm in the following way:

$$P'_{MUSIC} = \frac{1}{M - \sum_{i=1}^{p} |e_f^H w_i|^2}$$

where p is the frequency number, w_i is the i-th weight vector of the PCA NN after the learning, and e_f^H is the sinusoidal vector:

$$e_f^H = [1, e_f^{j2\pi f t_0}, \ldots, e_f^{j2\pi f t_{(L-1)}}]^H$$

where $\{t_0, t_1, ..., t_{(L-1)}\}$ are the first L components of the temporal coordinates of the uneven signal.

Furthermore, to optimize the performance of the PCA neural networks, we stop the learning process when $\sum_{i=1}^{p} |e_f^H w_i|^2 > M \; \forall f$, so avoiding overfitting problems.

These algorithms have been successfully applied to stratigraphic sequences derived from geological analysis of mountains [39] and to unevenly sampled data of light curves of stars [37].

6.2   The NExt system

The detection and classification of the objects in astronomical images are a multi-step task accomplished by our NN's based system Next ([1] and [2]) (see Fig. 4). After the standard preprocessing of the data [1] we perform the following steps:

1) we first run a 3x3 or 5x5 window on the image in order to determine the value of the central pixel;

2) we then use Robust PCA NNs to reduce to three the dimensionality of the input space. Principal vectors of the PCA are computed by the NN on a portion of the whole image. The values of the pixels in the transformed 3-dimensional eigen-space obtained via the principal vectors of the PCA NN are then used as inputs to unsupervised NN's to classify pixels in few classes.

3) Since supervised NN's need a large amount of labeled data to obtain a good classification, we use unsupervised NN's to segment the pixels into few

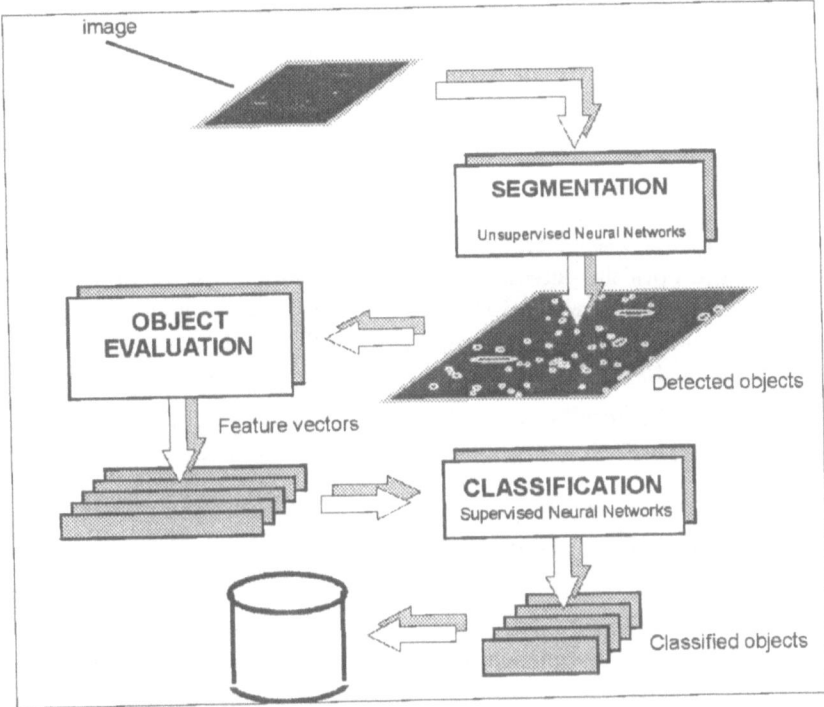

Fig. 4. An overview of the NExt processing steps

classes (one for the background and the others for the objects). The adopted NN is a hierarchical unsupervised NN [36], i.e. we never feed into the detection algorithm any a priori definition of what an object is, and we leave it free to find its own object definition. It turns out that image pixels are split in six classes, one coincident with what astronomers call background and five for the objects (in the astronomical sense). Afterwards, the class containing the background pixels is kept separated from the other classes which are instead merged together. Therefore, as final output, the pixels in the image are divided in "object" or "background".

4) Since objects are seldom isolated in the sky, we need a method to recognise overlapping objects and deblend them. We then adopt a generalisation of the method used by Focas [14] to isolate the objects against the noisy background.

5) Due to the noise, object edges are quite irregular. We therefore apply a contour regularisation to the edges of the objects in order to improve the following star/galaxy classification step.

5) We define and measure the features used, or suitable, for the star/galaxy classification, then we choose the best performing features for the classification step, through the sequential backward elimination strategy [7].

5) We then use a subset of the catalog [13] to learn, validate and test the classification performed by NExt on our images. The training set was used to train the optimized MLP NN, while the validation was used for model selection, i.e. to select the most performing parameters using an independent data set. As template classifier, we used SEx [6], whose classifier is also based on NN's.

The detection and classification performances of our algorithm were then compared with those of traditional algorithms, such as SEx. We wish to stress that in both the detection and classification phases, we were not interested in knowing how well NExt can reproduce SEx or the astronomer's eye performances, but rather to see whether the SEx and NExt catalogs are or are not similar to the "true", represented in our case by the catalog [13].

The experimental results, showing objects detection from background and star/galaxy separation are reported in [2] and therein discussed.

7   Conclusions and perspectives

In this paper we have given some main ideas about NN's in the area of Statistical Pattern Recognition and briefly shown two applications of NN's, the former to signal processing and the latter to image processing and objects classification.

Other researches are in progress in the Neural computing group of the Soft-computing Lab, regarding the use of Fuzzy NN's for classification, diagnostics and function approximation ([26] and [38]), the study of Fuzzy Neural Networks Based on Fuzzy Logic Algebras Valued Relations [8], and the design of NN's for the signal processing of the VIRGO Project for the detection of gravitational waves ([4] and [5]).

Acknowledgements

The author wishes to thank Professors Bruno D'Argenio, Giuseppe Longo, Leopoldo Milano and Fabrizio Barone for their help to begin and continue his applied research in Astronomy, Physics and Geology. He also acknowledges Dr.s Stefano Andreon, Angelo Ciaramella, Antonio Eleuteri, Giorgio Gargiulo and Nicola Pelosi for their contribution in the NN's applications to signal and image processing. Without their contribution this paper would not be the same.

This paper is based on a lecture given in 1999 and on the experience acquired by the author during his courses in Neural Networks hold at the University of Salerno from 1996 to 2000.

References

1. Andreon, S., Gargiulo, G., Longo, G., Tagliaferri, R., Capuano N. (1999) Neural Networks and Star/Galaxy Separation in Wide Field Astronomical Images. Proceedings of the 1999 International Joint Conference on Neural Networks IJCNN'99, Washington, Dc, July 10-16

2. Andreon, S., Gargiulo, G., Longo, G., Tagliaferri, R., Capuano, N. (2000) Wide Field Imaging. I. Applications of Neural Networks to object detection and star/galaxy classification. Submitted to MNRAS.

3. Baldi, P., Hornik, K. (1989) Neural networks for principal component analysis: learning from examples without local minima. Neural Networks 2, 7, 53–58.

4. Barone, F., De Rosa, R., Eleuteri, A., Garufi, F., Milano, L., Tagliaferri, R. (1999) A Neural Network-based ARX Model of Virgo Noise. Proceedings of the 11-th Italian Workshop on Neural Nets WIRN Vietri '99, M. Marinaro and R. Tagliaferri Ed.s, 171–183, Springer-Verlag, London

5. Barone, F., Garufi, F., Milano, L., Ciaramella, A., Eleuteri, A., Tagliaferri, R. (1999) A Neural Network approach for the Noise Identification and Data Quality of the VIRGO antenna. Proceedings of the 1999 Edoardo AMALDI Conference on Gravitational Waves, California Institute of Technology, Pasadena, CA, USA, July 12-16, in press.

6. Bertin, E., Arnouts, S., 1996, AAS **117**, 393.

7. Bishop, C. M. (1995) Neural Networks for Pattern Recognition. Oxford University Press, Oxford UK

8. Di Nola, A., Tagliaferri, R., Bělohlávek, R., Ciaramella, A. (2000) Fuzzy Neural Networks Based on Fuzzy Logic Algebras Valued Relations. DMI Soft-computing Lab, Int. Report.

9. Everitt B. (1977) Cluster Analysis Social Science Research Council. Heinemann Educational Books, London

10. Fisher, R. A. (1936) The use of multiple measurements in taxonomic problems. Annals of Eugenica **7**, 179–188

11. Fritzke, B. (1994) Growing Cell Structures - A Self-Organizing Network for Unsupervised and Supervised Learning. Neural Networks, **7**, 9, 1441-1460

12. Girosi, F., Poggio, T. (1990) Neural Networks and the best approximation property. Biological Cybernetics **63**, 169–176

13. Infante, L., Pritchet, C. (1992) ApJS, **83**, 237

14. Jarvis, J. F., Tyson, J. A. (1981) AJ, **86**, 476.

15. Jollife, I. T. (1986) Principal Component Analysis. Springer–Verlag, New York

16. Jutten, C., Herault, J. (1991) Blind separation of sources, partI: an adaptive algorithm based on neuromimetic architecture. Signal Processing, **24**, 1, 1–10

17. Karhunen, J., Joutsensalo, J. (1994) Representation and separation of signals using nonlinear PCA type learning. Neural Networks **7**, 113–127

18. Karhunen, J., Joutsensalo, J. (1995) Generalization of principal component analysis, optimization problems, and neural networks. Neural Networks **8**, 549–562

19. Kay, S. M. (1988) Modern spectral estimation: theory and application. Englewood Cliffs: Prentice-Hall

20. Koh, J., Suk, M., Bhandarkar, S. (1995) A Multilayer Self-Organizing Feature Map for Range Image Segmentation. Neural Networks **8**, 1, 67–86.

21. Kohonen, T. (1982) Self-organized formation of topologically correct feature maps. Biological Cybernetics **43**, 59–69
22. Kohonen, T. (1988) Self-organization and associative memory (2nd edition). Springer-Verlag, Berlin
23. Lloyd, S. (1982) Least squares quantization in PCM. IEEE Transactions on Information Theory IT-**28**, 2
24. Martinetz, T., Berkovich, S., Schulten, K. (1993) Neural-Gas Network for Vector Quantization and its Application to Time-Series Prediction. IEEE Transactions on Neural Networks **4**, 4, 558–568
25. Marple, S. L. (1987) Digital spectral analysis with applications. Prentice-Hall, Englewood Cliffs, N. Y.
26. Meneganti, M., Saviello, F., Tagliaferri, R. (1998) Fuzzy Neural Networks for Classification and Detection of Anomalies. IEEE Transactions on Neural Networks special issue on "Neural Networks and Hybrid Intelligent Models: Foundations, Theory and Applications" **9**, 5, 848-861
27. Narendra, P. M., Fukunaga, K. (1977) A branch and bound algorithm for feature subset selection. IEEE Trans. on Computers **26**, 9, 917–922
28. Oja, E. (1982) A simplified neuron model as a principal component analyzer. Journal of Mathematical Biology **15**, 267–273
29. Oja, E., Ogawa, H., Wangviwattana, J. (1991) Learning in nonlinear constrained Hebbian network. Artificial neural networks, T. Kohonen et al. (Eds.), 385–390, North-Holland, Amsterdam
30. Oja, E., Karhunen, J., Wang, L., Vigario, R. (1996) Principal and independent components in neural networks - recent developments. In Seventh Italian Workshop on Neural Networks, WIRN Vietri'95, M.Marinaro and R.Tagliaferri Ed.s, 16–35, World Scientific Pu. Singapore
31. Plumbley, M. (1993) Hebbian/anti Hebbian network which optimizes information capacity by orthonormalizing the principal subspace. Proc. IEE Conf. on Artificial Neural Networks, Brighton, UK, 86–90
32. Rose, K., Gurewitz, F., Fox, G. (1990) Statistical mechanics and phase transition in clustering. Phys. Rev. Lett. **65**, 945–948
33. Rumelhart, D. E., Hinton, G. E., Williams, R. J. (1986) Learnig internal representation by error propagation, in Parallel Distributed Processing: Explorations in the Microstructure of of Cognition, Volume 1: Foundations, D. E. Rumelhart, J. L. McClelland and the PDP group Ed.s, 318–362, MIT Press, Cambridge, MA
34. Sanger, T. D. (1989) Optimal unsupervised learning in a single-layer linear feedforward network. Neural Network, **2**, 459-473
35. Tagliaferri, R., Longo, G., Andreon, S., Zaggia, S., Capuano, N., Gargiulo, G. (1998) Astronomical Object Recognition by Means of Neural Networks. Proceedings of the 10-th Italian Workshop on Neural Nets WIRN, Marinaro M. and Tagliaferri R. Ed.s, 169–178, Springer-Verlag, London
36. Tagliaferri, R., Capuano, N., Gargiulo, G. (1999) Automated Labeling for Unsupervised Neural Networks: A Hierarchical Approach. IEEE Transactions on Neural Networks **10**, 199–203
37. Tagliaferri, R., Ciaramella, A., Milano, L., Barone, F., Longo, G. (1999) Spectral Analysis of Stellar Light Curves by Means of Neural Networks. Astronomy and Astrophysics Supplement Series, **137**, 391-405
38. Tagliaferri, R., Eleuteri, A., Meneganti, M., Barone, F. (2000) Fuzzy Min-Max Neural Networks: from Classification to Regression. Soft Computing, in press

39. Tagliaferri, R., Pelosi, N., Ciaramella, A., Barone, F., Milano, M. (2000) Soft Computing Methodologies for Spectral Analysis in Cyclostratigraphy. Computers and Geosciences, in press

Computational Aspects of Probability Logics

Sauro Tulipani

Dipartimento di Matematica, Univeristá di Perugia
Via Vanvitelli 1, 06123 Perugia, Italy - tulipani@dipmat.unipg.it

Abstract. This paper gives an overview of recent results about the computational complexity of the problem of checking the coherence for a partial probability assessment. Moreover, a decision procedure which works directly on the Boolean relations by simplification rules and by elimination of Boolean variables is presented.

keywords Coherent probability assessments–PSAT problem–NP-complete problems –elimination of Boolean variables

1 Introduction

In the last part of 20th century several authors ([17], [6], [18]) have proposed logics concerning probabilities. Such logics are useful for dealing with the probabilistic entailment as well for reasoning about the behaviour of the programs in computer science and for reasoning under uncertainty in knowledge-based systems.

The consistency problem and the entailment problem of such logics include two old problems investigated by Boole in the book *"An investigation of the laws of thought"* and called *"conditions of possible experience"* and *"general problem"*, respectively. The latter was formulated (see [13]) by: *Given the probabilities of any events of whatever kind, to find the probability of some other event connected with them"*. The former problem, which is very related, concerns the concistency of an assessment of probabilities to events represented as Boolean propositional sentences. It can be formulated mathematically as it follows.

A *probability assessment* is a map

$$\begin{cases} E_1 \mapsto p_1 \\ \ldots \mapsto \ldots \\ E_m \mapsto p_m \end{cases} \tag{1}$$

where $p_1, \ldots, p_m$ are given real numbers in the interval $[0, 1]$ and $E_1, \ldots, E_m$ are events presented as propositional logical sentences in some set of Boolean variables $\{X_1, \ldots, X_n\}$.

Boole's problem asks for any given probability assessment if there exists a probability distribution $P : \mathcal{F}(X_1, \ldots, X_n) \longrightarrow [0, 1]$ defined on the Boolean algebra of the propositional sentences (up to logical equivalence) built on the propositional variables $X_1, \ldots, X_n$, such that

$$P(E_i) = p_i, \qquad i = 1, \ldots, m. \tag{2}$$

Recently, this problem was investigated (in [9]) also from the complexity point of view and its restriction to events represented by Boolean clauses was called PSAT.

We observe also that Boole's problem can be reformulated as a satisfiability problem of the (non-truth functional) propositional fuzzy logic (see [10], [2]). According to Pavelka [19] a fuzzy set of formulas is a map $s : \mathbb{F} \longrightarrow [0,1]$ where $\mathbb{F}$ is the set of all propositional formulas. The semantics of the envelopes [2] is given by the set $\mathcal{M}$ of all fuzzy sets $P : \mathbb{F} \longrightarrow [0,1]$ which are probabilities up to logical equivalence of the formulas in $\mathbb{F}$. So, P satisfies s if and only if s is a fuzzy subset of P, i.e. $s \leq P$ pointwise. Now, let s be the extension to $\mathbb{F}$ of a map as in 1 where s takes the value 0 out of $\{E_1, \ldots, E_m, \neg E_1, \ldots, \neg E_m\}$ and $s(E_i) = p_i$, $s(\neg E_i) = 1 - p_i$, for $i = 1, \ldots, m$. Thus, s is satisfiable in the logic of envelopes if and only if there exists a probability distribution $P : \mathbb{F} \longrightarrow [0,1]$ such that $P(E_i) \geq p_i$, $P(\neg E_i) \geq 1 - p_i$, for $i = 1, \ldots, m$. This happens when the Boole's problem defined by 1 has a positive answer.

Boole gave a solution to the problem of consistency and a similar solution to the general problem by the following simple observation. Let $\alpha_1, \ldots, \alpha_{2^n}$ be the atoms of the free Boolean algebra $\mathcal{F}(X_1, \ldots, X_n)$. Then, we must have that

$$\sum_{j=1}^{2^n} P(E_i \wedge \alpha_j) = P(E_i) = p_i, \quad \text{for } i = 1, \ldots, m. \tag{3}$$

since the atoms form a partition of unity. So, the problem has a positive answer if and only if there is a nonnegative solution to the linear system

$$\begin{cases} \sum_{1 \leq j \leq 2^n} \xi_j = 1 \\ \sum_{1 \leq j \leq 2^n} a_{ij}\xi_j = p_i \end{cases} \qquad i = 1, \ldots, m \tag{4}$$

in the unknowns $\xi_j = P(\alpha_j)$, for $j = 1, \ldots, 2^n$, where

$$a_{ij} = \begin{cases} 1 & \text{if } \alpha_j \longrightarrow E_i \text{ is true} \\ 0 & \text{othewise.} \end{cases} \tag{5}$$

Boole's problem was re-discovered and refreshed by Halperin [12] who started the computational approach based on linear programming, see also [14] [15].

Boole's problem and its dual version, in the sense of linear programming, came up already in de Finetti work [5], in a more general setting, as a basic notion for the subjective probabilistic approach under the *"bet interpretation"* (see [4] [11] for recent developments).

Given events $E_1, \ldots, E_m$ with the relations that they satisfy, denote by $v_1, \ldots, v_s$, where $s \leq 2^n$, all possible outcomes of such events, in other words all possible truth-assignement $v_j : \{X_1, \ldots, X_n\} \longrightarrow \{0,1\}$ which are compatible with the given relations. Moreover, let $r_1, \ldots, r_m$ be costs of bets on such events. Then, the gain of the outcome v_j is defined as the value

$$\sum_{i=1}^{m} r_i \left(v_j(E_i) - p_i \right) \tag{6}$$

A probability assessment is called *coherent* and the associated bet game is called *reasonable* if and only if, for any given sequence of real numbers $r_1, \ldots, r_m$, the gains are not all positive and not all negative. This means that

$$\min_{1 \leq j \leq s} \sum_{i=1}^{m} r_i \left(v_j(E_i) \right) \leq \sum_{i=1}^{m} r_i p_i \leq \max_{1 \leq j \leq s} \sum_{i=1}^{m} r_i \left(v_j(E_i) \right) \tag{7}$$

Now, we denote by A the $(m+1) \times s$ matrix of the system in 4 where we assume that the lines are numbered from 0 to m; so, we have that $a_{0j} = 1$ for all $j = 1, \ldots, s$. Moreover, let p be the column vector whose entries are $1, p_1, \ldots, p_m$, let ξ and z be column vectors of s and $m + 1$ unknowns, respectively. Then, the de Finetti result can be stated along the linear programming theory as it follows.

Theorem 1. *The following are equivalent.*

(i) $\exists \xi \in \mathbf{R}^s$ *such that* $A\xi = p$, $\quad \xi$ *non negative.*
(ii) *The assessment is coherent, i.e. the game is reasonable.*
(ii) $\exists z \in \mathbf{R}^{m+1}$ $\quad {}^t z A \leq 0$, ${}^t z p \geq 1$, *where* ${}^t z$ *means transpose.*

Let A, ξ be as before, denote by $A_0, \ldots, A_m$ the $m+1$ rows of the $(m+1) \times s$ matrix A and by p^- the column vector whose entries are $1, p_1, \ldots, p_{m-1}$. Then, the following theorem by de Finetti is what is called the Fundamental Theorem of Probability (see [5]).

Theorem 2. *Assume that the assessment* $E_1 \mapsto p_1, \ldots, E_m \mapsto p_{m-1}$ *be coherent. Then the additional assessment* $E_m \mapsto p_m$ *preserves coherence if and only if*

$$p' \leq p_m \leq p''$$

where

$$p' = \min\{A_m \xi : \quad A_0 \xi + \ldots + A_{m-1} \xi = p^-\}$$
$$p'' = \max\{A_m \xi : \quad A_0 \xi + \ldots + A_{m-1} \xi = p^-\}$$

Theorem 1 gives a solution to the decision problem of coherence and hence to the Boole's consistency problem. Theorem 2 gives a solution to the Boole's more general problem.

In Section 2 we discuss some recent results about the computational complexity of the assessment consistency decision problem, in particular PSAT. This turns out to be a special case of the satisfiability problem for a logic which was proposed, in [6], for reasoning about probabilities.

In Section 3 we present a decision problem, called CPA, which is a reformulation of the problem PSAT when the Boolean relation among the events are explicitly known. Then, we describe a decision procedure, given in [1], which works directly on the Boolean relations by simplification rules and by elimination of Boolean variables.

2   Computational complexity of the consistency decision problem and satisfiability problems

2.1   The poblem PSAT

In order to present the decision problem for the consistency of a probability assessment as a concrete computational problem we require in 1 that the elements $p_1, \ldots, p_m$ be rational numbers. The problem, after the solution given by Theorem 1 seems to be inherently intractable having exponentially many unknowns. However, by a basic result in linear programming, a system $A\xi = p$ has a nonnegative soution if and only if there is a nonnegative ξ where at most $m + 1$ components are non null. This means that a coherent assessment extends to a probability distribution null but on $m + 1$ atoms. Moreover, there is a solution where such atoms have rational probability of length bounded by $m||p|| + m \log m$ where $||p||$ is the maximum length of the rational in $p_1, \ldots, p_m$. Such a solution can be used as a short certificate to witness the coherence. So, we have the following.

Theorem 3. *([9] 1988). The computational problem PSAT is in NP, hence it is NP-complete since it contains the satisfiability problem for propositional logic, called SAT.*

In [9] it is proved that the problem PSAT remains NP-complete even if restricted to the special case of clauses with at most 2 literals, called 2PSAT, when the ordinary 2SAT is, instead, polynomially solvable.

Moreover, following the same line of the proof of Theorem 1, the author prove that PSAT restricted to some class of clauses is polynomial-time reducible to MAXSAT on the same class of clauses. The problem MAXSAT is another generalization of SAT; an instance of it is given by an integer b and a set of Boolean clauses any of which is associated to a weight, the problem asks to find if there is a truth assignement such that the sum of the weights of all satisfied clauses is greater than the given bound b.

The reduction of 2PSAT to 2MAXSAT allows the authors to prove that significant subclasses of instances of 2PSAT are polynomially solvable (see [9], Section 4).

2.2 Satisfiability problems

We have already seen in Section 1 that the consistency problem for a probability assessment is substantially the satisfiability problem for the propositional fuzzy logic. This problem can be also formulated as a satisfiability problem for a logic which was proposed by Fagin, Halperin and Megiddo [6] for reasoning about probabilities.

Formulas of this logic are Boolean combinations of *basic weight formulas*. These are expressions of the form

$$a_1\, w(\varphi_1) + \ldots + a_k\, w(\varphi_k) \geq c \tag{8}$$

where $a_1, \ldots, a_k$, c are integers, $\varphi_1, \ldots, \varphi_k$ are propositional formulas and w is a symbol which is intended to be interpreted in a probability or more generally in a inner measure.

The semantics can be given by models which are structures

$$\mathcal{M} = (S, \mathcal{B}, \mu, \pi) \tag{9}$$

where S is a set, $\mathcal{B}$ is a subalgebra of the Boolean algebra $P(S)$ of the subsets of S, μ is an inner measure on $\mathcal{B}$ and π selects, for every element of S, a truth assignment for the propositional variables. So, the satisfation is defined by

$$\mathcal{M} \models a_1\, w(\varphi_1) + \ldots + a_k\, w(\varphi_k) \geq c \quad \text{iff} \quad a_1\, \mu(\varphi_1^{\mathcal{M}}) + \ldots + a_k\, \mu(\varphi_k^{\mathcal{M}}) \geq c \tag{10}$$

where $\varphi^{\mathcal{M}}$ is the interpretation entailed by

$$X^{\mathcal{M}} = \{s: \quad \pi(s)\,(X) = true\}$$

for every propositional variable X.

The authors provide axioms $(Ax)_{meas}$ for the semantics of *measurable spaces*. They are *sound* and *complete* with respect to the formal system based on the inference rule modus ponens. Moreover, it is proved that the satisfiability problem can be reduced to the solvability of a system like 4 using the same tecnique as in 3. So, the standard results in linear programming, discussed before Theorem 3, take to the proof of a small model theorem.

Theorem 4. *Assume Φ be a formula satisfiable in some measurable probability structure. Then, Φ is satisfiable in some $(S, \mathcal{B}, \mu, \pi)$ where*

- $|S| \leq m + 1$ *(m is the number of propositional variables in Φ);*

- *the probability size of each atom is not greater than* $m\|\Phi\| + m\log m$,
  ($\|\Phi\|$ *is the maximum length of the integers in* Φ *).*

From Theorem 4 it follows that the satisfiability problem for the logic of the measurable spaces is in NP. Hence, the satisfiability problem of this logic is NP-complete since, clearly, this problem contains PSAT. In fact, the satisfiability of the assessment $E \mapsto p$, when p is the rational number r/s, is equivalent to the satisfiability of the basic weight formulas $s\,w(E) \geq r$, $-s\,w(E) \geq -r$.

Then, the authors prove in the same paper also that the non-measurable semantics, where the weight w is interpreted in a inner measure, is axiomatizable. Moreover, the satisfiability problem is still NP-complete.

The complexity of satisfiability for the logic of polynomial weight-formulas is higher. A polynomial basic formula includes the product of weights; a typical formula is

$$2\,w(\varphi_1)\cdot w(\varphi_2) - 4\,w(\varphi_3)^2 + w(\varphi_1 \wedge \varphi_3) \geq 2$$

The satisfiability problem of this logic is in PSPACE.

The complexity of the decision problem when the logic is extended to allow first-order quantification over the reals is worse; there is an exponential space decision procedure.

3   Elimination of Boolean variables

Here we consider the probabilistic satisfiability problem for events $E_1, \ldots, E_m$ and assessments $E_i \mapsto p_i$, $i = 1, \ldots, m$, when the events are presented by means of relations that they satisfy. More precisely, it is given also a set $D_1, \ldots, D_k$ of Boolean relations that we can take to be dual clauses, i.e. conjunction of literals denoted as

$$X_{i_1}^{\varepsilon_1} X_{i_2}^{\varepsilon_2} \ldots X_{i_r}^{\varepsilon_r} \tag{11}$$

where the variables belong to $\{X_1, \ldots, X_m\}$. So, we think of $E_1, \ldots, E_m$ as generators of a Boolean algebra $\mathcal{B}$ and of $\{D_1, \ldots, D_k\}$ to be generators of the kernel of the morphism

$$d : \mathcal{F}(X_1, \ldots, X_m) \longrightarrow \mathcal{B}$$

on the free Boolean algebra induced by the map $X_i \mapsto E_i$, $i = 1, \ldots, m$.

Then, the problem CPA asks for

a probability distribution $Q : \mathcal{B} \longrightarrow [0,1]$ such that $Q(E_i) = p_i$, for $i = 1, \ldots, m$.

Note that the existence of Q is equivalent to the existence of

$$P : \mathcal{F}(X_1, \ldots, X_m) \longrightarrow [0, 1]$$

such that $Q \circ d = P$.

$$\mathcal{F}(X_1, \ldots, X_m) \xrightarrow{d} \mathcal{B}$$
$$\overset{P}{\searrow} \quad \downarrow Q$$
$$[0, 1]$$

Hence the CPA problem can be seen as the PSAT problem for the clauses

$$X_1, \ldots, X_m, D_1, \ldots, D_k$$

with the following probability assessment

$$p_1, \ldots, p_m, 0, \ldots, 0$$

Note also that the problem PSAT restricted to the values $p_i \in \{0, 1\}$ is equivalent to SAT, nevertheless CPA restricted to the range $\{0, 1\}$ is simply an evaluation of the clauses $D_1, \ldots, D_k$ for a truth-assignment to the variables $X_1, \ldots, X_m$. This could lead to think that CPA be in general easier. However, we have

Proposition 1. 2CPA *is NP-complete.*

The proof is quite the same as the one provided for 2PSAT in [9]. In fact, there is a polynomial reduction of the 3-colorability problem for graphs to 2CPA.

3.1   Simplification rules of instances of CPA

According to the discussion above we may represent an instance of CPA as a pair

$$\begin{cases} X_1 \mapsto p_1 \\ \ldots \ldots \ldots \\ X_m \mapsto p_m \end{cases} \qquad \mathcal{C} = \{D_1, \ldots, D_k\} \qquad (12)$$

where $\mathcal{C}$ is a set of dual clauses in the variables $\{X_1, \ldots, X_m\}$.

Now, we present some simplification rules for instances of CPA; each rule may introduce some constraints on the probability values $p_1, \ldots, p_m$ which we think as parameters. Each clause D is an expression as in 11, sometimes to remember its interpretation we write D as $D = 0$. The literal X' denotes the negation of the variable X; we represent a literal also as X^ε where $X^\varepsilon = X$ if $\varepsilon = 1$ and $X^\varepsilon = X'$ if $\varepsilon = 0$.

r1. Unitary clause rule: If $X_i = 0 \in C$ then

- delete all clauses containing X_i;
- delete the literal X_i' in each clause where it appears.

The rule produces the constraint $p_i = 0$.

Note that we may have a similar rule whenever $X_i X_j = 0$, $X_i X_j' = 0 \in C$ for some j.

r2. Subsumption rule: If D_i, $D_j \in C$ and $D_i \subseteq D_j$ as inclusion of literals, then delete D_j.

r3. Inclusion relations rule: Assume $C = C^- \cup C^+$ where $var(C^-) = \{X_1, \ldots, X_r\}$, $var(C^+) = \{X_{r+1}, \ldots, X_m\}$ and C^+ is a set of binary clauses where each clause means "inclusion of variables" (note that $XY' = 0$ if and only if $X \subseteq Y$). Then,

- delete C^+;
- delete the assessment $X_{r+1} \mapsto p_{r+1}, \ldots, X_m \mapsto p_m$;
- add the constraints $p_i \leq p_j$ for all i, j such that $X_i X_j'$ is in C^+.

r4. Weak pure literal rule: If $X_i X_j^\varepsilon = 0 \in C$ and X_i appears only once, then

- delete the clause;
- add the constraint $p_i + p_j \leq 1$ if $\varepsilon = 1$ and $p_i \leq p_j$ if $\varepsilon = 0$.

r5. Weak resolution rule: Assume $X_i' X_l^\varepsilon = 0$, $X_i X_r^\eta = 0 \in C$ and X_i does not appear in other clauses in C. Then

- replace both clauses by $X_l^\varepsilon X_r^\eta = 0$;
- add the constraints $p_l \leq p_i$, $p_i + p_r \leq 1$ when $\varepsilon = 1$, $\eta = 1$; analogously for other instances of ε, η.

3.2 The splitting rule

All the given rules, except r3, are analogous to the homonimous rules of Davis-Putnam. It is not difficult to prove that they are all correct with respect to the satisfiability of instances of CPA. However, in our framework we have to take care of the satisfiability of the constraints introduced by the rules.

As in Davis-Putnam, the crucial rule is the splitting rule which in our context introduces new constraints variables. Moreover, if we split an instance with respect to a given variable, say X_1, then we have to satisfy two new instances defined on the Boolean algebras of the traces on X_1 and X_1', respectively. Thus, the distribution probability on the new instances are not normalized since their maxima are p_1 and $1 - p_1$, respectively. Owing to this, we have to work with systems which are triple

$$S = (\mathcal{C}, p, t) \tag{13}$$

where $\mathcal{C}$ is a set of clauses in a set $\{X_1, \ldots, X_m\}$ of Boolean variables, $p = (p_1, \ldots, p_m)$ is a list of probability assessment for the variables and t represents the maximum for the range of the measure $\mu : \mathcal{F}(X_1, \ldots, X_m) \longrightarrow [0, t]$ we look for. As before, because of the constraints introduction, we think $p_1, \ldots, p_m, t$ as parameters. In fact, more generally they are represented as linear polynomials with rational coefficients, practically the coefficients of the variables can be $0, -1, 1$.

Now, we may state the splitting rule with respect to any given variable; to simplify notation we assume that the splitting variable be X_1.

r6. Splitting rule: Given a system $S = (\mathcal{C}, p, t)$ as in 13, then split it into two systems $S^\varepsilon = (\mathcal{C}^\varepsilon, p^\varepsilon, t^\varepsilon)$, for $\varepsilon = 0, 1$, where p^1 and p^0 are the follwing assessment, respectively.

$$\begin{cases} X_2 \mapsto z_2 \\ \ldots \quad \ldots \ldots \\ X_m \mapsto z_m \end{cases} \qquad \begin{cases} X_2 \mapsto p_2 - z_2 \\ \ldots \quad \ldots \ldots \\ X_m \mapsto p_m - z_m \end{cases}$$

where $z_2, \ldots, z_m$ are new real variables. Moreover, $t^\varepsilon = p_1$ if $\varepsilon = 1$, $t^\varepsilon = t - p_1$ if $\varepsilon = 0$ and $\mathcal{C}^\varepsilon = \mathcal{C}\{X_1 \leftarrow \varepsilon\}$ where $\{X_1 \leftarrow \varepsilon\}$ means the substitution of the variable X_1 for the truth value ε. In other words $\mathcal{C}^1$ is obtained by deleting in $\mathcal{C}$ all the clauses containing X_1' and by deleting the variable X_1 in the clauses where it appears; $\mathcal{C}^0$ is obatined dually. Finally, the replacement of S with S^1 and S^0 adds the constraints

$$0 \le z_i \le p_1, \qquad 0 \le p_i - z_i \le t - p_1 \qquad \text{for } i = 2, \ldots, m. \tag{14}$$

Definition 1. Let $S = (\mathcal{C}, p, t)$ be a system as in 13 which is associated to a set of constraints $\mathcal{U}$. Assume that z be all the variables appearing in $\mathcal{U}$ or in t or in $p = (p_1, \ldots, p_m)$. Then, we say that $S; \mathcal{U}$ is satisfiable if there is an assignment a of real numbers for the list of variables z and there is a measure $\mu : \mathcal{F}(X_1, \ldots, X_m) \longrightarrow [0, t(a)]$ such that

- $\mu(X_i) = p_i(a), \qquad i = 1, \ldots, m;$

- $\mu(\mathbb{1}) = t(a)$;
- $\mu(D) = 0$, for every $D \in \mathcal{C}$;
- the real contraints in $\mathcal{U}$ are satisfied by the substitution of z for a.

Now, let $\mathcal{U}^+$ be obtained by $\mathcal{U}$ with the addition of the constraints in 14 generated by the splitting rule. Then, we have the correctness.

Proposition 2. *$S;\mathcal{U}$ is satisfiable if and only if $S^1;\mathcal{U}^+$ and $S^0;\mathcal{U}^+$ are simultaneously satisfiable with the same substitution for the constraint variables in $\mathcal{U}^+$.*

Note that the splitting rule is in a sense dual respect to the homonymous Davis-Putnam rule. The Proposition 2 indicates that our splitting rule matches Davis-Putnam rule if one changes the goal of "satisfiability" with the goal of "unsatisfiability".

3.3   A procedure for satisfiability

The rules r1–r6 can be used to build a procedure for checking the satisfiability of instances for CPA. Observe in fact that when S has no clause at all then $S;\mathcal{U}$ is satisfiable if and only if the set of real contraints $\mathcal{U}$ is satifiable. So, the procedure eliminates the Boolean variables by building a tree by means of the rules. In the benefit of efficiency it is better to use, while it is possible, the non splitting rules and to explore the tree depth-first.

Finally, note that, because of the rules r3, r4, r5 the procedure seems to be more qualified to solve instances of 2CPA.

References

1. Baioletti M, Capotorti A, Tulipani S, Vantaggi B (2000) Elimination of Boolean variables for probabilistic coherence. Soft Computing, vol. 4 N.2, 81–88.
2. Biacino L (1993) Generated envelopes. J. Math. Anal. Appl. 172: 179-190.
3. Boole G (1854) An investigation of the laws of thought. London, Walton and Maberley, (reprint New York, Dover 1958).
4. Coletti G. (1994) Coherent Numerical and Ordinal Probabilistic Assessments. IEEE Trans. on Systems, Man, and Cybernetics, 24(12): 1747-1754.
5. De Finetti B (1970) Teoria della probabilitá vol.I,II. Torino: Einaudi (Engl. Transl. (1974) Theory of probability vol.I,II, London: Wiley & Sons).
6. Fagin R, Halperin J Y, Megiddo N (1990) A Logic for Reasoning about Probabilities. Information and Computation, 87: 78-128.
7. Frisch a M, Haddawy P (1994) Anytime Deduction for Probabilistic Logic. Artificial Intelligence, 69: 93-122.
8. Garey M R, Johnson D S (1979) Computers and Intractability. San Francisco: Freeman and Company.

9. Georgakopoulos G, Kavvadias D, Papadimitriou C H (1988) Probabilistic Satisfiability. Journal of Complexity, 4: 1-11.

10. Gerla G (1994) Inferences in probability logic. Artificial Intelligence 70: 33-52.

11. Gilio A (1993) Probabilistic Consistency of Knowledge Bases in Inference Systems. Lecture Notes in Computer Science, In: Clarke, M., Kruse, R., Moral, S., vol. 747, pp. 160-167. Springer-Verlag.

12. Hailperin T (1965) Best Possible Inequalities for the Probability of a Logical Function of Events. American Mathematical Monthly, 72: 343-359.

13. Hailperin T (1986) Boole's logic and probability, studies in logic and the foundations of mathematics 85. New York, North-Holland.

14. Hansen P, Jaumard B (1990) Algorithms for the Maximum Satisfiability Problem. Computing, 44: 279-303.

15. Hansen P, Jaumard B, Poggi de Aragão M (1995) Boole's Conditions of Possible Experience and Reasoning Under Uncertainty. Discrete Applied Mathematics, 60: 181-193.

16. Jaumard B, Hansen P, Poggi de Aragão M (1991) Column Generation Methods for Probabilistic Logic. ORSA Journal on Computing, 3: 135-148.

17. Nilson N J (1986) Probabilistic Logic. Artificial Intelligence 28: 71–87.

18. Nilson N J (1993) Probabilistic Logic Revisited. Artificial Intelligence 59: 39–42.

19. Pavelka J (1979) On fuzzy logic I: many valued rules of inference. Z. Math. Logik Grundl. Math. 25: 45-52.

Survey of Theory and Applications of Lukasiewicz-Pavelka Fuzzy Logic

Esko Turunen

Tampere University of Technology
P.O. Box 692, FIN-33101 Tampere, Finland - esko.turunen@cc.tut.fi

Abstract. We demonstrate how approximate reasoning, many classification tasks, case-based reasoning, etc. can be viewed as applications of many valued similarity and, thus Lukasiewicz-Pavelka logic.

1   Introduction

Two valued logic fits well to mathematical reasoning, indeed, it is the 'metamathematics of mathematics'. Applying two valued logic outside mathematics, however, rises anomalies that we cannot accept as they contradict our everyday experiences. An alternative approach to avoid paradoxes rising from seeing real world's phenomena only black or white is to accept more than two truth values *true* and *false*. This is the starting point of many-valued logics and fuzzy logic. Following Lotfi Zadeh [12], the inventor of Fuzzy Set theory, we distinguish fuzzy logic in *broad sense* i.e. everything concerning vagueness and fuzziness, from fuzzy logic in *narrow sense*, i.e. the formal logical calculus of fuzziness. In this survey we focus on the latter.

Jan Lukasiewicz [8] was the first to investigate systematically many-valued logics in 1920's. In 1935, Morchaj Wajsberg showed[1] that infinite valued sentential logic was complete with respect to the axioms conjectured by Lukasiewicz. Twenty-three years later, in 1958, C.C. Chang introduced MV-algebras, which allowed him to give another completeness proof for Lukasiewicz logic. For decades many-valued logic was far from the mainstreams of mathematical research, it was only after the 'fuzzy boom' started in 1965 with Zadeh's seminal paper 'Fuzzy Sets' that the situation has changed a bit. In 1979 Jan Pavelka [10] published a paper entitled 'On fuzzy logic' in which he generalized Lukasiewicz's logic by introducing fuzzy consequence operations, general fuzzy rules of inference, fuzzy proofs, etc. Pavelka studied the real unit interval valued fuzzy sentential logic and proved that necessary and sufficient condition for the completeness of his logic is the continuity of the implication operation. In this survey we recall an outline of Lukasiewicz-Pavelka fuzzy logic.

---

[1] Unfortunately Wajsberg's proof was never published

The traditional Greek notion of *analogia*, meaning 'proportion' is usually taken to assert the similarity, or partial identity of two objects. Following Leibniz, the founder of mathematical logic, the identity of two objects A and B means that they share all their properties, hence, objects A and B may be said to be partially identical if they share some (or most) of their properties. Immanuel Kant's Logik [7] formulated the problem on analogy in the following terms: 'Analogy concludes from partial similarity of two things to total similarity according to the principle of specification: Things of one genus, which we know to degree in much, also agree in the remainder as we know it is some of the genus but do not perceive it in others'. According to Niiniluoto [9], 'The real challenge · · · is that we have to extend our treatment from simple analogy to multiple analogy'.

We shall see how generalizing equivalence relation can solve this challenge on the basis of Łukasiewicz-Pavelka many-valued logic, and applying the generated many-valued equivalence on Zadeh's fuzzy sets. We introduce an algorithm to construct fuzzy IF-THEN inference systems having a special feature that whenever the output would not be unique the final decision should be left to human experts. Thus, as much as possible the intelligence relies on a real controller, and technical defuzzification methods are not needed. We demonstrate how approximate reasoning, many classification tasks, case-based reasoning, etc. can be viewed as applications of many-valued similarity and, thus, Łukasiewicz-Pavelka logic.

2   The Algebra of Łukasiewicz-Pavelka Logic

An axiom system of a logic generates an algebraic structure. In classical logic this structure is Boolean algebra, while infinite valued Łukasiewicz-Pavelka logic, being an extension of two-valued logic, generates a more general algebraic structure, called *MV-algebra* [1],[2]. To minimize the axioms we first set, however, the following

Definition 1. Let L be a non-void set, $1 \in L$ and $\rightarrow$, * be a binary and a unary operation, respectively, defined on L such that, for each $x, y, z \in L$,

$$1 \rightarrow x = x, \tag{1}$$

$$(x \rightarrow y) \rightarrow [(y \rightarrow z) \rightarrow (x \rightarrow z)] = 1, \tag{2}$$

$$(x \rightarrow y) \rightarrow y = (y \rightarrow x) \rightarrow x, \tag{3}$$

$$(x^* \rightarrow y^*) \rightarrow (y \rightarrow x) = 1. \tag{4}$$

Then the system $\langle L, \rightarrow, {}^*, 1 \rangle$ is called a Wajsberg algebra.

Now define on a Wajsberg algebra $\langle L, \rightarrow, {}^*, 1 \rangle$ a binary relation $\leq$ by

$$x \leq y \text{ iff } x \rightarrow y = 1. \tag{5}$$

Then $\leq$ is an order relation on L and $\mathbf{1}$ is the greatest element in L and, by
defining two binary operations $\wedge$ and $\vee$ on a Wajsberg algebra L via

$$x \vee y = (x \rightarrow y) \rightarrow y, \tag{6}$$
$$x \wedge y = (x^* \vee y^*)^*. \tag{7}$$

we obtain a lattice $\langle L, \leq, \wedge, \vee \rangle$. Moreover, by defining on a Wajsberg algebra
L two binary operation $\odot$, for each $x, y \in L$, via

$$x \odot y = (x \rightarrow y^*)^*. \tag{8}$$
$$x \oplus y = x^* \rightarrow y, \tag{9}$$

the following equations - the original MV-algebra axioms by Chang [1]- are
satisfied in every Wajsberg algebra.

$$x \oplus y = y \oplus x \ , \ x \odot y = y \odot x, \tag{10}$$
$$x \oplus (y \oplus z) = (x \oplus y) \oplus z \ , \ x \odot (y \odot z) = (x \odot y) \odot z, \tag{11}$$
$$x \oplus x^* = \mathbf{1} \ , \ x \odot x^* = \mathbf{0}, \tag{12}$$
$$x \oplus \mathbf{1} = \mathbf{1} \ , \ x \odot \mathbf{0} = \mathbf{0}, \tag{13}$$
$$x \oplus \mathbf{0} = x \ , \ x \odot \mathbf{1} = x, \tag{14}$$
$$(x \oplus y)^* = x^* \odot y^* \ , \ (x \odot y)^* = x^* \oplus y^*, \tag{15}$$
$$x^{**} = x \ , \ \mathbf{1}^* = \mathbf{0}, \tag{16}$$
$$x \vee y = y \vee x \ , \ x \wedge y = y \wedge x, \tag{17}$$
$$x \vee (y \vee z) = (x \vee y) \vee z \ , \ x \wedge (y \wedge z) = (x \wedge y) \wedge z, \tag{18}$$
$$x \oplus (y \wedge z) = (x \oplus y) \wedge (x \oplus z) \ , \ x \odot (y \vee z) = (x \odot y) \vee (x \odot z). \tag{19}$$

Conversely, given an MV-algebra $\langle L, \oplus, \odot, ^*, \mathbf{0}, \mathbf{1} \rangle$, we can define an operation
$\rightarrow$ for all $x, y \in L$ by $x \rightarrow y = x^* \oplus y$. Then the Wajsberg algebra axioms
hold. Thus, there is a one-to-one correspondence between MV-algebras and
Wajsberg algebras. An MV-algebra is called *complete* if it contains lowest
upper bound and least lower bound of any of its subset $\{x_i \mid i \in \Gamma\}$, denoted
by $\bigvee\{x_i \mid i \in \Gamma\}$ and $\bigwedge\{x_i \mid i \in \Gamma\}$, respectively.

In the real unit interval, the most used MV-algebra structure called Łukasie-
wicz algebra, too, is obtained by setting $\mathbf{1} = 1, \mathbf{0} = 0$ and, for all $x, y \in [0, 1]$,

$x \rightarrow y = \min\{1 - x + y, 1\}$, $x^* = 1 - x$
$x \odot y = \max\{x + y - 1, 0\}$, $x \oplus y = \min\{x + y, 1\}$
$x \wedge y = \min\{x, y\}$, $x \vee y = \max\{x, y\}$.

These algebraic operations will offer us an elegant tool to interpret the logical
connectives implication, negation and two kinds of conjunction and disjunction, respectively. Moreover, for equivalence in Łukasiewicz-Pavelka logic we
have the operation $x \leftrightarrow y = 1 - |x - y|$ which, in general setting, is defined
by

$$x \leftrightarrow y = (x \rightarrow y) \wedge (y \rightarrow x).$$

3   Łukasiewicz-Pavelka Logic

Now we start to develop logic, which allows more truth values than only 'false' and 'true'. We assume all the time that the set L of values of truth forms a complete MV-algebra. The formalized language of this logic is composed of four kinds of building blocks:
(i) The set of *propositions* is an infinite set $L = \{p_i; i \in \mathcal{N}\}$. Propositions are sometimes denoted by p, q, r, s, t, w, too.
(ii) For any element $a \in L$ there is an *inner truth value* **a** in the language.
(iii) The logical connectives are imp (read 'implies') and **and** (read 'and' or 'conjunction').
(iv) There are auxiliary symbols $\},],), (, [, \{$ in the language of the logic.

Inner truth-values and propositional variables are *atomic formulas*.

Definition 2. The set $\mathcal{F}$ of well-formed formulas is constructed in the following way:
(i) atomic formulas are in $\mathcal{F}$,
(ii) if α and β are in $\mathcal{F}$ then $(\alpha \text{ imp } \beta)$ and $(\alpha \text{ and } \beta)$ are in $\mathcal{F}$.

Propositional variables correspond statements like *it is raining*, etc. The main difference between two-valued sentential logic and Lukasiewicz-Pavelka logic are the inner truth-values, which can be regarded as generalizations of the falsum sign $\perp$ of classical logic. We follow the idea of intuitionistic logic and abbreviate $(\alpha \text{ imp } \mathbf{0})$ by $(\text{non} - \alpha)$. The logical connective **non** (read 'non') is called *negation*. Giving *semantic interpretation* to a formula $\alpha \in \mathcal{F}$ means we associate a value of truth $v(\alpha) \in L$ to α, in other words, we define *truth value function* $v : \mathcal{F} \searrow L$ by setting

Definition 3. A function $v : \mathcal{F} \searrow L$ such that, for any inner truth value **a** and for any formulas α, β,

$$v(\mathbf{a}) = a, \qquad (20)$$
$$v(\alpha \text{ imp } \beta) = v(\alpha) \rightarrow v(\beta), \qquad (21)$$
$$v(\alpha \text{ and } \beta) = v(\alpha) \odot v(\beta), \qquad (22)$$

is called (fuzzy) valuation or (fuzzy) truth value function.

We introduce a logical connective **or** (read 'or' or 'disjunction') as an abbreviation

$$(\alpha \text{ or } \beta) = [\text{non-}(\text{non-}\alpha \text{ and } \text{non-}\beta)].$$

This generalizes the state of affairs in classical sentential logic and makes the formalized language of many-valued logic easier to read.

Remark 1. For any valuation v, any $\alpha, \beta \in \mathcal{F}$,

$$v(\text{non-}\alpha) = v(\alpha)^*, \tag{23}$$
$$v(\alpha \text{ or } \beta) = v(\alpha) \oplus v(\beta). \tag{24}$$

Generally, $v(\alpha \text{ and } \beta) < v(\alpha) \wedge v(\beta)$ and $v(\alpha) \vee v(\beta) < v(\alpha \text{ or } \beta)$. In application we may need, however, disjunctive and conjunctive connectives, denote them by $\overline{\text{and}}$ and $\overline{\text{or}}$, respectively, such that

$$v(\alpha \overline{\text{and}} \beta) = v(\alpha) \wedge v(\beta), \ v(\alpha \overline{\text{or}} \beta) = v(\alpha) \vee v(\beta).$$

Abbreviating can do this

$$(\alpha \overline{\text{or}} \beta) = [(\alpha \text{ imp } \beta) \text{ imp } \beta)]$$
$$(\alpha \overline{\text{and}} \beta) = [\text{non-}(\text{non-}\alpha \overline{\text{or}} \text{non-}\beta)],$$

even if the abbreviation of the logical connective $\overline{\text{or}}$ is far from being obvious. We also introduce a logical connective **equiv** by abbreviating

$$(\alpha \text{ equiv } \beta) = [(\alpha \text{ imp } \beta) \overline{\text{and}} [(\beta \text{ imp } \alpha)],$$

thus generalizing the situation in classical logic. Then we have, for any valuation v, any formulas $\alpha, \beta \in \mathcal{F}$,

$$v(\alpha \text{ equiv } \beta) = v(\alpha) \leftrightarrow v(\beta).$$

In classical logic truth value functions $v : \mathcal{F} \searrow \{0,1\}$ satisfy the truth tables

$v(\alpha \text{ or } \beta)$	$v(\beta) = 1$	$v(\beta) = 0$
$v(\alpha) = 1$	1	1
$v(\alpha) = 0$	1	0

$v(\alpha \text{ and } \beta)$	$v(\beta) = 1$	$v(\beta) = 0$
$v(\alpha) = 1$	1	0
$v(\alpha) = 0$	0	0

$v(\alpha \text{ imp } \beta)$	$v(\beta) = 1$	$v(\beta) = 0$
$v(\alpha) = 1$	1	0
$v(\alpha) = 0$	1	1

$v(\text{non-}\alpha)$	$v(\alpha) = 1$	$v(\alpha) = 0$
	0	1

It is easy to verify that fuzzy valuations satisfy these tables and that in two valued case the truth value tables of the logical connectives $\overline{\text{and}}$ and and as well as $\overline{\text{or}}$ and or coincide. Complete truth tables are not, of course, possible in logic with infinite many values of truth. We may, however, write instances of them.

As an example, we calculate that if $v(\alpha) = 0.2$, $v(\beta) = 0.6$ and $v(\gamma) = 0.9$ then

$$v([\alpha \text{ imp } (\text{non-}\beta\text{or non-}\gamma)] \text{ imp } \gamma) = 0.9$$

In logic we are interested in the logical consequences of given statements. From semantic point of view this raises a question

Associating fixed values of truth to a set of well-formed formulas $T \subseteq \mathcal{F}$, what is the least degree of truth, or greatest lower bound of such degrees, of an arbitrary formula $\alpha \in \mathcal{F}$ with respect to T?

This leads us to the following semantic definitions

Definition 4. A fuzzy set T of formulas is a function $T : \mathcal{F} \searrow L$. A truth value function $v : \mathcal{F} \searrow L$ satisfies T if $T(\alpha) \leq v(\alpha)$ for any formula $\alpha \in \mathcal{F}$. If there exists a valuation v such that v satisfies T the T is called satisfiable.

The void set $\emptyset$ can be regarded as a fuzzy set of formulas by defining $\emptyset(\alpha) = 0$ for all formulas $\alpha \in \mathcal{F}$. The void set is of course satisfiable.

Definition 5. The degree of validity of a formula $\alpha \in \mathcal{F}$ (with respect to a fuzzy set of formulas T) is a value

$$C^{\text{sem}}(T)(\alpha) = \bigwedge \{v(\alpha) \mid v \text{ satisfies } T\}. \tag{25}$$

In particular, if T is the void set we define the degree of tautology of a formula α by

$$C^{\text{sem}}(\alpha) = \bigwedge \{v(\alpha) \mid v \text{ is a valuation }\}. \tag{26}$$

This definition is very natural and generalizes the concept of tautology in classical logic. If $C^{\text{sem}}(T)(\alpha) = a$ we write $T \models_a \alpha$, in particular $\models_a \alpha$ if T is the void set. Of special interest will be formulas α such that $\models_1 \alpha$. Evidently, if $\models_1 \alpha$, then $T \models_1 \alpha$ for any fuzzy set T of formulas (even for those T which are not satisfied by any valuation v!)

Proposition 1. *Let α, β, γ, α_1, β_1, α_2, β_2 be formulas and c any inner truth value. Then the following forms of formulas are universally valid at the degree 1, except for, of course, the inner truth value c, which is universally* *valid at the degree c, i.e.*

$$\models_1 \alpha \text{ imp } \alpha, \quad (27)$$

$$\models_1 (\alpha \text{ imp } \beta) \text{ imp } [(\beta \text{ imp } \gamma) \text{ imp } (\alpha \text{ imp } \gamma)], \quad (28)$$

$$\models_1 (\alpha_1 \text{imp} \beta_1) \text{ imp } \{(\beta_2 \text{imp} \alpha_2) \text{ imp } [(\beta_1 \text{imp} \beta_2) \text{ imp } (\alpha_1 \text{imp} \alpha_2)]\}, \quad (29)$$

$$\models_1 \alpha \text{ imp } 1, \quad (30)$$

$$\models_1 0 \text{ imp } \alpha, \quad (31)$$

$$\models_1 (\alpha \text{ and } \text{non-}\alpha) \text{ imp } 0, \quad (32)$$

$$\models_c c, \quad (33)$$

$$\models_1 \alpha \text{ imp } (\beta \text{ imp } \alpha), \quad (34)$$

$$\models_1 (1 \text{ imp } \alpha) \text{ imp } \alpha, \quad (35)$$

$$\models_1 [(\alpha \text{ imp } \beta) \text{ imp } \beta] \text{ imp } [(\beta \text{ imp } \alpha) \text{ imp } \alpha], \quad (36)$$

$$\models_1 (\text{non-}\alpha \text{ imp } \text{non-}\beta) \text{ imp } (\beta \text{ imp } \alpha). \quad (37)$$

The definitions of valuation and the degree of validity of formula α are natural and relatively easy generalizations of the corresponding concepts in two valued logic. Now we consider the following related non-trivial problem

Knowing that a formula α is valid at a certain degree, do there exist a fuzzy set of axioms and fuzzy rules of inference by which we can infer α at the same degree?

In other words, is Łukasiewicz-Pavelka logic axiomatizable? To find an answer to this question, we start by defining on what we mean by fuzzy axiom, fuzzy rule of inference, fuzzy proof, etc.

A *rule of inference* in Classical Propositional Logic is an n-ary operation on the set of well-formed formulas which with a finite sequence of formulas $\alpha_1, \cdots, \alpha_n$ $(1 \leq n)$ in a formalized language associates another formula β in this language in such a way that β is a logical consequence of the formulas $\alpha_1, \cdots, \alpha_n$. This fact is usually denoted as follows

$$\frac{\alpha_1, \cdots, \alpha_n}{\beta}$$

Formulas $\alpha_1, \cdots, \alpha_1$ are called *premises* and β the *conclusion* of this rule of inference. For example,

$$\frac{\text{non} - (\text{non} - \alpha)}{\alpha}$$

and

$$\frac{\alpha, (\alpha \text{ imp } \beta)}{\beta}$$

are rules of inference in Classical Propositional Logic, called *Rule of Double Negation* and *Modus Ponens*, respectively. By saying that a formula β is a *logical consequence* of a set S of formulas we mean that if every formula α belonging to S is acknowledged to be true, then β must be accepted as true. Thus, the most important property of rule of inference is soundness, i.e. rule of inference preserves truth.

We define a fuzzy rule of inference as consisting of two components. The first component operates on formulas and is, in fact, a rule of inference in the usual sense; the second component operates on truth values and says how the truth value of the conclusion is to be computed from the truth-values of the premises such that the degree of truth is preserved. More accurately, we set

Definition 6. An n-ary fuzzy rule of inference *is a scheme*

$$R \quad : \quad \frac{\alpha_1, \cdots, \alpha_n}{r^{syn}(\alpha_1, \cdots, \alpha_n)} \;,\quad \frac{a_1, \cdots, a_n}{r^{sem}(a_1, \cdots, a_n)}$$

where the well-formed formulae $\alpha_1, \cdots, \alpha_n$ *are the* premises *and the well-formed formula* $r^{syn}(\alpha_1, \cdots, \alpha_n)$ *is the* conclusion.

The values $a_1, \cdots, a_n$, $r^{sem}(a_1, \cdots, a_n) \in L$ *are the corresponding truth-values. The mapping* $r^{sem} : L^n \searrow L$ *is semi-continuous on each variable, i.e. it holds always that*

$$r^{sem}(a_1, \cdots, \bigvee_{j \in \Gamma} a_{k_j}, \cdots, a_n) = \bigvee_{j \in \Gamma} r^{sem}(a_1, \cdots, a_{k_j}, \cdots, a_n), 1 \leq k \leq n.$$

We assume the fuzzy rule of inference is sound, *i.e. for each valuation* v *holds*

$$r^{sem}(v(\alpha_1), \cdots, v(\alpha_n)) \leq v(r^{syn}(\alpha_1, \cdots, \alpha_n)).$$

Proposition 2. *The following schemes are fuzzy rules of inference in Łukasiewicz-Pavelka logic*

Generalized Modus Ponens:

$$R_{GMP} \quad : \quad \frac{\alpha, (\alpha \text{ imp } \beta)}{\beta} \;,\quad \frac{a, b}{a \odot b}$$

a-Consistency-testing rules:

$$R_{a-CTR} \quad : \quad \frac{\mathbf{a}}{\mathbf{0}} \;,\quad \frac{b}{c}$$

where **a** *is an inner truth value, and* $c = 0$ *if* $b \leq a$ *and* $c = 1$ *elsewhere.*
a-Lifting Rules:

$$R_{a-LR} \quad : \quad \frac{\alpha \qquad b}{(\text{a imp } \alpha) \quad a \to b},$$

where **a** *is an inner truth value.*
Rule of Bold Conjunction:

$$R_{RBC} \quad : \quad \frac{\alpha, \beta \qquad a, b}{(\alpha \text{ and } \beta) \quad a \odot b},$$

Definition 7. *A fuzzy set* A *of logical axioms is a finite set of forms of formulas each being an inner truth value* **a**, *then* $A(\mathbf{a}) = a$, *or a tautology* α *at the degree 1, then* $A(\alpha) = 1$. *Elsewhere* $A(\alpha) = 0$.

For example, the following set of forms of formulas is, by (27)-(37), a set of logical axioms:

(Ax.1) α imp α,
(Ax.2) $(\alpha$ imp $\beta)$ imp $[(\beta$ imp $\gamma)$ imp $(\alpha$ imp $\gamma)]$,
(Ax.3) $(\alpha_1 \text{imp} \beta_1)$ imp $\{(\beta_2 \text{imp} \alpha_2)$ imp $[(\beta_1 \text{imp} \beta_2)$ imp $(\alpha_1 \text{imp} \alpha_2)]\}$,
(Ax.4) α imp 1,
(Ax.5) 0 imp α,
(Ax.6) $[(\alpha$ and non-$\alpha)$ imp 0,
(Ax.7)   **a**,
(Ax.8) α imp $(\beta$ imp $\alpha)$,
(Ax.9) $(1$ imp $\alpha)$ imp α,
(Ax.10) $[(\alpha$ imp $\beta)$ imp $\beta]$ imp $[(\beta$ imp $\alpha)$ imp $\alpha]$,
(Ax.11) $(\text{non-}\alpha$ imp non-$\beta)$ imp $(\beta$ imp $\alpha)$.
where $\alpha, \beta, \gamma, \alpha_1, \alpha_2, \beta_1, \beta_2$ are well-formed formulas and **a** any inner truth-value. Values $A(\delta)$ are obvious.

Definition 8. *Let* A *be a fixed set of logical axioms,* R *a fixed finite set of fuzzy rules of inference and* T *fuzzy set of formulas called* non-logical *axioms. Then a (zero-order) fuzzy theory is a triplet* $\langle A, R, T \rangle$. *In particular, if the set of logical axioms* A *is composed of Ax.1 - Ax.11, and the set of fuzzy rules of inference* R *contains* R_{GMP}, R_{a-CTR}, R_{a-LR}, R_{RBC}, *we denote a fuzzy theory simply by* T, *and if* T *is the void set we talk about* Fuzzy Propositional Calculus.

A *metaproof* of a well-formed formula α in a fuzzy theory $\langle A, R, T \rangle$, denoted by w, is a finite sequence

$$\alpha_1 \ , \ a_1$$
$$\vdots \quad \vdots$$
$$\alpha_m \ , \ a_m$$

of pairs $(\alpha_i, a_i) \in \mathcal{F} \times L$ such that the following holds: (i) $\alpha_m = \alpha$, (ii) for each $i, 1 \le i \le m, \alpha_i$ is a logical axiom, or α_i is a non-logical axiom, or there are a fuzzy rule of inference in R and formulas $\alpha_{i_1}, \cdots, \alpha_{i_n}$ with $i_1, \cdots, i_n < i$ such that $\alpha_i = r^{\mathbf{syn}}(\alpha_{i_1}, \cdots, \alpha_{i_n})$, (iii) for each $i, 1 \le i \le m$, the value a_i is given by

$$a_i = \begin{cases} a & \text{if } \alpha_i \text{ is the axiom } \mathbf{a} \\ 1 & \text{if } \alpha_i \text{ is some other logical axiom} \\ T(\alpha_i) & \text{if } \alpha_i \text{ is a non-logical axiom} \\ r^{\mathbf{sem}}(a_{i_1}, \cdots, a_{i_n}) & \text{if } \alpha_i = r^{\mathbf{syn}}(\alpha_{i_1}, \cdots, \alpha_{i_n}). \end{cases}$$

The value a_m is denoted by $Val_{\langle \mathbf{A},\mathbf{R}, T\rangle}(w)$ and is called the *degree* of the metaproof w. Because a formula α may have many metaproofs with different degrees, we define the *degree of deduction* of the formula α in fuzzy theory $\langle \mathbf{A}, \mathbf{R}, T\rangle$ by

$$C^{\mathbf{syn}\langle \mathbf{A},\mathbf{R}\rangle}(T)(\alpha) = \bigvee \{ Val_{\langle \mathbf{A},\mathbf{R}, T\rangle}(w) | w \text{ is a metaproof for } \alpha \text{ in } \langle \mathbf{A}, \mathbf{R}, T\rangle \}.$$

The case $C^{\mathbf{syn}\langle \mathbf{A},\mathbf{R}\rangle}(T)(\alpha) = a$ is denoted by $\langle \mathbf{A}, \mathbf{R}, T\rangle \vdash_a \alpha$, in particular, $\vdash_a \alpha$ if the set of logical axioms A is composed of Ax.1 - Ax.11, the set of fuzzy rules of inference R contains R_{GMP}, R_{a-CTR}, R_{a-LR}, R_{RBC} and T is the void set.

Let the set of logical axioms A be composed of Ax.1 - Ax.11, and the set of fuzzy rules of inference R contains the fuzzy rules of inference R_{GMP}, R_{a-CTR}, R_{a-LR}, R_{RBC}. Fuzzy theories are thus identified by means of their sets T of non-logical axioms. We will write $C^{\mathbf{syn}}$ instead of $C^{\mathbf{syn}\langle \mathbf{A},\mathbf{R}\rangle}$. Obviously, for any fuzzy theory T, if $\vdash_1 \alpha$ then $T \vdash_1 \alpha$, and by Ax.7, for any inner truth value a, $a \le C^{\mathbf{syn}}(T)(\mathbf{a})$. This leads us to the following

Definition 9. A fuzzy theory T is consistent *if, for any inner truth value* a, $a = C^{\mathbf{syn}}(T)(\mathbf{a})$, *and otherwise T is contradictory.*

Proposition 3. *A fuzzy theory T is contradictory iff $T \vdash_1 \alpha$ holds for each $\alpha \in \mathcal{F}$.*

Proof. Assume T is contradictory. Then there exists an inner truth value a such that $a \ne C^{\mathbf{syn}}(T)(\mathbf{a})$. If for each metaproof w for a holds $Val_T(w) \le a$, then $a \le C^{\mathbf{syn}}(T)(\mathbf{a}) \le a$, hence $C^{\mathbf{syn}}(T)(\mathbf{a}) = a$, which is not the case. Therefore there exists a metaproof w for a such that $Val_T(w) \not\le a$. For every formula $\alpha \in \mathcal{F}$, we have now the following metaproof:

$$
\begin{array}{llll}
\text{a} & , Val_T(w) , & assumption \\
0 & , \quad 1 & , & R_{a-CTR} \\
0 \text{ imp } \alpha , & 1 & , & Ax.4 \\
\alpha & , \quad 1 & , & R_{GMP}
\end{array}
$$

We conclude that $T \vdash_1 \alpha$ holds for each $\alpha \in \mathcal{F}$. Conversely, if $T \vdash_1 \alpha$ holds for each $\alpha \in \mathcal{F}$, then, in particular, $T \vdash_1 0$, i.e. $C^{syn}(T)(0) = 1 \neq 0$.

Proposition 4. *A fuzzy theory T is contradictory iff the following condition holds* (C):

> *There is a formula α and metaproofs w, w' for α, non-α, respectively, such that $Val_T(w) = a$, $Val_T(w') = b$ and $0 < a \odot b$.*

Let T be a fuzzy theory. The choice of the logical axioms Ax.1 - Ax.11 and soundness of fuzzy rules of inference guarantee, for each formula α, each metaproof w for α in T, each valuation v which satisfies T, that $Val_T(w) \leq v(\alpha)$. Thus,

$$
\bigvee \{Val_T(w) \mid w \text{ is a metaproof for } \alpha \text{ in } T\} \leq \bigwedge \{v(\alpha) \mid v \text{ satisfies } T\},
$$

by symbols, $C^{syn}(T)(\alpha) \leq C^{sem}(T)(\alpha)$. (This (in-)equality holds even if T is not satisfiable as $\bigwedge \{\emptyset\} = 1$.) We write

Theorem 1. (Soundness Theorem for Fuzzy Propositional Calculus) *Let T be a fuzzy theory. For each formula α, if $T \vdash_a \alpha$, $T \models_b \alpha$, then $a \leq b$.*

Corollary 1. *Any satisfiable fuzzy theory T is consistent.*

The most important theoretical result concerning Łukasiewicz-Pavelka logic is the *Completeness Theorem* of Fuzzy Propositional Calculus; for any fuzzy theory T, for each formula $\alpha \in \mathcal{F}$ and for any value $a \in L$, holds

$$
T \vdash_a \alpha \text{ if, and only if } T \models_a \alpha.
$$

The rather long proof of this fact is, however, omitted. We conclude this section by giving an easy example, which should illustrate a possible application of Łukasiewicz-Pavelka logic.

Example Assume p stands for *It is raining enough* and q stands for *Potato is growing fast*. We study a fuzzy theory T such that $T(\text{non}-p \text{ imp non}-q) = 1$ standing for *If it is not raining enough then potato is not growing fast* and $T(q) = 0.7$ standing loosely for *Potato is growing more or less fast*. Now we are interested in the degree of deduction of p. We find the following metaproof for p:

$$(\text{non} - p \text{ imp non} - q) \text{ imp } (q \text{ imp } p) \ , \ 1 \ , \qquad Ax.11$$
$$(\text{non} - p \text{ imp non} - q) \qquad\qquad , \ 1 \ , \ non\text{-}logical \ axiom$$
$$(q \text{ imp } p) \qquad\qquad\qquad\qquad , \ 1 \ , \qquad R_{GMP}$$
$$q \qquad\qquad\qquad\qquad\qquad , \ 0.7 \ , \ non\text{-}logical \ axiom$$
$$p \qquad\qquad\qquad\qquad\qquad , \ 0.7 \ , \qquad R_{GMP}$$

Therefore $0.7 \leq C^{syn}(p)$. Since a valuation v such that $v(p) = v(q) = 0.7$ satisfies T, we have, by Completeness Theorem, $0.7 \leq C^{syn}(T)(p) = C^{sem}(T)(p) \leq 0.7$. Thus, the degree of deduction of p is 0.7. Freely speaking, *Is is raining more or less enough.*

4   Similarity-based Reasoning

The objective in approximate reasoning is to draw conclusions from partially true premises. Our idea is to look for the most similar premise, the IF-part, and fire the corresponding conciliation, the THEN-part. Moreover, the degree of similarity may be composed of various partial similarities. Lukasiewicz-Pavelka logic provides a reasonable method to do this task, in a sense Lukasiewicz-Pavelka logic is the only many-valued logic to come off this challenge as we shall now see.

Recall a binary operation $\odot : [0,1]^2 \searrow [0,1]$ is called *t-norm* if, for all elements $x, y, z \in [0,1]$, (i) if $x \leq y$, then $x \odot z \leq y \odot z$, (ii) $x \odot y = y \odot x$, (iii) $x \odot 1 = x$, (iv) $x \odot (y \odot z) = (x \odot y) \odot z$, (v) $x \odot 0 = 0$.

In particular, *continuous t-norms* $\odot$ and their residua $\rightarrow$ play a fundamental role in fuzzy logic. The most frequently used continuos *t*-norms in various fuzzy inference systems generate the following algebraic structures
Gödel algebra:

$$x \odot y = \min\{x, y\}, \ x \rightarrow y = \begin{cases} 1 & \text{if } x \leq y \\ y & \text{otherwise.} \end{cases}$$

Product t-algebra:

$$x \odot y = xy, \ x \rightarrow y = \begin{cases} 1 & \text{if } x \leq y \\ \frac{y}{x} & \text{otherwise.} \end{cases}$$

Lukasiewicz algebra:

$$x \odot y = \max\{0, x + y - 1\}, \ x \rightarrow y = \begin{cases} 1 & \text{if } x \leq y \\ 1 - x + y & \text{otherwise.} \end{cases}$$

These three examples are fundamental since, in a certain sense, they characterize all possible continuos *t*-norms (for details, see [3],[4]). They are the generators of all *BL-algebras* of the real unit interval, too; by fixing a continuous t-norm we fix a *Basic Logic*, a well-defined many-valued logic modeling mathematically fuzzy reasoning. Łukasiewicz-Pavelka logic is, in particular, a Basic Logic. The operations $\odot$ and $\rightarrow$ are the algebraic counterparts of the logical connectives conjunction and implication, respectively. In particular, the complement x^* of an element $x \in [0,1]$ defined by $x^* = x \rightarrow 0$, is the algebraic counterpart of negation, while many-valued equivalence is interpreted algebraically by *bi-residuum* defined, for all $x, y \in [0,1]$, via

$$x \leftrightarrow y = \min\{(x \rightarrow y), (y \rightarrow x)\}.$$

In any *BL*-algebra, a bi-residuum $\leftrightarrow$ has the following properties (cf. [11])

$$x \leftrightarrow x = 1, \tag{38}$$

$$x \leftrightarrow y = y \leftrightarrow x, \tag{39}$$

$$(x \leftrightarrow y) \odot (y \leftrightarrow z) \leq x \leftrightarrow z, \tag{40}$$

$$x \leftrightarrow 1 = x. \tag{41}$$

Now we can set the following [13] important

Definition 10. Let A be a non-void set and $\odot$ a continuos t-norm. Then a fuzzy similarity S on A is such a binary fuzzy relation that, for each $x, y, z \in A$,
(i) $S\langle x, x \rangle = 1$ (everything is similar to itself),
(ii) $S\langle x, y \rangle = S\langle y, x \rangle$ (fuzzy similarity is symmetric),
(iii) $S\langle x, y \rangle \odot S\langle y, z \rangle \leq S\langle x, z \rangle$ (fuzzy similarity is weakly transitive).

Trivially, fuzzy similarity is a generalization of classical equivalence relation, thus called *many-valued equivalence*, too. Notice that, by weak transitivity, partial similarity of x and y, and y and z imply only a lower bound for the degree of similarity of x and z.

Recall a *fuzzy set* X is an ordered couple (A, μ_X), where the *reference set* A is a non-void set and the *membership function* $\mu_X : A \searrow [0,1]$ tells the degree to which an element $a \in A$ belongs to the fuzzy set X.

Theorem 2. *Any fuzzy set* (A, μ_X) *on a reference set A generates a fuzzy similarity S on A, defined by*

$$S(x, y) = \mu_X(x) \leftrightarrow \mu_X(y), \text{ where } x, y \text{ are elements of A.}$$

Moreover,

$$\text{if } \mu_X(y) = 1 \text{ then } S(x, y) = \mu_X(x).$$

Proof. By (38)-(41).

It is worth noting that, in Łukasiewicz-Pavelka logic, 'the negation of equivalence is distance'. Indeed, for all $a, b \in [0, 1]$,

$$(a \leftrightarrow b)^* = 1 - [1 - |a - b|] = |a - b|,$$

the Euclidean distance between a and b.

Theorem 3. *Consider n Łukasiewicz valued fuzzy similarities S_i, $i = 1, \cdots, n$ on a set X. Then*

$$S\langle x, y \rangle = \frac{1}{n} \Sigma_{i=1}^n S_i \langle x, y \rangle$$

is a Łukasiewicz valued fuzzy similarity on X. More generally, the weighted mean

$$S\langle x, y \rangle = \frac{1}{M} \Sigma_{i=1}^n m_i \cdot S_i \langle x, y \rangle,$$

where $M = \Sigma_{i=1}^n m_i, m_i \in \mathcal{N}$, is again a Łukasiewicz valued fuzzy similarity on X, called total fuzzy similarity relation.

Proof. Since all S_i, $i = 1, \cdots, n$ are reflexive and symmetric so is S. The weak transitivity of S can be seen in the following way. Let $A = S\langle x, y \rangle \odot S\langle y, z \rangle$. If $A = 0$, then trivially $A \leq S\langle x, z \rangle$, therefore assume $A > 0$. Then

$$\begin{aligned}
A &= (\tfrac{1}{n} \Sigma_{i=1}^n S_i \langle x, y \rangle) \odot (\tfrac{1}{n} \Sigma_{i=1}^n S_i \langle y, z \rangle) \\
&= \tfrac{1}{n} (\Sigma_{i=1}^n S_i \langle x, y \rangle + \Sigma_{i=1}^n S_i \langle y, z \rangle - n) \\
&= \tfrac{1}{n} [(S_1 \langle x, y \rangle + S_1 \langle y, z \rangle - 1) + \cdots + (S_n \langle x, y \rangle + S_n \langle y, z \rangle - 1)] \\
&\leq \tfrac{1}{n} (S_1 \langle x, z \rangle + \cdots + (S_n \langle x, z \rangle) \\
&= S\langle x, z \rangle
\end{aligned}$$

thus, S is weakly transitive, and therefore a Łukasiewicz valued fuzzy similarity on X. The other part is now an easy generalization of this result. The proof is complete.

Theorem 2 does not hold for other BL-algebras than Łukasiewicz algebra. Indeed, consider the following two fuzzy similarities S_1 and S_2 on a set $\{a, b, c\}$ (with respect to any BL-algebra on the real unit interval!), defined by

S_1	a	b	c
a	1	1	0
b	1	1	0
c	0	0	1

and

S_2	a	b	c
a	1	0	0
b	0	1	1
c	0	1	1

The combined fuzzy relation is not a fuzzy similarity on the set $\{a, b, c\}$ if one uses Gödel algebra or Product t-algebra; in general, weak transitivity does hold.

4.1   Algorithm to Construct Fuzzy Inference Systems

A control situation comprehends a system S, an input universe of discourse X, the IF-parts, and an output universe of discourse Y, the THEN-parts. We assume there are n input variables and one output variable. The dynamics of S are characterized by a finite collection of IF-THEN-rules; e.g.

Rule 1 IF x is A_1 and y is B_1 and z is C_1 THEN w is D_1
Rule 2 IF x is A_2 and y is B_2 and z is C_2 THEN w is D_2

$$\vdots \qquad\qquad\qquad\qquad\qquad \vdots$$

Rule k IF x is A_k and y is B_k and z is C_k THEN w is D_k
where $A_1, \cdots, D_k$ are fuzzy sets of height 1, that is, in each fuzzy set there is at least one element that obtains the membership degree 1. Generally, the output fuzzy sets $D_1, \cdots, D_k$ should obtain all the same values $\in [0,1]$ the input fuzzy sets $A_1, \cdots, C_k$ do, however, the outputs can be crisp actions, too. All these fuzzy sets are to be specified by the fuzzy control engineer. We avoid disjunction between the rules by allowing some of the output fuzzy sets D_i and $D_j, i \neq j$, be possibly equal . Thus, a fixed THEN-part can follow various IF-parts. Some of the input fuzzy sets may be equal, too (e.g. $B_i = B_j$ for some $i \neq j$). However, the rule base should be consistent; a fixed IF-part precedes a unique THEN-part. Moreover, the rule base can be incomplete; if an expert is not able to define the THEN-part of some combination 'IF x is A_i and y is B_i and z is C_i' then the rule should be skipped.

Now we are in the position to formulate an algorithm a fuzzy control engineer has to perform to construct a total fuzzy similarity based inference system.

Step 1. Create the dynamics of S, i.e. define the IF-THEN rules, give the shapes of the input fuzzy sets (e.g. $A_1, \cdots, C_k$) and the shapes of the output fuzzy sets (e.g. $D_1, \cdots, D_k$).

Step 2. Give weights to various parts of the input fuzzy sets (e.g. to A_i.s, B_i.s and C_i.s) to emphasize the mutual importance of the corresponding input variables.

Step 3. Put the IF-THEN-rules in a linear order with respect to their mutual importance, or give some criteria on how this can be done when necessary.

Step 4. For each THEN-part i, give a criteria on how to distinguish outputs with equal degree on membership (e.g. w_0 and v_0 such that $\mu_{D_i}(w_0) = \mu_{D_i}(v_0)$, $w_0 \neq v_0$).

A general framework for the inference system is now ready. Assume then that we have actual input values, e.g. (x_0, y_0, z_0). The corresponding output value w_0 is found in the following way.

Step 5. Consider each IF-part of the rule base as a crisp case, and compare the actual input values separately with each IF-part, in other words, count total fuzzy similarities between the actual inputs and each IF-part of the rule base; by the above Theorems, this is equivalent to counting weighted means, e.g.

$$m_1\mu_{A_1}(x_0) + m_2\mu_{B_1}(y_0) + m_3\mu_{C_1}(z_0) = \text{Similarity(actual,Rule 1)}$$
$$m_1\mu_{A_2}(x_0) + m_2\mu_{B_2}(y_0) + m_3\mu_{C_2}(z_0) = \text{Similarity(actual,Rule 2)}$$
$$\vdots \qquad\qquad\qquad\qquad \vdots$$
$$m_1\mu_{A_k}(x_0) + m_2\mu_{B_k}(y_0) + m_3\mu_{C_k}(z_0) = \text{Similarity(actual,Rule } k)$$

where m_1, m_2 and m_3 are the weights given in Step 2.

Step 6. Fire an output value w_0 such that

$$\mu_{D_i}(w_0) = \text{Similarity(actual,Rule } i)$$

corresponding to the maximal total fuzzy similarity Similarity(actual,Rule i), if such Rule i is not unique, use the mutual order given in Step 3, and if there are several such output values w_0 utilize the criteria given in Step 4.

Of course, we can specify our algorithm by putting extra demands, for example, in some cases the degree of total fuzzy similarity of the best alternative should be greater than some fixed value $\alpha \in [0,1]$, sometimes all the alternatives possessing the highest fuzzy similarity should be indicated, or the difference between the best candidate and second one should be larger than a fixed value $\beta \in [0,1]$. All this depends on an expert's choice.

4.2   Multi-phase Vehicle Control

The first example to illustrate the Algorithm origins from traffic signal control. Consider a T-junction[2] (Figure 1), where traffic flow on the main street (phase A) is assumed to be from two to ten times more intensive than traffic flow from the other direction. Normally the green traffic signal phase order is A-B-C-A, however, if there is low request, i.e. very few or no vehicles in the next phase B or C, then this phase can be skipped. Thus, the order can be e.g. A-C-A-B-C or A-B-A-B-C. The task is to determine the right phase order; fuzzy phase selector - imitating traffic policeman's action - decides the next signal group.

---

[2] This traffic signal control system is operating in Kontula, Helsinki

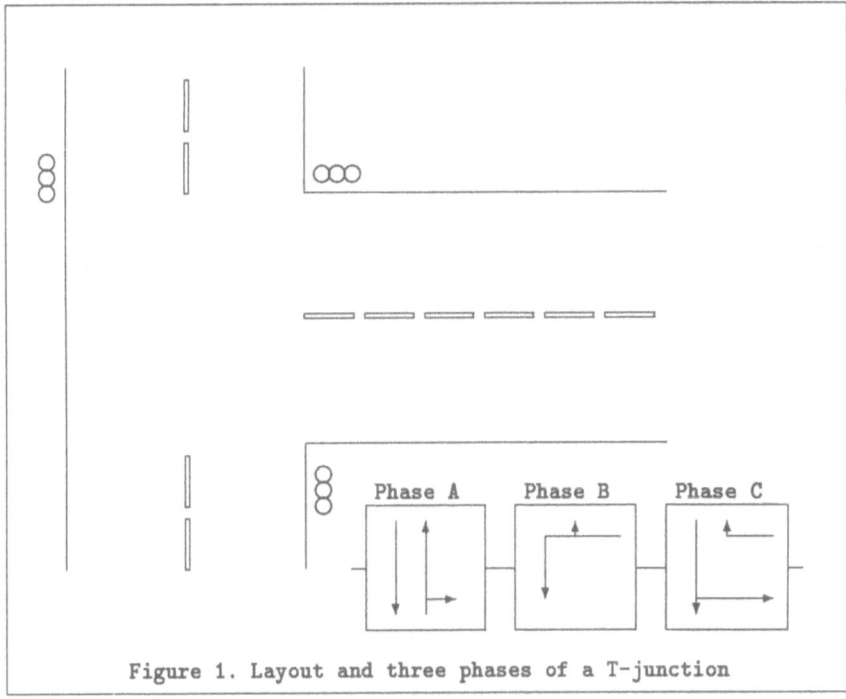

Figure 1. Layout and three phases of a T-junction

The basic principle is that phase B can be skipped if there is no request or if total waiting time of vehicles V(B) in phase B is low, and similarly, phase C can be skipped if there is no request or if total waiting time of vehicles V(C) in phase C is low. Therefore, after phase B the next phase is C or A, and after phase C the next phase is A. In details, the dynamics of the inference is the following. After phase A,

IF V(B) is high   AND V(C) is any   THEN phase is B
IF V(B) is medium   AND V(C) is over saturated   THEN phase is C
IF V(B) is low   AND V(C) is more than medium THEN phase is C
IF V(B) is less than low AND V(C) is more than medium THEN phase is C

The corresponding fuzzy sets are defined by the following membership functions

	Total wait time V(B) [10 sec]															
	0	1	2	3	4	5	6	7	8	9	10	11	12	13	14	15
less than low	1.0	0.1	0.0	0.0	0.0	0.0	0.0	0.0	0.0	0.0	0.0	0.0	0.0	0.0	0.0	0.0
low	0.0	0.3	0.7	1.0	0.7	0.3	0.0	0.0	0.0	0.0	0.0	0.0	0.0	0.0	0.0	0.0
medium	0.0	0.0	0.0	0.0	0.5	1.0	1.0	1.0	1.0	1.0	0.5	0.0	0.0	0.0	0.0	0.0
high	0.0	0.0	0.0	0.0	0.0	0.0	0.0	0.2	0.3	0.5	0.7	0.8	1.0	1.0	1.0	1.0

	Total wait time V(C) [10 sec]															
	0	1	2	3	4	5	6	7	8	9	10	11	12	13	14	15
over saturated	0.0	0.0	0.0	0.0	0.0	0.0	0.0	0.0	0.0	0.0	0.0	0.5	1.0	1.0	1.0	1.0
more than medi	0.0	0.0	0.0	0.0	0.0	0.0	0.0	0.0	0.0	0.0	0.5	1.0	1.0	1.0	1.0	1.0
any	1.0	1.0	1.0	1.0	1.0	1.0	1.0	1.0	1.0	1.0	1.0	1.0	1.0	1.0	1.0	1.0

Corresponding to Step 3 of the Algorithm, if the maximal total similarity is not unique, the phase with the longest waiting time will be fired, or in the worst case, the next phase will not be skipped. The performance of the fuzzy phase control is now straightforward; for example, after phase A, if there are 7 vehicles in phase B and 3 vehicles in phase C, then the next phase will be B.

HUTSIM traffic simulator simulated the performance of this control system, constructed at Helsinki University of Technology, and the results were compared to those determined by a fuzzy phase control based on a Mamdani-style fuzzy controller and a non-fuzzy control algorithm. The average waiting time per vehicle turned out to be the shortest in fuzzy similarity based control as can be seen in Table 1, Table 2 and Table 3.

veh/hour	200	400	600	800	1000	1200	1400	1600
Tot.sim	12.2	12.2	12.6	13.1	12.9	13.8	14.6	15.5
Mamdani	12.1	13.0	12.7	14.9	14.4	15.7	17.2	17.6
Non-fuz	12.1	12.9	13.5	13.9	14.6	16.1	17.2	17.5

Table 1. Average waiting time/vehicle. Vehicle flow ratio 10:1.

veh/hour	200	400	600	800	1000	1200	1400	1600
Tot.sim	12.9	12.7	13.2	12.7	12.8	13.9	15.8	17.2
Mamdani	12.7	13.0	13.7	14.0	14.2	15.3	16.8	21.5
Non-fuz	13.7	12.8	13.3	13.1	14.3	15.3	17.8	19.9

Table 2. Average waiting time/vehicle. Vehicle flow ratio 10:2.

veh/hour	200	400	600	800	1000	1200	1400	1600
Tot.sim	11.2	11.9	12.4	13.4	14.1	18.0	17.4	20.5
Mamdani	11.7	13.2	13.5	13.9	14.0	18.1	18.1	22.0
Non-fuz	12.4	11.9	13.4	14.1	14.2	17.6	18.3	77.6

Table 3. Average waiting time/vehicle. Vehicle flow ratio 10:5.

4.3   Determining Athlete's Anaerobic Thresholds

The second example on how to utilize the Algorithm takes us to the realm of sports medicine. The maximal performance capacity is essential in many sports like football, while in some other sports like long distance cycle racing the submaximal endurance capacity play a more important role. At low exercise levels energy is yielded mostly aerobically, but when approaching maximal exercise level, the aerobic process with increasing lactate production start to play a more perceptible role. To guide successfully athlete's training programs, it is therefore of importance to be able to identify his aerobic and anaerobic thresholds, which are functions of blood lactate, ventilation and oxygen uptake. The test protocol of a continuous incremental exercise, which is performed e.g. by bicycle ergometer, starts with a 3 minutes warm up, then the load is increased every second minute and blood lactate, ventilation and oxygen uptake are measured. The planned duration of the test is about 20-25 minutes, the test is carried out until volitional exhaustion so usually there are $x_1, \cdots , x_n$ measurements, where $n = 10 \cdots 12$. Here is a part of a possible test result:

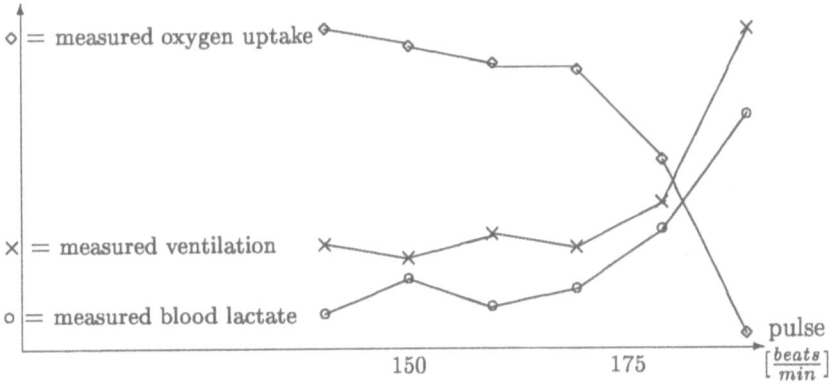

According to skilled sports medicine specialists, the *anaerobic threshold* is such a pulse (beats/min) that

• the amount of blood lactate is increasing rapidly (more that 0.2 mmol/l)

• ventilation is increasing clearly

• oxygen uptake is decreasing

• pulse is 15-25 (+-5) beats/min less than maximal pulse.

For example, in the case above, the anaerobic threshold would be about 170 beats/min. We construct[3] an expert system based on total fuzzy similarity to imitate sports medicine specialists' reasoning. First we connect the measured values x_i with lines as done above (in fact, they are first order spline functions!). Then, corresponding to each test, we create a fuzzy set called

Ventilation is increasing clearly
such that the x_i possessing the absolutely highest positive change of ventilation will have the membership degree 1, the x_j with the absolutely lowest positive change or non-positive change will have the membership degree 0 and all the other x_i:s with positive change will have a linearly scaled degree of membership in this fuzzy set. Due to the spline function, also the values in between measured ones will have a reasonable degree of membership. In a similar manner we define another fuzzy set called

Oxygen uptake is decreasing.
The last fuzzy set we need is called

The amount of blood lactate is increasing rapidly
and is context independent. It has the following shape

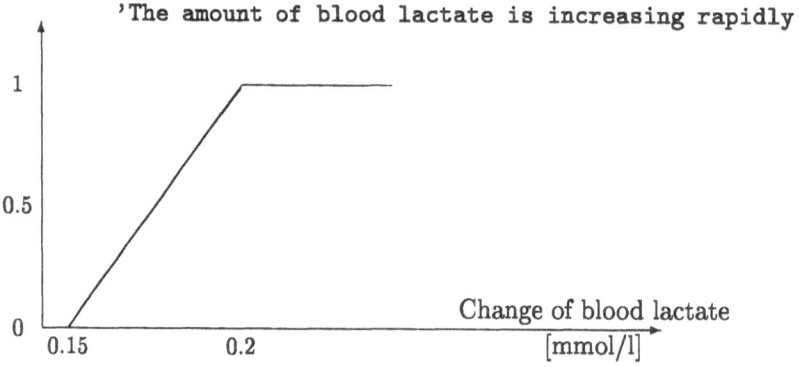

We assume there is an 'ideal object' a which belongs to all these fuzzy sets at the highest degree; we compare each measured value x and the values obtained by the spline functions with a; the one(s) which has (have) the highest degree of total similarity will be the anaerobic threshold(s) if it (they) fulfils the last crisp criteria 'pulse is 10 to 30 beats/min less than maximal pulse'. It may happen, however, that even the found highest degree of total fuzzy similarity is too low, or that there are too many possible anaerobic thresholds to draw a reliable conclusion. This fact is informed to the user with a recommendation that something went wrong and the test should be repeated.

---

[3] A prototype can be found in [5]

Example Having the following test results, define the anaerobic threshold

Pulse	Δ blood lactate	Δ ventilation	Δ oxygen uptake
142	0.14	+21	+0.1
147	0.13	-3	+0.1
156	0.19	-2	-0.1
161	0.20	+1	-0.2
169	0.29	+27	-0.7
174	0.30	+26	-0.7
181	0.35	+26	-0.5
186	0.37	+27	-0.6

For simplicity, we consider only the measured values and calculate the corresponding fuzzy sets and the degrees of total fuzzy similarity. In this test the measured maximal pulse is 186 beats/min so, to fulfill the crisp rule 'pulse is 10 to 30 beats/min less than maximal pulse', we do not have to consider all the cases. We obtain the following fuzzy sets and results

Pulse	Blood Lac.inc.	Vent.inc.	Oxygen Up.inc.	Tot.sim.
156	0.8	0	0	0.27
161	1	0.04	0.17	0.40
169	1	1	1	1
174	1	0.96	1	0.97

The maximal fuzzy similarity is obtained at the value 169 beats/min, while the value 174 beats/min is quite good, too. Since there is always uncertainty involved in the measurements, sports medicine specialists would drive into a conclusion: Anaerobic threshold is close to 170 beats/min.

4.4   Classification Tasks

Theory of total fuzzy similarity offers us a powerful tool to handle with partial similar objects. For example, consider the following table of information which was collected from The World Almanac and Book of Facts 1998.

State	GDT	area	pop	bir	mor	life	65	lite	tel	car
Finland	18.2	130,1	5.1	11	5	74	14	100	1.8	2.7
Denmark	21.7	16,6	5.3	12	5	74	15	100	1.6	3.1
Belgium	19.5	11.8	10.2	12	6	74	16	99	2.2	2.4
France	18.7	210	58.0	11	6	75	16	99	1.8	2.4
Italy	18.7	116.3	57.5	10	7	75	17	97	2.3	1.9
Spain	14.3	195.4	39.2	10	6	75	16	97	2.6	2.8
Slovakia	7.2	18.9	5.4	13	11	69	11	100	4.8	5.4
Bulgaria	4.9	42.9	8.7	8	15	67	16	98	3.3	5.4
Romania	4.6	92.0	21.4	10	23	66	13	97	7.6	10.7
Colombia	5.3	440.8	37.4	21	25	70	5	91	10.0	32.5
Tanzania	0.8	364.0	29.5	41	105	40	3	68	328	589
Nepal	1.2	56.8	22.6	37	77	54	3	28	276	-

GDP = per capita GDP $ 1000   area = area sq.mi
pop = population 10^6 bir = births per 1000 pop
mor = infant mortality/1000 live births life = life expect. at birth
65 = age distrib. % 65+   lite = literacy%
tel = 1 telephone per x persons   car = 1 car per x persons

We may express the information of this table by fuzzy set in various ways. For example, corresponding to the first column, we may construct a fuzzy set High GDT by scaling, i.e.

$$\mu_{\text{High GDT}}(\text{state x}) = \frac{(\text{GDT of state x}) - (\text{Lowes GDT})}{(\text{Highest GDT}) - (\text{Lowest GDT})}$$

In a similar manner we construct the fuzzy sets

                                        Huge Area
          High Population   High Amount of Births
          High Infant Mortality Long Life Expectancy
          Many Old Aged   High Literacy

corresponding to the columns 2-8. For the last two columns it is more reasonable to define a fuzzy set Telephone Per Person by

$$\mu_{\text{Telephone Per Person}}(\text{state x}) = \frac{1}{1 \text{ telephone per x persons in state x}}$$

and similarly a fuzzy set Car Per Person. This results

State	GDT	area	pop	bir	mor	life	65	lite	tel	car
Finland	0.82	0.27	0	0.09	0	0.97	0.79	1	0.56	0.37
Denmark	1	0.01	0	0.12	0	0.97	0.86	1	0.63	0.32
Belgium	0.89	0	0.1	0.12	0.01	0.97	0.93	0.99	0.45	0.42
France	0.86	0.46	1	0.09	0.01	1	0.93	0.99	0.56	0.42
Italy	0.86	0.24	0.99	0.06	0.02	1	1	0.96	0.43	0.53
Spain	0.65	0.43	0.64	0.06	0.01	1	0.93	0.96	0.38	0.36
Slovakia	0.31	0.02	0.01	0.15	0.06	0.83	0.57	1	0.21	0.19
Bulgaria	0.20	0.07	0.07	0	0.10	0.77	0.93	0.97	0.30	0.19
Romania	0.18	0.19	0.31	0.06	0.18	0.74	0.71	0.96	0.13	0.09
Colombia	0.22	1	0.61	0.39	0.20	0.86	0.14	0.88	0.10	0.03
Tanzania	0	0.82	0.46	1	1	0	0	0.56	0	0
Nepal	0.02	0.11	0.33	0.88	0.72	0.40	0	0	0	0

Example 1 Clearly, Romania, Bulgaria and Slovakia are countries mutually equal with respect to the above information. By maximal fuzzy similarities generated by the above fuzzy sets we calculate Similar⟨Slovakia,Bulgaria⟩ = 0.9050, Similar⟨Slovakia,Romania⟩ = 0.8380 and Similar⟨Romania,Bulgaria⟩ = 0.8420.

Example 2 Typical features of an *undeveloped country* are low GDP per capita, high rate of birth, high infant mortality, short life expectancy at birth and low literacy percentage. In the light of above facts, which are the three most undeveloped countries? It is reasonable to use the following fuzzy sets

Low GDP = (High GDP)*

High Amount of Births

High Infant Mortality

Short Life Expectancy = (Long Life Expectancy)*

Low Literacy =(High Literacy)*

and to assume that a 'typical undeveloped country' belongs to each fuzzy set at the degree 1. Comparing now each state with such a typical country results the following degrees of total fuzzy similarity: Finland 0.06, Denmark 0.03, Belgium 0.06, France 0.05, Italy 0.05, Spain 0.09, Slovakia 0.21, Bulgaria 0.23, Romania 0.27, Colombia 0.33, Tanzania 0.89 and Nepal 0.84.

The result is right, however, by utilizing more specified fuzzy sets than those obtained by simple scaling would result even more divergence in the group.

4.5   Case-based Reasoning

Many-valued similarity can be used to model Case-based Reasoning, too. Having n model cases, we want to consider a new one similar enough to be compared with the cases in the database. We illustrate the idea by an example, which is again from medicine.

Maximal heartbeat rate, HRmax, is a good measure of cardiorespritory fitness, indeed, population studies have shown that low fit individuals have a significantly greater risk of all-cause mortality compared with high fit subjects, indicating health-related validity of this measure. Direct measurements of HRmax, however, require expensive apparatus and laboratory facilities and the test protocols require the individual to exercise to the exhaustion. Submaximal test protocols provide inexpensive, safe and feasible way of testing healthy adults. Several HRmax prediction methods, based on sub-maximal exercise tests are already available. The prediction accuracy and their ability to classify fitness are acceptable on group level, but not so for individuals.

At The Tampere Research Center of Sports Medicine the following preliminary research was carried out. 57 healthy females of age 34 to 51 years got through a 2 kilometers walk test including a test for sub-maximal heartbeat rate HRsub and another test for HRmax, and resulting the following information

	age	W-ind	mlk	HR_{sub}	ww%	time	weight	length	HR_{max}
C1	34	23.39	36.4	174	75.2	15.7	68.4	171	197
C2	34	25.16	33.5	165	73.8	16.8	72.7	170	192
C3	35	21.77	38.4	150	68.7	16.1	58.9	165	192
⋮									
C57	51	18.48	39.7	140	69.0	15.7	53.4	170	173

The range of each variable was calculated first, for e.g. Body Mass Index (W-ind) it was from 18.48 to 32.66, and on this bases eight scaled fuzzy sets were greaten. Each of them generated a fuzzy similarity relation, and an experienced medical doctor weighted each factor by weights 5, 7, 4, 20, 50, 7, 5 and 2, respectively. This yield 57 total similarity relations corresponding to each case. The relevance of such treatment was tested by calculating the internal dependence of the cases. It turned out that whenever total fuzzy similarity of two different cases was greater than 0.925, then the corresponding difference in HRmax values was less or equal to 10 beats/min. Such a result is superior to other methods. The database can be used to predict HRmax for objects that go through a 2 kilometers walk test.

References

1. C. C. Chang, Algebraic analysis of many-valued logics. Trans. Amer. Math. Soc. 88 (1958), 467-490.
2. R. Cignoli, M.L. D'Ottaviano, D. Mundici, Algebraic Foundations of many valued Reasoning. Kluwer Acad. Publ. 1999.
3. R. Cignoli, F. Esteva, L. Godo, A. Torres, Basic Fuzzy Logic is the logic of continuous t-norms and their residua, submitted.

4. P. Hájek, Metamathematics of fuzzy logic, Kluwer Acad. Publishers, Dordrecht, 1998.
5. M. Hempilä, Defining an athlete's aerobic and anaerobic thresholds with fuzzy total similarity. *Diploma Thesis*. Lappeenranta University of Technology, 1998.
6. U. Höhle, On the Fundamentals of Fuzzy Set Theory. J. of Math. Anal. and Appl. 201 (1996) 786-826.
7. I. Kant, Logic, New York: Bobbs-Merrill. 1974..
8. J. Łukasiewicz, Selected works. Cambridge Univ. Press. 1970.
9. I. Niiniluoto, Analogy and Similarity in Scientific Reasoning D. H. Helman (ed.) Analogical Reasoning. Kluwer Acad. Publ. 1988. 271-298.
10. J. Pavelka, On fuzzy logic, I,II,III Zeitsch. f. Math. Logik. 25 (1979), 45-52, 119-134, 447-464.
11. E. Turunen, Mathematics behind Fuzzy Logic, Advances in Soft Computing, Physica-Verlag, Heidelberg, 1999.
12. L. Zadeh, Fuzzy Sets, Information and Control 8 (1965) 338-353.
13. L. Zadeh, Similarity Relations and Fuzzy Orderings, Information Sciences 3 (1971) 177-200.